T0269267

Astronomers' Universe

More information about this series at http://www.springer.com/series/6960

David S. Stevenson

The Exo-Weather Report

Exploring Diverse Atmospheric
Phenomena Around the Universe

 Springer

David S. Stevenson
Carlton le Willows Academy
Nottingham, UK

ISSN 1614-659X   ISSN 2197-6651   (electronic)
Astronomers' Universe
ISBN 978-3-319-25677-1   ISBN 978-3-319-25679-5   (eBook)
DOI 10.1007/978-3-319-25679-5

Library of Congress Control Number: 2016943436

Printed on acid-free paper

This Springer imprint is published by Springer Nature
The registered company is Springer International Publishing AG Switzerland

This book is dedicated to my sisters Avril Stevenson-Davies, Karen Suzuki, and Mairi Allardice and their families; and to my cousin Lesley Duncan.

Preface

My father fought in both World Wars. In 1915, aged 16 he joined the British Navy, serving on HMS Orion. Training in communications, he ultimately took the surrender of the German Fleet, passing the terms of surrender to the German High Command (Fig. 1). During the Second World War, my father served in the merchant fleet and took part in the evacuation of Dunkirk. The science of meteorology took off during the First World War, because it became understood that the weather was a key variable in winning battles. There were obvious patterns, such as the "lowering of the sky" ahead of a rain, but the term front had yet to be coined, except outside the unpleasant confines of the Trenches. Changes in wind direction and cloud cover were noted as rain bands came and went as well as the turning of the seasons (Fig. 1).

However, it wasn't until the late nineteenth century before much of the underlying science was known and not until the 1920s that our understanding of frontal boundaries emerged. Later still, in the 1930s and 1940s the jet streams were discovered and the driving force behind much of the movement of our weather became apparent. My father taught me the nature of frontal systems that he had learnt while at sea, along with the names of all of the clouds. However, the idea that other planets might have weather was still in its infancy.

Until the Soviet Venera probes landed on Venus, it was thought rather likely Venus was a rather pleasant tropical world. That Venus could experience a runaway greenhouse effect, driven by carbon dioxide, was simply not understood. John Tyndall may have done the first experiments with various gases as early as 1850, but the implications of such discoveries were not realized until much later. Indeed, the idea that altering the concentration of greenhouse gases can have an impact on terrestrial climate is still rather contentious in some circles.

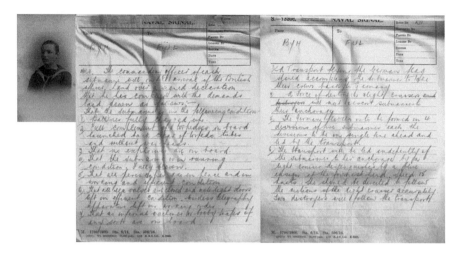

Fig. 1 The original transcript of the terms of surrender of the German fleet in 1918, taken by my father David Stevenson. My father rather mischievously kept a copy of the transcript he recorded to be passed onto the fleet admiral. In amongst the tragedy of the First World War, and in large part because of it, the science of meteorology really took off. Respective Navies and fledgling air forces needed to know how the weather would impact their activities

We are at a point that we have gathered a lot of data on the atmospheres of the planets in our solar system and are beginning to gather information on planets hundreds or thousands of light years away. Through a greater understanding of our planet, we can see how the climate of other planets is affected by the composition of their atmosphere; the influence of their parent star and the proximity of such worlds to their stars. Take Kepler 452b, for example. This world is a fairly good match for the Earth a billion years into our future. As its Sun-like star slowly advances its years, Kepler 452b is being stranded on the hot side of its star's habitable zone. Our observations of this world will paint a more accurate picture of how our world will eventually find its habitability coming to an end.

Meteorology is a broad physical science, but this breadth is its strength. It encompasses simple, generic observational skills, through to complex mathematical modeling of fluids. Models are best tested by comparison with observation and it often turns out that simple observations can provide some very useful truths. In Chap. 1 I describe a workable model for predicting winter weather in the UK several months in advance. This is testable, which is the hallmark of good science. However, it can be underpinned by solid mathematics, something I could with brushing up on.

I've arranged the chapters in increasingly hierarchical but overlapping structures that provide clear links between phenomena on all the worlds described. Lightning is introduced in Chap. 4, but links to all of the subsequent chapters (bar Mars) on the worlds of our solar system. The greenhouse effect emerges in Chap. 3 then links to the others, while terrestrial monsoons are described in Chap. 2 before linking through to the climate of Mars and extrasolar worlds. Hopefully, this approach provides clear links and shows the interconnectedness of the underlying science.

Meteorology and climate science are truly fascinating aspects of planetary science where we humble earthlings get to experience firsthand how the physics of other worlds plays out.

Nottingham, UK   David S. Stevenson

Contents

xvi   Contents

1. What We Know About the Weather on Earth

Why Do We Have Weather?

Weather is such a ubiquitous part of our lives that most of us take it completely for granted. Yet it controls nearly every aspect of our life, from our route to and from work to the food we eat and more. The weather is the talking point for the random meeting, or the casual chat on the Metro. The weather is something that we all have an opinion on; most notably the accuracy, or perceived lack-thereof, of TV weather forecasts, or the dress sense of their presenters. Yet, most have little understanding of why weather happens at all.

Weather, for all its daily and geographical complexity, can be broken down to one simple statement: it is the transfer of energy from one area to another. Weather is a manifestation of our parent star's inability to heat the Earth's surface evenly. A round, three dimensional planet is notoriously difficult to deliver energy to in any kind of even-handed manner. Areas under more direct sunlight, such as at the equator, warm most strongly as heating is more intense when the Sun is overhead. This is simply because the energy is delivered to a smaller surface area than if it were illuminated at an angle.

Worse still, the presence of an atmosphere ensures that energy from our star is diluted or diverted on its multimillion kilometer passage from star to planet. Much of the energy is reflected by clouds or particles in the air, or absorbed by elements and compounds that swirl within it (Fig. 1.1).

If the Earth were more like our Moon, with little or no atmosphere, then the daytime temperatures would soar to nearly 175 °C, while at night we would languish at –125 °C: a range of 300 °C. The atmosphere cushions the rise and fall of temperature,

© Springer International Publishing Switzerland 2016
D.S. Stevenson, *The Exo-Weather Report*, Astronomers' Universe,
DOI 10.1007/978-3-319-25679-5_1

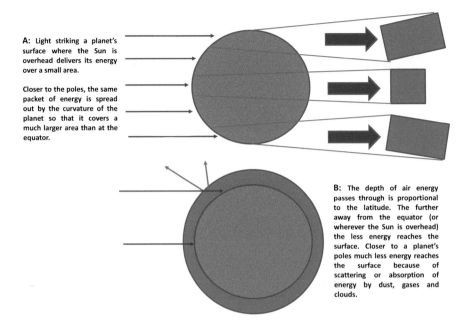

A: Light striking a planet's surface where the Sun is overhead delivers its energy over a small area.

Closer to the poles, the same packet of energy is spread out by the curvature of the planet so that it covers a much larger area than at the equator.

B: The depth of air energy passes through is proportional to the latitude. The further away from the equator (or wherever the Sun is overhead) the less energy reaches the surface. Closer to a planet's poles much less energy reaches the surface because of scattering or absorption of energy by dust, gases and clouds.

FIG. 1.1 The effect of the curvature of a planet and any atmosphere on the amount of heating different parts of its surface will experience. Were the Earth a pancake facing directly into the sunlight all of it would be heated to the same extent and weather, although it would exist, would be profoundly dull. (**a**) Light striking a planet's surface where the Sun is overhead delivers its energy over a small area. Closer to the poles, the same packet of energy is spread out by the curvature of the planet so that it covers a much larger area than at the equator. (**b**) The depth of air energy passes through is proportional to the latitude. The further away from the equator (or wherever the Sun is overhead) the less energy reaches the surface. Closer to a planet's poles much less energy reaches the surface because of scattering or absorption of energy by dust, gases and clouds

in part by absorbing and retaining some of the Sun's energy, but also by effectively transporting it from place to place. This transfer of energy we call wind.

On any object with significant gravity energy can be transported vertically from the bottom to the top of the atmosphere by convection. Advection is the transport of energy from place to place in a (roughly) horizontal direction. This can happen in response to convection pulling and pushing air around or, more directly, in response to temperature differences and the spin of our planet. On the Earth, convection dominates the transport of energy at the hottest and (in our case) most equatorial regions, but is prevalent elsewhere where conditions are suitable, particularly in the summer months when heating is strongest. By contrast, advection dominates energy transport between the tropics and the poles.

The Highs and Lows of Meteorology

Where air rises, it exerts less pressure on the underlying layers or the surface: this creates an area of lower pressure that draws in air from the surroundings. Where the air is moist, clouds can form and precipitation falls. Essentially: what goes up must come down. Otherwise, when air is sufficiently cold and dense, it descends under its own weight. As it falls it compresses the air underneath, generating an area of higher pressure—an anticyclone—and warms somewhat as it compresses under its own weight. In addition, pressure will rise and fall if it is forced to do so. Pressure falls where air is forced to rise over hills and mountains, and conversely in the extreme environs of tornadoes and hurricanes, as the air circulates rapidly around the central low pressure core, inflowing air cannot keep pace with the air that is being sucked out of the storm's top. Some air then is forced to descend inside the center of the storm: the eye of the hurricane in particular is a region of descending, warming air that breathes a brief window of calm in the otherwise violent storm.

In addition to this simple relationship overall, warm air exerts a higher pressure than colder air. Thus, if you move vertically in a column of warm air pressure falls more slowly with height than if you are moving upwards through a column of cold air. This has some interesting effects. Over Siberia (and North America) during the winter, the air becomes extremely cold. This generates an intense area of high pressure over Siberia. Yet, if you rise above 3000 m the high pressure area has been replaced by a finger-shaped area of low pressure—a trough. Conversely, Scandinavia is commonly the location of what is known as blocking anticyclones. These are areas of high pressure, with a warm core, which extend through the full height of the troposphere. Although the air may be very cold at the surface, you don't have to travel far upwards to encounter air that is much warmer than you would expect for that latitude and altitude in the winter. Blocking anticyclones are also common over the North Pacific in the winter, particularly in La Niña years, and over the same region in the summer of El Niño years (Chap. 2). It is the presence of the core of warm and relatively high pressure air that makes these atmospheric features so stubbornly difficult to move. Indeed, the current climatic tribulations taking Alaska and California by force, are largely down to the persistence of a blocking anticyclone over the northern Pacific; this is one block that has lasted, with only minor interruptions, for several years.

Likewise, low pressure areas tend to become colder with height. The major low pressure areas of the mid and high latitudes develop cold cores as they deepen. However, in the summer, and over the Tropics, many low pressure areas, including hurricanes, are warm-cored throughout. Within these thermal lows pressure is low at the surface but soon morphs into an area of high pressure at greater heights, due to warmer air exerting a higher pressure than cold air.

Wind Direction: Waterwheels and a Suspect Tale from the Front

Air flow on any rotating planet is driven by two opposing forces: pressure and spin. Pressure is the simplest to understand. Air of lower pressure effectively draws air into it from regions with higher pressure. Strictly speaking a force is exerted by the high pressure air, which directs gases towards the regions with lower pressure. This is the pressure force. Intimately linked to this idea is temperature: air of a higher temperature has a higher pressure. There are three fundamental gas laws describing this relationship of pressure, temperature, volume and amount of gas. Boyle's Law dictates that the volume of gas increases as the pressure decreases: gas under higher pressure has a smaller volume, if you keep the temperature and mass of gas the same. Charles' Law requires that the volume of gas increases as the temperature increases. And Avogadro's Law tells us that equal volumes of all gases, at the same temperature and pressure, have the same number of molecules: for a given mass of an ideal gas, the volume and amount (moles) of the gas are directly proportional if the temperature and pressure are constant. Together, these laws guide much of modern science and are integral to our understanding of how atmospheres work.

Charles's law, incidentally, is often misattributed as Gay-Lussac's Law, as it was published in 1802 by the French chemist, Joseph Louis Gay-Lussac. However, the publication followed a century after from the work of Guillaume Amontons and Jacque Charles, whose work Gay-Lussac extended. Amonton's work languished for a century before Gay-Lussac revamped it, along with many unpublished results from Charles. Gay-Lussac's triumph lay in the use of individual gases, such as oxygen and nitrogen, unavailable in Amontons's day and his ability to interpret, unite and extend the work of Charles and Amontons, producing two of the three core gas laws biologists, chemists and physicists use today.

If we look simply at pressure and temperature, particles of the various gases that make up the air will exert a higher pressure at a higher temperature because they move faster (they have greater kinetic energy). Collisions between the particles and any surface will, therefore, carry more energy and thus exert a greater force than those in a colder and denser gas. Imagine a baseball hitting a wall at 10 km per hour versus 80 km per hour. This idea has important ramifications for the structure of the atmosphere as we shall see later in this chapter.

Spin is a more complex concept. The Earth rotates on its axis once per day. That means that a piece of soil at the equator has to move at nearly 1700 km per hour to make a full rotation in this interval. Conversely, standing at the Earth's rotation poles, you don't rotate at all (except slowly around your middle, of course). That means that any stray packet of air trundling north or south towards the poles will find itself, increasingly, moving faster than the ground underneath; and this means that, relative to the surface, it will steadily bend towards the east as it travels away from the equator. Similarly, but in the opposite direction, air moving from either pole towards the equator will curve towards the west

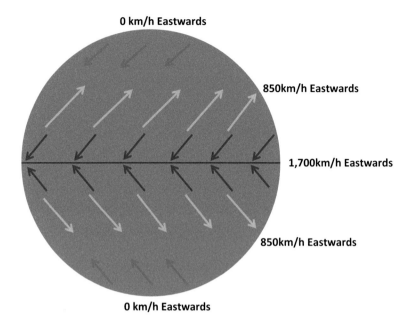

0 km/h Eastwards

850km/h Eastwards

1,700km/h Eastwards

850km/h Eastwards

0 km/h Eastwards

Fɪɢ. **1.2** The effect of a spinning Earth on the movement of air to and from the equator. Air moving away pole-wards is deflected eastwards while air moving towards the equator moves to the west. Numbers indicate the rate of rotation of the surface around the polar axis

because the ground underneath it is moving more swiftly towards the east than it is (Fig. 1.2). A similar thing happens when a person steps off a moving bus or train—their speed, doesn't match that of the ground underneath.

Known as the Coriolis Effect, this made its presence known during the latter part of World War I. In 1918, German troops positioned a large gun 120 km to the northeast of Paris. The gun was used to shell Paris from behind the German lines with the aim of terrorizing the population rather than inflicting significant damage. Because the distance was so far, the Earth's rotation had an impact on the shell's trajectory; they reached higher than any other projectiles up to that time, and their path was therefore subject to forces that had theretofore unaccounted for. Initially most of the shells fell to the west of the French capital, because when they began their journey they were moving eastwards at a slower speed than Paris, which lay to the southwest and thus had further to travel in its rotation around the Earth's axis (Fig. 1.3).

Fig. 1.3 The 1918 "Paris gun". Shells were fired at Paris from 120 km to the north-east of the capital. Shells curved to the west of the city as a result of the Coriolis Effect (*red arrow*). The gunners soon corrected for this and repeatedly shelled Paris. French map courtesy of Wikipedia Commons

Unfortunately, after some tweaking of the trajectory, Paris was positioned firmly in the gun's sights and was shelled up to 20 times a day, killing 250 people. The "Paris Gun" was notable for one other meteorological first. The gun fired the first man-made object 42.3 km into the stratosphere; a feat not beaten until Germany launched the infamous V2 rockets in 1944. The sheer height of the trajectory combined with the distance brought Gustav Coriolis's effect into play.

Otherwise, the Coriolis Effect is mostly experienced by air or liquids that are moving northwards or southwards on any rotating body. Its discoverer, Gaspard–Gustave de Coriolis, first described the Coriolis Effect while thinking about machinery. Coriolis was interested in *work*, or more fully *work done*, a definition of energy: the ability of a force to move a mass over a distance. He investigated "work" in relation to objects that were rotating, in particular water wheels, which were very much in demand to drive the hardware of the early industrial revolution. Coriolis had no interest in the dynamics of the Earth's atmosphere, or indeed any planetary atmosphere, although the title of his 1832 paper was inadvertently suggestive of this: "On the equations of relative motion of a system of bodies." In this work Coriolis detailed various forces which afflict rotating bodies, including planets, but it would take nearly 70 years before Coriolis' name became clearly associated with the meteorological effect he inadvertently described.

The Coriolis Effect is proportional to the speed that the body spins and its overall diameter. Thus a small planet has a weak effect, as does a planet that rotates slowly. Therefore, both Mars (small) and Venus (slow) have weak Coriolis Effects, while mighty Jupiter, with its 17 h rotation has an effect much stronger (about three times) than that found on the Earth.

This effect works at the surface and throughout the bulk of the atmosphere. On the Earth hot air rises above the Equator (or more precisely where the Sun is overhead) and sinks at the poles. You might then expect a simple pattern of air flow from pole to equator. Easterly winds would dominate the planet[1] because these would be returning cold air from the poles towards the equator; indeed, this pattern is evident on Venus (Chap. 5). However, this is not seen in the Earth's atmosphere, because the Coriolis Effect prevents air from flowing in such a simple pattern. The issue is the air rising above the equator. This air rises over 10 km through the lowest layers of the atmosphere, until it hits a wall, called the tropopause, the nature of which we shall return to later. At the tropopause the air is diverted to the north and south and the Coriolis Effect begins to work its magic.

---

[1] An interesting consequence of this would be that the Earth's spin would decelerate. Winds would be blowing in the opposite direction to its spin and thus exert a frictional force on its surface.

High above the band of rising air, air flows north and south. The Coriolis force diverts this towards the east, in a westerly flow. Initially, the winds move northeastwards or southeastwards towards the pole, but by the time the air has reached roughly 30° N or 30° S the air has been turned right angles to its original direction and flows strongly to the east. There is no further north or south movement. Were this all that happened then the tropics would be completely isolated from the poles and the only means by which heat could be transported further north or south would be via ocean currents. However, the planet has a trick to play with this air. The air steadily cools as it initially rises near to the equator. By the time it has traveled 30° north or south it is dense enough to sink towards the ground once more. So down it falls towards the ground in a steady stream, pouring over the Tropics of Cancer and Capricorn. The planet's steadiest belt of weather, the dry Horse Latitudes, are created from this pattern, called Hadley cells. Past the 30th parallel most of the air returns to the equator, while the rest streams northwards or southwards towards the poles, entraining air from outside the Hadley cells.

Once more the Coriolis Effect comes into play as air moving towards the North Pole is again bent towards the east, producing the mid-latitude westerly winds. The southern hemisphere mirrors this with the winds arcing round to form the roaring forties and even more descriptively screaming fifties. The Coriolis Effect keeps the air turning towards the east so once again it fails to reach either pole and the atmosphere comes to the rescue. Cold air streaming away from either pole, towards the west, collides with the westerly belt and directs much of it upwards towards the tropopause.

Finally, after a fairly exhausting trip a portion streams northwards (in the Northern Hemisphere) and southwards (in the Southern Hemisphere) to reach the Polar Regions. The circuit completes when this now profoundly chilly air sinks under its own weight towards the surface. From here the only way is away from the pole to complete the loop, reuniting the air of our planet in three great circulations between pole and equator. This is illustrated in Fig. 1.4. It must be emphasized that this pattern is broadly symmetrical around the equator, meaning that the same overall pattern of air flow is seen in the Southern Hemisphere.

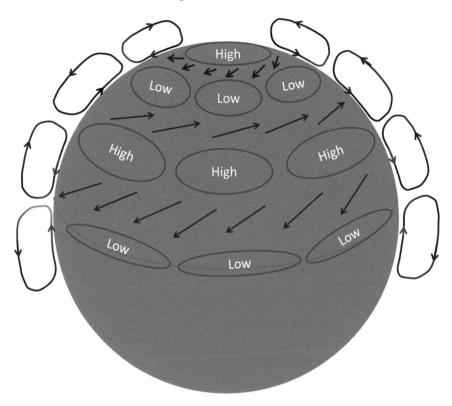

Fɪɢ. 1.4 The general pattern of wind flow and pressure on the Earth in the Northern Hemisphere. Geographical features such as oceans and mountains strongly affect this generic pattern; as does the tilt of the Earth throughout the seasons

The Vertical Structure of the Earth's Atmosphere

As well as horizontal transitions across the globe at different latitudes, the atmosphere can be broken into different layers as it increases in altitude. In each layer temperature and pressure vary in predictable ways.

Extending through the first few kilometers of the atmosphere is the region of densest and most turbulent air, known as the troposphere. This is thickest (17 km) over the Tropics where the air is warmest and exerts the greatest pressure. The depth of the troposphere decreases in a series of steps as you go further towards each pole. Each step is marked by a current of fast moving air known as a jet stream (returned to later in this chapter). Thus, where as above the equator the top of the troposphere is 17 km high, it is only 10 km or so above the ground over Spain, the bulk of the US and southern Australia. Even closer to the poles and it drops even closer towards the surface, ultimately lying 7–8 km above the ground at the Poles.

Throughout the troposphere temperatures fall as altitude rises—indeed this is a defining feature of this layer. This can make the air unstable and prone to convect. However, there is such variability in this region that simply describing it as a simple layer would do it a great disservice. The troposphere is broken horizontally into different regions that move under the influence of pressure and the Coriolis Effect, leading to a very dynamic and complex zone. Fortunately, as you go higher the complexities subside and things become somewhat simpler to understand.

Above the troposphere lies the stratosphere. As the name implies, air in this layer is stratified, displaying relatively little vertical movement. The dividing line between the troposphere and the stratosphere is a boundary called the tropopause. Here temperatures stop falling with increasing altitude and begin to rise. This change causes the air to stop convecting as from this point upwards the air is warmer the higher you go. The reason the temperatures rise is down to one gas: ozone. Ultraviolet light is absorbed by oxygen, which is found in the form of a molecule (O_2). In this diatomic (two atom) form, the bonds can be broken and then reassembled into ozone (O_3). Ozone is superbly efficient at absorbing ultraviolet light, particularly that in the range of longer wavelengths closest to the visible part of the spectrum. It is these wavelengths that are most harmful to life, thus ozone helps ensure complex life can populate the solid surface of the Earth. Without it, life would be confined to the ocean depths where UV is unable to penetrate deeply (Fig. 1.5).

As you progress upwards through the atmosphere, the density of gases steadily diminishes. The stratosphere begins where the atmospheric pressure is approximately one fifth that found at the surface (200–300 mb) and ends with a pressure less than one tenth this value. Although very dense compared to the vacuum of space, the air is so rarified that any living organism (for example a stray microbe) would be rapidly freeze-dried.

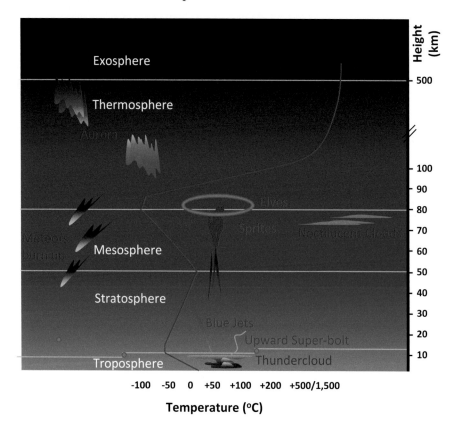

FIG. 1.5 The overall structure of the Earths Atmosphere. The different layers are separated by *white lines*. The thickness of the troposphere varies with latitude and is thickest nearest the equator. Jet streams (*small, blue circles*) mark the position of the jumps. The *red line* indicates how temperature changes with height. The temperature of the thermosphere varies from 500 °C to over 1000 °C when the Sun is most active. Various phenomena are indicated. Sprites, Elves and Blue Jets that are associated with thunderstorms are described in Chap. 4

Above the ozone layer, the mesosphere begins and the temperature falls once more from around zero Celsius to around –100 °C. The air is very dry but does hold a small amount of water vapor—just enough to form rare noctilucent clouds out of water ice crystal. These form at an altitude of around 70–80 km, predominantly in the early spring at the poles where the air is coldest. Their frequency is increasing, which suggests that more moisture is escaping through the stably stratified layer below from the moist troposphere at the atmosphere's base. This may be a sign that rising global temperatures (Chap. 2) are driving more water vapor into the higher atmosphere.

As you ascend through the mesosphere the abundance of molecules falls as ultraviolet light and increasing amounts of x-rays and gamma rays break them apart. Atomic oxygen, nitrogen, and the molecule hydroxyl, which is unstable at the Earth's surface, dominate the composition of the gases in the mesosphere. Hydroxyl is produced when water vapor is split by UV releasing hydrogen to space. The left-over oxygen and hydrogen remain bound together, but are ultimately split up at higher altitudes. The atmospheric pressure is less than 1000th that at the surface making the mesosphere a fairly good analog for the atmosphere that rests on the surface of Mars.

The mesosphere is also home to some enigmatic electrical phenomena known as sprites, elves and tendrils. These microsecond long features appear above some, but not all thunderstorms and appear to form electrical connections between the turbulent troposphere and the thermosphere above (see Chap. 4).

Additionally, the mesosphere is also the region where most meteorites are vaporized. This leads to a regional enhancement in the abundance of silicates and metals, particularly iron and nickel. All told, the mesosphere is one of the atmosphere's most inscrutable regions. It is too high for balloons and jet craft to probe, yet too low for satellites to sample. Measurements are always taken remotely from the surface or from space and are as yet relatively piecemeal in nature. Some of these recent observations are described in Chap. 10.

Above 80 km in height, the temperature begins to rise once more. Gases are increasingly ionized forming bands of rarified gases at different heights. These ionized regions are important for ground-based radio communication as they efficiently reflect electromagnetic waves to allow humans to communicate with one another around the curved surface of the planet. These gases also give this layer its name, the ionosphere. Technically, the ionosphere encompasses both the mesosphere and the ionized layer above which we now call the thermosphere. Whereas the amount of ionization is limited in the mesosphere, all of the gases are ionized to varying extents in the thermosphere. Temperatures rise to over 500 °C in this region. Not that you would feel it hot were you able to experience it: the concentration of gases is so low that you would still freeze solid if out of the glare of the Sun. The maximum temperature can exceed 1000 °C when the Sun is active and the Earth is subjected to the highest intensity of ultraviolet and x-radiation.

Towards the top of the thermosphere you might bump into passing orbiting craft, including the International Space Station which orbits at 400 km altitude. As you went higher you would enter the magnetosphere. In its regions above the equator the Earth's magnetic field guides the vapors and the solar wind into warped donuts of plasma. Here the density of gas is trillions upon trillions of times lower than the density found at the Earth's surface. It is effectively an extension of the near vacuum of space, where hydrogen and helium pause on their way out into the Solar Wind. The hydrogen comes mostly from the break-up of water vapor that results from ultraviolet light splitting the molecule up, while the helium comes from the Earth's crust and interior. Helium is synthesized by the radioactive decay of heavy elements, such as uranium, or was originally trapped there when the Earth formed. In addition, small amounts of oxygen and nitrogen are drawn outwards into space. This is partly as a result of their entrapment within the flow of escaping hydrogen and helium, something called hydrodynamic escape. Alternatively, oxygen and nitrogen escape directly from the effect of ultraviolet light bombarding these gases and energizing them sufficiently to escape in their own right.

This region where gases escape from the thermosphere to the magnetosphere or the vacuum of space is known as the exosphere. The exosphere is less of a region of atmosphere as much as it is a transition from atmosphere to space. The density of gases isn't so much measurable in millibars rather in terms of individual particles per cubic centimeter. The only thing that would exert noticeable pressure upon you would be the impact of a passing meteor or a man-made satellite.

As far as weather and climate are concerned most of the action occurs in the troposphere, with its upper surface, the tropopause, keeping a lid on most meteorological activity. However, within the stratosphere and mesosphere, meteorological phenomena do occur that are at least visible from, if not having a direct effect upon, the surface below. With our sortie of the atmosphere complete we return to the troposphere to engage in some atmospheric hand-to-hand combat. However, in Chap. 4 we will look again at some of the more exotic atmospheric phenomena that stir the layers above.

The Language of War-Fronts

During the First World War, and in it immediate aftermath, meteorologists were trying to understand patterns within the flow of air and the link to the weather in general. A group of scientists working within the Bergen School of Meteorology harnessed the language of the Great War to describe the observed patterns of temperature and wind flow. The term "front" became established as a generic term that described the boundary between two *competing* air masses. In the Bergen scenario, known as the "Norwegian Cyclone Model", a low pressure first develops as a disturbance along the so-called polar-front: this is the boundary between tropical air that is moving pole-wards from the equator and polar air that is moving towards the equator. A wave develops along the polar front as warm air nudges into it. Along this boundary, the overriding warm air condenses as clouds and precipitation intensifies. As it does so pressure falls along this portion of the boundary (Fig. 1.6).

To the rear of this surface feature, cold air begins to undercut the warm air, again driving the warm air upwards as a second band of precipitation develops. This region will become the cold front. Between the two fronts is a region of warm air that has intruded into the colder polar air: this is the warm sector. The idea of fronts is most closely tied to Jacob Bjerknes at the Bergen School. Bjerknes often referred to the warm front as the steering line, while the cold front became known as the squall line (Fig. 1.7). Although these terms are still often in use in the US they are generally not used in Europe. Here, the terms cold and warm fronts have stuck and most actively describe the changes in conditions associated with the passage of each frontal boundary.

In the Bergen model, the low pressure moves along the frontal zone as an intensifying wave. As it does so, the warm front advances effectively at the same rate as the low pressure as a whole, while the cold front progressively undercuts the warm air to the rear.

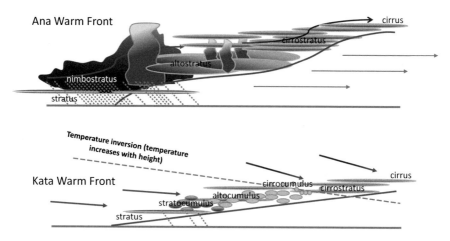

FIG. 1.6 Idealized warm fronts. In the *top diagram*, the warm air advances quickly and is unstable, meaning that it is able to rise by convection, as well as rise because it is being forced over the wedge of colder air. This "Ana warm front" brings extensive precipitation, a strong rise in temperature and a significant change in wind direction. On occasion thunderstorm cells may be embedded (*grey*). Above the "kata front" air is descending from higher up in the atmosphere. This may be associated with frontal boundaries in the upper air. Descending air causes it to warm and evaporate much of the cloud layer. Clouds are typically more broken and precipitation lighter (more *dispersed blue dots*). *Red, blue* and *graded arrows* indicate overall direction of air flow. The vertical scale is exaggerated relative to the horizontal scale. The frontal surfaces are typically only a few degrees to the horizontal

After anything from one to several days, the cold front advances so far that it undercuts the warm front, forming an occluded front (Fig. 1.8). The low pressure center, which is often referred to as an extra-tropical cyclone, then typically becomes isolated within the cold air to the rear of the cold front. Denied access to the energy contained in the warm air, the low pressure progressively fills in and decays.

The Bergen Model was based exclusively on surface observations of clouds, pressure, temperature and wind direction. There was little access to data from greater elevations; therefore, the link to processes happening further up in the atmosphere was not understood. Nonetheless, the model has worked extremely well and accurately describes the processes occurring in frontal low pressure areas. For most scenarios involving the formation and development of low pressure areas, the Bergen model works more than adequately.

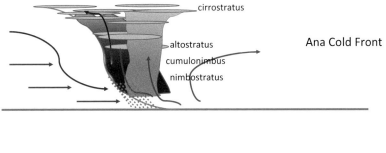

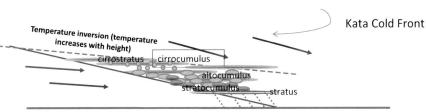

FIG. 1.7 Cross sections through idealized cold fronts. The *upper fig-ure* shows the stereotypical "Ana cold front" with strongly converging air and strong uplift along the front. Rain typically falls immediately after the cold front has passed from thick nimbostratus and occasion-ally cumulonimbus. In the lower half of the figure the kata cold front has air subsiding and warming aloft. This evaporates clouds, leading to lighter precipitation. Kata fronts often form when they are overrun aloft by upper level cold fronts (Fig. 1.13). However, they can form more gener-ally when front meander into areas of subsiding air. The vertical scale is exaggerated in each diagram with the slope of each front being at most 5–10° and usually much less

The frequently strong differences in temperature, humidity and pressure associated with the air in front of and behind fron-tal boundaries led to the concept of air masses. This idea is still taught today and is, generally, a worthwhile method of defining both different regional air types, as well as providing a predictive measure of the weather that will follow when one of these air masses advances overhead. Figures 1.9 and 1.10 show two other types of atmospheric front. Figure 1.9 shows an upper level front where changes in temperatures and pressure are only found above 5–10,000 feet. Many of these are formed when stratospheric air is sucked downwards into the troposphere. Figure 1.10 shows the "dry line": a boundary between dry continental air and moist mar-itime air. Such fronts often encourage the formation of vigorous thunderstorms during the spring in the US that are often associ-ated with tornadoes.

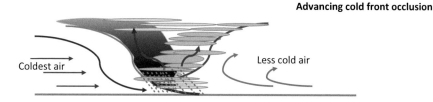

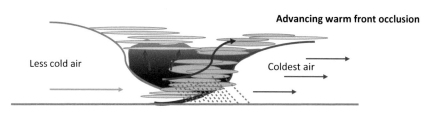

FIG. **1.8** The occluded front. The cold front almost always moves faster than the warm front, which it follows. This means that sooner or later the cold front will catch the warm front and lift the entire intervening warm sector aloft—and with it the rain bearing cloud. When the advancing cold air behind the cold front is the coldest air in the storm system the warm front and the cold air ahead of it are lifted en mass. The coldest air follows the front so this is called a cold front occlusion. Precipitation tends to be more intense and shorter in duration than in the opposite case when the coldest air lies ahead of the warm front. In the winter in the western States and the UK, air from the Pacific or Atlantic, respectively, tends to be warmer than the air ahead of it so occlusions tend to be of the warmer variety. Nimbostratus and altostratus dominate the rain-bearing clouds

The air mass concept has its limitations, however. In particular, the system of fixed air masses needs tweaking to account for the effects of direction of passage of each air mass and whether or not the air mass has passed over water with temperatures that differ from the bulk air mass. For example, dry continental polar air can only reach the UK via the North Sea. In winter the sea is considerably warmer than the air mass, so much like the passage of continental polar air over the Great Lakes, this dry cold air mass is heated from below and picks up considerable moisture. By the time it reaches the UK's eastern shores, it frequently drops showers of rain, sleet, hail or snow, depending on how much it has been modified by its passage over water. Similarly, maritime polar air, which starts out near Iceland, may take a fairly circuitous route to the west of the UK (or the west coast of North America from a starting point south of Alaska). If it travels far to the south before

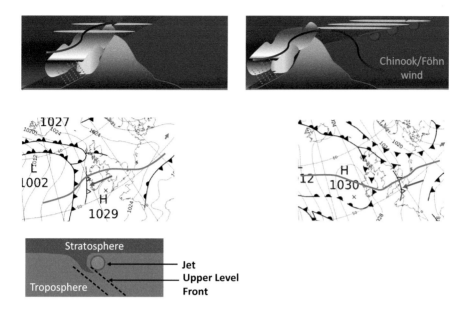

Fɪɢ. 1.9 Upper level fronts. Warm air flows over the mountains and with sufficient momentum it continues onwards faster at height than at the surface. This upper level front can continue forward while the remainder of the air mass is modified as it pours through valleys and over mountain tops. Air cools at a slower rate when moist than when it is dry, ensuring that it is far warmer in the lee of mountains than on the windward side. Such warm air is known as Föhn in Europe and Chinook in Canada and the far north west US states. Upper level fronts can also form when stratospheric air enters the troposphere at a jet stream. The synoptic charts show an upper level cold front (arrowed) crossing the UK bringing thunderstorms as it swept over the warm sector

arcing back northwards, the air is generally warmer than the waters over which it travels. Subsequently, the air mass is cooled at its base and becomes fairly stable. Rather than producing copious showers, such drab air brings banks of broken low cloud and only slight precipitation.

The Jet Stream

Between 7 and 16 km above the surface of the planet lie narrow bands of fast moving air. These air currents had been suspected as early as 1883 when Krakatau erupted, throwing several cubic kilometers of ash and sulfur dioxide into the air. It was apparent to observers that the ash was initially carried quickly from east to west along a narrow, equatorial band, referred to this as the

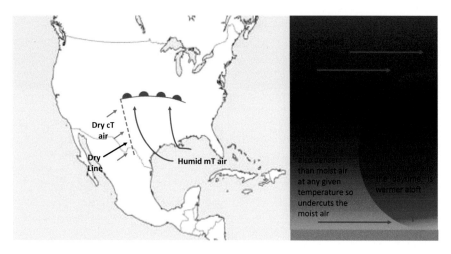

Fig. 1.10 A different kind of front. The *dry line* separates air with differing humidity. In the US, Argentina and India the dry line is responsible for the formation of severe thunderstorms in the Spring and Summer months. Dry continental (cT) air to the west undercuts and overrides warm moist air from the south. The undercutting action lifts and heats the air at the surface and generates instability. Because temperatures fall at different rates in dry and humid air any storm that is developing near the boundary may penetrate the overlying dry layer. As temperatures fall more rapidly with height (in the daytime) convection can rapidly intensify once the boundary is breached and warm moist air invades the cooler air above. North American map courtesy of http://www.vectors4all.net/vectors/north-america-map-vector

"equatorial smoke band". The significance of this band of rapidly moving air was not understood until the 1920s when Japanese meteorologist Wasaburo Oishi detected another strong band of air which swept westwards high above Mount Fuji during the winter months. Oishi released a flotilla of balloons, known colloquially as Pibals. These were allowed to drift upwards into the prevailing winds. During the winter months some were carried towards the United States.

Oishi's work fell victim to his desire to promote Esperanto and unwisely Oishi published his many studies in this artificial language. Unfortunately, few shared his passion for Esperanto and his work languished, with few readers outside of Japan. Consequently, it is the American pilot Wiley Post who is normally credited with the discovery of the Pacific jet stream. Wiley had invented a pressurized suit that allowed him to fly above 6200 m. During many such flights Wiley observed that he was zipping merrily across the surface of the planet at speeds far in excess of his air speed. Clearly, something was giving his aircraft a boost. The transient nature of the effect implied that these high winds were confined to narrow bands only a kilometer or two thick and less than a couple of hundred wide. In 1939 the term *jet stream* came into play, owing its origin to Heinrich Seilkopf who coined the term *Strahlströmung;* the literal translation of which means "jet streaming".

Air Masses

Air masses are defined as any large body of air with a specific temperature, pressure and humidity. In the UK we are afflicted by five main types; in the US six or more. These can be described as follows:

Maritime Tropical (mT) air: warm and humid air from the Azores (UK), Caribbean (US) or the Indian and Pacific Ocean (Australia and New Zealand). In the UK this air mass brings relatively warm summer weather, but with moderate to high humidity. In the winter, it usually brings cloud, damp conditions with above normal temperatures.

Continental Tropical (cT) air is dry and warm in the winter or hot in the summer. This is a rare but often welcome visitor to the UK in the winter when it brings clear, mild conditions. In the summer it is somewhat more prevalent and brings our hottest weather from Spain and North Africa. In the US and Australia this air mass originates over the continental interior, bringing the hottest and driest conditions.

Maritime Polar (mP) air: generally humid air that is cool in the summer and relatively cold in the winter. Usually associated with variable cloud and showery weather, this often follows the passage of the cold front. This air mass is most common in northwest North America, and Western Europe, as well as southern Australia and New Zealand. The Atlantic, Pacific and circum-Antarctic region sources this air mass.

Maritime Arctic (mA): a rare visitor to the UK in the summer, this brings cold, clear but showery conditions, along with some of the coldest, most snowy weather in the winter and spring. In the US this originates over the Arctic Ocean or the northern north Pacific and Atlantic oceans in the winter, along with air blowing across the sea from Greenland. In the southern hemisphere, maritime (ant)arctic air rarely afflicts Australia and New Zealand, being largely confined to the regions closest to Antarctica.

Continental Polar (cP) air: this is a dry air mass that originates in the continental interiors or Europe and North America. In the summer it brings warm, dry weather, but in the winter it can bring a deep dry cold. On the windward side of lakes (or in the UK along the east coast) cP air can bring extensive snow showers. cP air is responsible for extensive snows along the eastern shores of the Great Lakes, particularly in the early winter.

Continental Arctic (cA) air. Largely unknown in the UK, but a common part of the winter scene in Canada and the northern States, this largely seasonal air mass brings a deep and largely dry cold. Like cP it can bring snow along the windward sides of lakes, if there is sufficient warmth in the water to generate showers. Similarly, in continental Europe cA air brings deep, dry cold from Russia during the winter.

Initially, Oishi and Post's work implied one jet stream, located over the mid-latitudes, with the Krakatau eruption cloud suggesting another easterly jet. However, during the Second World War it soon became clear that there were numerous, distinct jet streams lying at different latitudes. Fighter pilots, undertaking long distance flights—particularly between the US and the UK—encountered various bands of high velocity winds which varied in position but were encountered most frequently at particular latitudes (Fig. 1.11).

These newly discovered belts of fast winds were employed to various effects during the Second World War. Aside from pilots trimming their fuel consumption, the most notorious use of the jet stream was Japan's aerial bombing of the Pacific northwest coast of the US between November 1944 and April 1945. Using Wasaburo's studies, at least 9000 incendiary bombs were

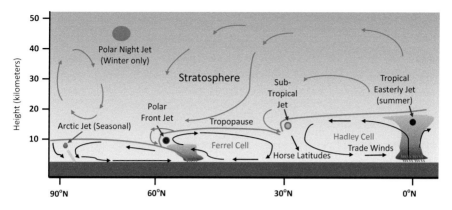

FIG. 1.11 General circulation in the atmosphere of the Earth below 50 km. The tropopause marks the boundary between the increasingly warm stratosphere and the unstable troposphere, which chills with height. *Black arrows* indicate the general direction of winds in the troposphere, while *blue arrows* indicate air flow in the stratosphere. At two locations along the tropopause the level drops sharply. Here, the greatest temperature differences are found and it is here that the two strongest jet streams—the polar front and sub-tropical jets—are located. Above the rising band of air nearest the equator a third seasonal easterly jet is found, while a fourth seasonal jet separates Arctic (or Antarctic) air from warmer polar air. Higher up in the stratosphere a fifth, seasonal, jet is located. Dry air can enter the troposphere from the stratosphere near the sub-tropical and polar front jets, giving rise to upper level fronts (Fig. 1.12). Not shown is air that leaks directly northwards, or southwards from the Tropical Hadley Cells to the Polar Cells

dispatched on stratospheric balloons. The westerly jet stream was used to guide these onto the forests of California and Oregon. Despite the prescient work of Wasaburo, his measurements of the westerly jet stream were too imprecise to allow an intensive campaign of bombardment. Rather than incinerate the western States, most of the devices fell short of their targets having taken longer than expected to reach the US coastline. Nonetheless, some of the devices reached the US mainland. This was to be the last international terror and it was to be the last international terror attack on the US mainland until September 11th 2001.

Over time the true extent and nature of the jet streams has come into focus, and with this improved resolution, a clearer understanding of the role these ribbons of air play in determining the weather (and climate) at the surface of the Earth. However, jet streams are ubiquitous in the atmospheres of all rotating planets and they even appear to have analogs inside many stars, including our Sun. It's worth asking why planets generate these peculiar high velocity streams of air, since they run counter to most of the other observed air flows. However, a closer look at the movement of air reveals that jet streams are an inevitable consequence on living on a rotating body.

Rotation, Rotation, Rotation

Jet streams are powered by two competing forces: pressure and temperature combined with the Coriolis Effect. The first of these returns us to the work of the French physicists Guillaume Amontons, Jacques Charles and Joseph Louis Gay-Lussac. Together they showed how air of different temperatures occupies different volumes and has different pressures. For any fixed mass of gas the pressure and volume will be greater at greater temperatures. So, as we observe, warm air expands because it particles are moving more quickly and tend to spread out. This gives the warm air a greater pressure than the same mass colder air. Therefore, pressure exists between the warmer tropics on the Earth and the cold Polar Regions. This force drives air from warm to cold areas.

Much as the air wants to move in this direction, the Coriolis Effect tends to block it, causing instead the air to move with the Earth's rotation. The Coriolis Effect overpowers the movement of warm air away from the equator, directing it into two broad streams at about 30° N and 30° S of the equator: these are the sub-tropical jet streams that overly the belt of high pressure areas known as the Horse Latitudes that were mentioned above. The Horse latitudes contain typically calm, hot and dry weather. It is said that the weather was so calm that the Spanish galleons on route to the New World were left with dead or dying horses that the crew could not keep on board. In the Horse Latitudes, these sick or dead animals would be tossed overboard. Vivid though this imagery is, a more likely origin of the term relates to the financial state of the ship crews early in their voyages. Sailors were not adept at managing finances and they would soon end up indebted to their ship's paymasters. The period over which this debt was repaid became known as the "dead horse time". This period, lasting one to two months, tended to coincide with the passage of the ship over the 30th parallel, hence the association. Regardless of the origin of the name, the weather in the Horse Latitudes has an unflinching banality that is only punctuated in some locations, at certain times, by the occasional tropical storm…

North of the Horse Latitudes, the winds blow northwards at the surface and are directed by the Coriolis Effect to the east. Where these winds encounter the cold polar air streaming away from the Arctic and Antarctic, there is a sharp difference in air pressure because of the sharp difference in temperature. It is this sharp transition in air temperature and pressure that generates the planets strongest jets: the polar-front Arctic or Antarctic jets. The rapid change in pressure causes air to move rapidly and the Coriolis Effect directs this rapidly moving air into bands of high velocity westerly winds. It is these jets that Oishi and Post encountered.

In general the Polar Front jet stream lies at a height of about 7–12 km above the Earth's surface, while the sub-tropical jet lies somewhat higher at 10–16 km. The seasonal equatorial easterly jet is the highest jet stream of the atmosphere's lowest layer the troposphere, although other jets are recognizable at greater heights within the stratosphere (Fig. 1.11).

Rossby Waves

This is where things get a little bit more complicated, but also crucially important when we start to look at the atmospheres of other planets. Rather than thinking about altitude in terms of kilometers and miles, meteorologists use pressure as a proxy for it. For convenience, meteorologists pick a pressure value and then plot the height above the surface at which this pressure is found (Fig. 1.12).

Looking at the atmosphere this way provides a kind of mirror image of the differences in pressure that you would find if you were looking at a set altitude. For example the 500 mb level represents the surface below which half the mass of the atmosphere is found, while at 300 mb roughly two thirds of the atmosphere's mass is below you. In general, meteorologists prefer looking at three levels in the atmosphere at differing distances above the Earth's surface. At the surface the average atmospheric pressure is 1000 millibars (1000 mb or 1 bar). This is also the nominal

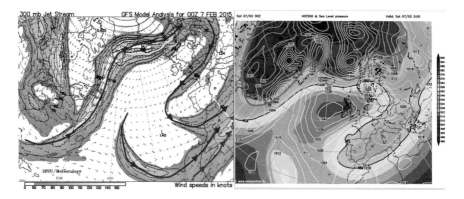

FIG. 1.12 Rossby Waves, jets and surface pressure. This view of the part of the Northern Hemisphere shows the "weather" at different heights in the atmosphere. Left, is a 300 mb view showing the jet stream—or more precisely jet streams—at the tropopause. The *light blue line* shows the Arctic jet, the purple shows a very convoluted Polar-Front jet and the red shows the sub-tropical jet just fringing the lower right of the image. Between the Arctic and Polar-front jets, over eastern Canada, is a fourth jet stream lying between both. The Arctic and Polar-front jets steer storm systems from west to east, while the sub-tropical jet lies above the belt of high pressure systems over the tropic of Cancer. *Arrowhead* indicate the direction of air flow along each jet. At right shows the 500 mb level and surface level pressure. Surface pressure features are shown in white, such as a large high pressure area, blocking flow across the Atlantic, while the 500 mb surface is shown in various colors. Warm (and raised high pressure) is shown in *orange* and *red*, while cold, and depressed surfaces are shown in *blue, purple* and *pink*. 300 mb chart courtesy of Californian Regional Weather Service; 500 mb chart courtesy of www.netweather.tv

"surface" used in extraterrestrial worlds that is used to describe their "sea level", particularly where there is no solid or liquid surface, such as in a gas giant (Chap. 7). Above the surface, meteorologists pick a second level at 850 millibars (850 mb). This pressure is found near 1500 m, or 5000 feet. At this level most of the effects of friction, such as buildings, hills and forests are lost and air is moving freely across most of the planet's surface. Next up is the 500 mb level, which corresponds to a height of roughly 5000 m, or 18000 feet. The 500 mb level represents the height at which air broadly flows along pressure lines, driven by differences in pressure and the Coriolis Effect: it is neither "diverging or converging" across these pressure lines. Air flow at this height also tends to be the driving force for most of the important surface features, such as storm systems at mid-latitudes such as

Europe, the bulk of Asia and North America. Similarly, in the Southern Hemisphere storm tracks across southern Australia, New Zealand, southernmost Africa and South America are driven by air movements at this altitude.

The final altitude that meteorologists concentrate on is found at 200 or 300 mb. This is the level of the polar front jet stream that organizes the flows beneath it and often drives the development and decay of storm systems. The jet lies nearer 200 mb in summer, when the air is warmer, and slightly lower at 300 mb in winter, when air is colder, denser and hugs more closely to the ground. This level is found at approximately 9300 m, or 30,000 feet (Fig. 1.13).

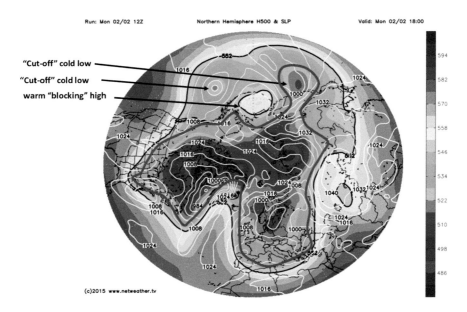

FIG. 1.13 A 500 mb chart for early February 2015. Rossby Waves are broadly defined by the edge of the *blue–purple–pink* regions centered on the Arctic Ocean (*outlined*, approximately in *red*, with *broader outline in black*). In this particular chart, three areas are identified that have been cut-off from their sources. In the Pacific there are two *blue* circulations (low pressure zones, sometimes called "cold-pools") and one, warm *yellow–orange* region—a blocking anticyclone. Elsewhere, note how the Rocky Mountains divert the flow of air from Seattle, south-westwards towards Florida before it heads back up the east coast of the US and out into the Atlantic. This is a characteristic feature of mid-latitude Rossby Waves on the Earth, where mountains divert and anchor the overall flow of air. This should also be true of other planets with varied surface terrain (Chap. 10). Modified chart courtesy of www.netweather.tv

The temperature of the air is critical in determining the pressure layer. Cold air tends to fall towards the surface because it is denser than warm air. Therefore, the 500 mb level is lower in cold air than it is in warm air. Warm air forms domes, while cold air forms troughs and depressions in the 500 mb surface. The air at 500 mb and above is clearly organized into a series of ridges and troughs which extend from the equator towards the Polar Regions and vice versa (Fig. 1.13). The troughs are colder at any given latitude than the ridges.

These troughs and ridges form features called Rossby Waves (Fig. 1.14). These waves can move westward or eastward, and the

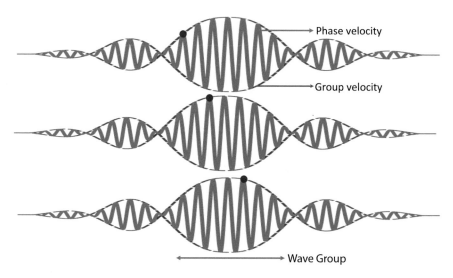

FIG. 1.14 Rossby Waves are a fairly complicated idea but one that we are all familiar with. They consist of waves of different wavelengths that can move in different directions. Imagine each blue wave is linked to its neighbor in a group (red dashed outline) then this group of waves will move together as a packet. This packet (or group) of waves can move towards the right (east) or towards the left (west). Larger groups of waves (those with the longest wavelengths) tend to move westward, while shorter wave groups move eastward. A low pressure area (indicated by a purple dot) moves with one wave inside the group. Rossby waves appear in the jet streams, as well as the Tropics and between the Tropics and the Polar Regions. Rossby waves are critical in organizing the atmospheric flow on all planets and will be encountered in chapters 5–10. In Chap. 10 we will encounter how Rossby waves drive a phenomenon known as super-rotation. Underlying wave image courtesy of Wikipedia Commons: Oleg Alexandrov and modified by book author

way they do this is determined by the spin of the planet, its geographical features and the temperature of the air below. Within the Earth's troposphere three types of Rossby waves are discernible. The most obvious waves lie along the Polar Front jet stream and organize surface low and high pressure areas into alternating east-west patterns. Another, very important set lie along either side of the equator and are visible as alternating bands of higher and lower pressure within the easterly trade winds. These extend in a north-south pattern that is paired across the equator. A final set of Rossby waves extends northwards across the pole. Both of these north-south trending waves are most obvious during El Niño weather phases (Chap. 2). A low pressure area over the central Pacific has a paired high pressure area to its northeast over the northeast Pacific (and also to the southeast of the equator). Continuing further northwards, over the North Polar Region, surface pressure tends to be low once more. This pattern was particularly evident during the 2015 El Niño.

Now, let's look just at the east–west trending waves. It has to be said that the idea that these waves can be moving east and west at the same time is probably confusing. However, things are not quite what they appear. To get a better idea of what's going on imagine throwing a rock into a pond. Instead of a single wave moving outward away from the impact point, you get a package of waves of different heights (amplitudes) moving outwards. Because these move together, they are called a wave group. At the front of the wave, smaller wavelets are moving forwards more slowly than the whole group of advancing waves. This means that they appear to move *backwards* into the body of the group, growing larger as they do so, while new small waves appear in their place at the front. Eventually the original forward-most wave passes to the rear of the group and grows smaller once more. Atmospheric Rossby waves are similar with one set of waves (those with the longest wavelength—well over 1000 km long) moving towards the west. Within these large westward moving waves are smaller waves (still a mighty 700–1200 km long wavelength) that move towards the east. Low pressure areas develop along their poleward face and move from west to east (purple dots in Fig. 1.14).

Although planetary Rossby Waves may be more complex in their structure and movement, the underlying rules apply, with smaller waves moving through the body of the group, growing first larger, then smaller as they traverse the group as a whole. In reality the longest wavelength Rossby waves tend to be locked by these geographical features and even without such barriers might not move westward or eastward. Meanwhile, the shortest wavelength features, associated with surface low pressure and high pressure regions, tend to propagate from west to east along the crests and troughs of the broader Rossby Waves. This is in the general direction of the air flow within the jet streams and is driven by them. However, on many planets the westerly-moving waves are free to propagate. On Earth the westerly movement is often hard to discern, except when the overall movement of surface lows is blocked by warm areas of high pressure.

If Rossby waves are aligned along the direction of wind flow then there isn't much of a problem as the mountains do not obstruct the path of the air. However, mountains, such as the Rockies, Andes and the Himalayas very much get in the path of the waves. As a result waves will become locked around these features. Perhaps most significant of these mountains are the Rockies. Air flowing upwards, across the mountains, bends towards the pole on the windward side, then bends towards the tropics, on the leeward side. This pattern is fairly weakly defined for much of the summer, when the continental interior is hot, although the jet still dips towards the south of the Great Lakes. However, during the winter the jet stream and the air that immediately underlies it clearly bows upwards towards the north of the Rockies, before bending south across the Great Plains and out towards the eastern seaboard.

A Method of Predicting UK Winters at the End of August

Every October—usually in the second week of the month—there is a "Coldest Winter Ever!" headline in a well-known British Tabloid. This is something of a British tradition, much like egg custard or coronary heart disease. Despite a dismal success rate the pattern persists year on year, much like the Asian Monsoon.

Most of the UK population have learnt to scoff at this, in part because it is invariably wrong; in part because it's an amusing tradition. However, I am fairly certain I know the origin of the Tabloid's erroneous system.

The system I use is well-trodden—or at least it should be—but there is little evidence it is put into practice (Fig. 1.15, below). I look at the pattern of Rossby Waves at the mid-tropospheric level of 500 mb in the summer or early autumn. By late August the circulation of air over the Arctic has formed patterns that is an excellent predictor of the pattern in the North Atlantic in the following winter. Where there is a well-established cold circulation over northern Canada and Greenland, the following winter is invariably dominated by westerly winds over Western Europe. The best recent example was the winter of 2013–2014, which brought a relentless tide of miserable wet weather to the UK, France and the Benelux countries. The UK suffered its wettest winter on record with severe flooding and extensive storm damage. The average pattern month on month during the autumn was of an increasingly well organized polar vortex. Conversely, in the preceding year in the summer of 2012 the polar vortex was poorly organized with warm blocking high pressure areas punctuating its flow. The winter of 2012–2013 was one of the UK's coldest in recent years with extensive snow through to late March. Other recent cold winters in the UK also follow the same pattern of air flow above the Arctic during the late summer and early autumn. Indeed, the more you look at the pattern the more straightforward the correlation appears. That is not to say that it has always held - only that it has over the last seven years - pretty much without fail (six years out of seven predicted correctly using this method)[2].

Does it apply to other regions? An obvious complication, for North America is the continental landmass which cools down dramatically in the autumn. While the UK, in particular England and Wales, drowned under the wettest winter

---

[2] A short paper entitled "An underlying predictability in the winter weather patterns in the North Atlantic Basin" will be published in the peer-review Journal "Hypothesis" by this author in the summer of 2016.

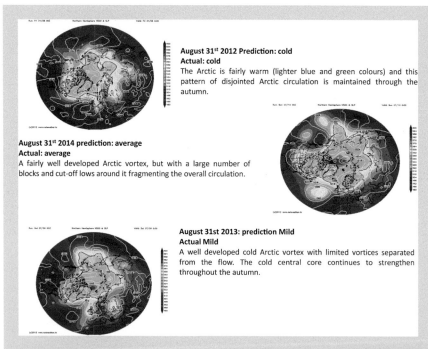

August 31st 2012 Prediction: cold
Actual: cold
The Arctic is fairly warm (lighter blue and green colours) and this pattern of disjointed Arctic circulation is maintained through the autumn.

August 31st 2014 prediction: average
Actual: average
A fairly well developed Arctic vortex, but with a large number of blocks and cut-off lows around it fragmenting the overall circulation.

August 31st 2013: prediction Mild
Actual Mild
A well developed cold Arctic vortex with limited vortices separated from the flow. The cold central core continues to strengthen throughout the autumn.

FIG. 1.15 A surprising link between the northern summer weather pattern in the Arctic and the ensuing winters in western Europe. See textbox for details

on record in 2013–2014, North America was periodically hit by a drifting Polar Vortex—the large pool of cold air normally centered over the Arctic. This brought temporary spells of severe cold and also helped pin the jet stream south of its normal position, which in turn fired storm systems directly at the UK, rather than allowing them to drive further north towards Iceland and north Scotland. Similarly, while the winter of 2014–2015 was downright average in the UK (a mixture of mild and cold), the eastern States were pummeled by a succession of severe nor'easters. The Arctic patterns should influence North America as it does here, but not necessarily in the same way. Undoubtedly, some sense can be made of the US patterns and link them to the summer pattern in the same manner, but not being as familiar with the minutiae of US meteorology I won't attempt to do that here!

So what of the Tabloid system, which appears to come from Exacta Weather? Is that just a fanciful guess that routinely is wrong or is there something to it? Well, it turns out there is probably some method there. If you look at the same system of Arctic air in October you get a picture which is similar to the one I use earlier in the season. It appeared that every October the Arctic goes a bit haywire for a while and appeared to show a jumbled circulation that earlier on is indicative of cold weather. Perhaps the Tabloid forecaster is looking at the pattern here and judging it the same way each year. Instead he or she should check the pattern in September and re-check in November when the October glitch clears...

Other websites make very strong claims about their success. One site claim to beat the UK Met Office. However, one of their June forecasts for July 2014 (http://www.weatheraction.com/resource/data/wact1/docs/BI1407JUL30dSlat9cWowContrastsProd27JunTypCor.pdf) (for the UK) painted a picture of "deluges", "tornadoes" and torrential hail. July 2014, in the UK, was a rather warm, dry month—something I got right with the method described above. Similarly, claims of an astounding 52 % success rate for gales (in 1 year) don't impress when by chance alone 50 % might be expected. Be very careful of cherry-picked data. I'll be clear I may get winters largely correct (6 out of 7) but my success rate for predicting UK summers is hardly spectacular (currently about 2 out of 3), so take my forecast for July 2014 with a pinch of salt; ditto that of others...

In the winter, when the continental interior is cold, the jet bends sharply to the Gulf of Mexico. It may break into one stream tracking back along the Appalachians, while another branch develops over the Gulf and tracks north eastward along the eastern seaboard. Thus the topography and temperature of the North American continent strongly anchor the upper level Rossby wave, producing either a mild trough in the summer or a much stronger trough during the winter.

Similarly, the Himalayas divert a branch of the westerly polar front jet stream to their south during the winter months. This helps direct cool, dry air over India and maintains the dry winter monsoon. When the spring comes and India heats up, this southerly jet stream is trapped and is eventually broken up. All the while a second branch tracks east, along the northern edge of Tibet, towards Japan, It was this jet stream that Oishi identified and, later, the Japanese army used to deliver bombs to California and Oregon. Free from a mountainous anchor, this northern branch is also free to move. Thus when Asia moves into its summer, the jet drifts northwards, eventually leaving Japan stranded in warm, tropical air.

In the Southern Hemisphere only the Andes really get in the way of the polar front jet. There is a greater effect from ocean currents, in particular the cold Humboldt Current that flows up the western side of South America. The polar front jet is diverted northwards towards the equator in the eastern Pacific before turning back southeastwards over the southern Andes and out into the southern Atlantic. In general it only clips the southernmost shores of Australia during the coldest months of the year, but it frequently swings north eastwards over the Tasman Sea to New Zealand, before heading out over the Pacific. This configuration favors the formation of storm systems to the west of New Zealand, during the winter.

Other Jets That Drive Terrestrial Weather

In addition to the broad, upper westerly jets and the Rossby waves that they contain, are a myriad of lower level and seasonal jet streams. These bands of concentrated airflow are defined as existing below 700 mb or about 12,000 feet (3500 m). The definition is a little wobbly as jets at 600 mb are well below the height of the main jets that blow at the top of the atmosphere's lowest layer, the tropopause. Perhaps these are mid-level jets?

Low level jets are found across the planet, particularly, but not exclusively in the tropics, where they strongly influence the seasonal patterns of rainfall. Normally, across the equatorial regions winds are deflected minimally by the Coriolis Effect, which means that there are only weak (horizontal) forces directing rainfall. However, because the Earth is tilted on its axis the area that is most strongly illuminated by the Sun moves north and south between the Tropics of Cancer and Capricorn. On June 22nd the Sun shines directly down on the Tropic of Cancer, leading to the strongest heating, here. Conversely, in the southern summer, centered on or around December 22nd, the Tropic of Capricorn is most thoroughly heated. Wherever the Sun shines most intensely, the belt of tropical thunderstorms follows: this is the inter-tropical convergence zone, or ITCZ. However, particularly in the Northern hemisphere, where there is a lot of land, there is a lag between the Sun being overhead and the highest temperatures on the ground. Therefore, instead of June being hottest in the Northern Summer (or December in the south), July (or January in the southern hemisphere) is the hottest.

Concentrating on the northern hemisphere, July sees the peak of the summer (wet) monsoon over southern Asia and over West Africa. It is also the season where most rain falls across southwestern North America. On each continent the arrival of a low level jet stream marks the onset of the peak rainy season.

As summer advances, the belt of equatorial easterlies drifts northwards, strengthens and begins to be influenced by the Coriolis Effect. At the surface the easterly winds begin to bend towards the west, arcing around the southern tip of India and northeastwards towards the southern Himalayas. However, that isn't all. The strength of the Asian monsoon is linked to the strength of another low level jet stream: the Somali jet. The jet is a consequence of the topography of eastern Africa. Alongside the Great Rift Valley lie the Virunga Mountains and the Ethiopian Highlands. As the sun moves north and winds cross the equator winds from the southern hemisphere initially move to the west south of the equator, before, as we have seen then moving north eastwards towards India. The mountains that flank the Rift Valley (in particular the Ethiopian Highlands) intercept the low level winds and channel them into a narrow (low level) jet stream. This, too, is diverted by the Coriolis Effect to the east as it continues to travel northwards towards India from the eastern coast of Somalia (Fig. 1.16).

High above all of these low level features a strong easterly jet forms over northern India. This jet acts like an exhaust system, removing the air that has been dragged across the surface towards the continent. The combination of the Somali low level jet and the upper easterly jet serve to guarantee the strength of the vital summer monsoon. Without the integrity of these jets, India would dry out and the rice paddies that fuel the country wither and die.

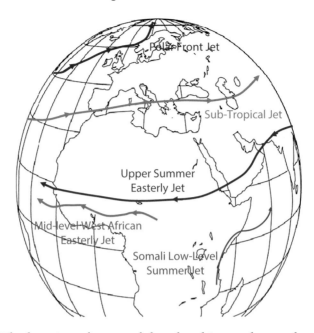

Fɪɢ. 1.16 The location of seasonal, low-level jets and upper level jets over Africa in the summer. These jets, perhaps more than any others affect the well-being of the greatest number of people on the planet—either directly over Africa, or across Asia where they influence the summer monsoon rainfall or in the Americas and Pacific where they influence the formation of hurricanes and typhoons. The underlying African map is courtesy of Bruce Jones Design Incorporated

Meanwhile, back over Africa to the north of the seasonal easterly jet, over the Sahara, and somewhat further north, is a seasonal westerly jet. Like the easterly jet over northern India this easterly jet appears once the African interior has heated up—particularly the southern Sahara. Air above this area begins to rise strongly, helping draw moisture northwards from the African coastline and into the Sahel. Here, it helps deliver welcome (if erratic) summer rains. Like the Indian easterly jet the seasonal African mid-level easterly jet and the more northerly westerly jet act like an exhaust system, carting away the warm air that has been dragged in across the surface. This westerly African jet ultimately directs moisture eastwards into the interior of Asia where it influences the summer weather of Russia.

Much like the Somali jet, another low level jet develops over southwestern North America during the summer months. Again, this jet—diverted this time by the Rockies—brings moisture to the south western States. In turn this helps drive the formation of thunderstorms and larger clusters of storms that go by the grand term, mesoscale convective complexes. Mesoscale systems simply refers to size: these complexes of storms vary in size from 100–500 km across and are variously found in the tropics—forming the cores of easterly wave ridges—or storm systems that spawn tornadoes in the Mid-Western States in the US.

To the east of India much of the summer rain is concentrated in a narrow band called the Mei-yu (China) or Baiu (Japan) front (Fig. 1.18). Here, another south westerly jet over southern China meets the westerly winds blowing out of central Asia—winds that may have started out with the monsoon rains over West Africa. As the two air flows meet, storms are generated which move towards the north east. Most of the heavy rain and thunderstorms lie to the south and east of the front where very humid tropical air is moving northwards. North of the front lie belts of low stratus cloud and fog, with only light precipitation. Although the Baiu front owes its origin to the tropics, it is structured very like a mid-latitude cold front, with colder air to the north of the front under-cutting the warmer air trying to advance from the south. Just like a cold front, waves develop along the front at regular intervals, probably initiated by waves running out of the tropical easterly flow. These are foci of more intense precipitation bringing regular pulses of heavy rain to northern China, southern Korea and Japan. The Mei-Yu front persists throughout the summer months before retreating back south towards Indonesia in the autumn, where it decays away.

Meanwhile, over western Africa an easterly jet develops that has a more insidious effect. As well as usefully directing moisture and storms from the Cameroon coast, northwestwards, this mid-level jet also directs the movement of squall lines called easterly waves. These begin as clusters of storms over West Africa, which are embedded in the broad easterly flow. As these continue to grow they are driven off the coast and out into the Tropical Atlantic. From here, some become sufficiently well organized that they develop into circular storm systems. It is these, from July to September (or occasionally later) that may develop into Atlantic Hurricanes. Thus there is a clear link between the success of the rainy season in West Africa and the strength of the Atlantic hurricane season.

Aside from bringing hurricanes, easterly waves bring the Amazon, the Caribbean and Florida another benefit. Easterly storms often kick up dust storms over the Sahel and southern Sahara. Until the advent of comprehensive satellite monitoring, it wasn't appreciated that these storms frequently carry Saharan sand all the way across the Atlantic before dumping it on the ocean's western shores. Without these easterly propagating storms, the beaches of Florida and the Caribbean—or, indeed, the lushness of the Amazonian rainforest, for which they fertilize—would be severely depleted (Fig. 1.17). One reason that these waves can reach

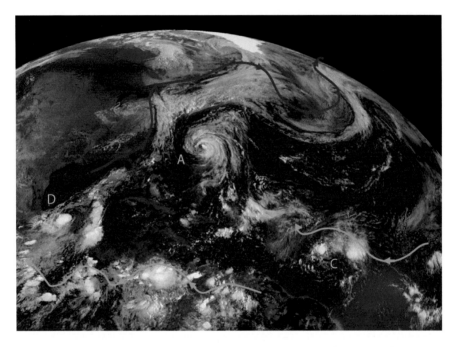

Fig. 1.17 NOAA's GOES-13 satellite took this image of 4 tropical systems in the Atlantic on September 8, 2011. Hurricane Katia in the western Atlantic between Bermuda and the U.S. East coast (A); Tropical Storm Lee's remnants affect the northeastern U.S. (B); Tropical Storm Maria in the central Atlantic (C); and newborn Tropical Storm Nate in the Bay of Campeche, Gulf of Mexico (D). Credit: NASA/NOAA GOES Project. The location of the polar front jet is also shown in purple and easterly waves in turquoise. You can see where tropical air escapes northward into the westerly flow near Nate in the Gulf and where Lee's remnants impact the polar front jet

across the full width of the Atlantic is the presence of another low level, seasonal jet. This one is very low level, indeed. Extending from the surface to around 1500 m this Caribbean Low Level jet directs easterly waves into the Caribbean where they bring summer rains—and less fortunately hurricanes. The jet also supplies extensive moisture towards the Caribbean coast of Central America, where the rains sustain the rain forests of Belize, Costa Rica and the Caribbean coast of Venezuela.

The easterly waves behave in much the same way as the Rossby waves further north. The moist air pouring in across Western, equatorial Africa during the summer monsoon season is much cooler than the hot Saharan air to the north. The difference in air temperatures generates a strong pressure difference and leads to the development of the easterly jet stream described above. This is because the air is moving towards the west as it crosses the equator and heads towards the center of the Sahara. A major difference between the westerly jets and this easterly jet is the pressure surface at which it occurs. The westerly jets occur at the 200–300 mb level—which lies between 10,000 and 13,000 m— or the tropopause. The equatorial easterly jet occurs at a lower elevation—approximately 4–5000 m above sea level. At this level the density of air is greater and there is a more direct drive on the storm systems that develop over Africa and have cumulonimbus towers that extend upwards of 8000 m.

Meanwhile in the Southern Hemisphere, during the summer months, a low level jet transfers moist, warm air from the Corral Sea across Australia towards low pressure areas over the continent's southwest flank. Again, this can drive the development of mesoscale storm systems, just as the analogous jet does in the southwest US.

Both the Somali jet and the US jet owe their existence to the presence of mountains. In each case the north–south aligned ranges intercept the east–west flow of tropical air and divert it northwards. Without the activity of these jets normally dry areas would become deserts and are otherwise sustained by the life-giving rains these low level jets bring. That said, the activity of the US jet has been marked by its absence (or at least its ineffectiveness) and a prolonged drought has afflicted the western States. This highly unfortunate state of affairs has been linked to a persistent wobble in the Polar Front jet to the north. Instead of migrating the overlying Rossby wave has led to the persistence of a ridge of high pressure over these States, which has served to maintain the dearth of rain.

How the Jet Stream Brought a Sting to the Bergen Frontal Model

At the time of its conception the Bergen frontal model was based exclusively on surface pressure, wind and temperature measurements. Although it was obvious such surface features were linked to processes occurring higher in the atmosphere, there was no means to determine what was going on at these upper levels. The jet stream, or rather the jet streams were unknown except in outline. Therefore, accurate though the Bergen model is, it does not fully describe meteorological phenomena, much like Newton's Law of Gravitation does not describe gravity in extreme environments.

Over the decades since the Bergen model was synthesized a number of storm systems have required meteorologists to re-think what is going on. In particular a class of mid-latitude storms is known that appear to "explosively" develop, with a rapid decrease in pressure and narrow walls of extreme winds that defy a straightforward explanation. Three of these are noteworthy, but by no means unusual: the 1979 Fastnet Storm; the 2011 New Zealand "Bomb"; and the 1993 US. "superstorm". Although quite distinct storm systems in their own rights, each system shared a common feature: a rapid drop in their central pressure alongside the development of hurricane-strength winds around the core of the low pressure core.

Fastnet

In August 1979, the international yachting race, the Admiral Cup, was underway in the English Channel. The 605-mile course ran from Cowes, direct to the Fastnet Rock and then west to Plymouth, via the Scilly Isles. The Fastnet stretch was the climax of the race and was, at the time, one of the most prestigious races of its kind.

On August 11th a small low pressure area developed to the south west of Ireland and began intensifying rapidly as it went north eastwards towards the Irish Sea. Winds of 50–60 km per hour had been forecast, significant in their own right, particularly for the month of August. However, by August 13th winds continued to strengthen to near hurricane force—storm force 10–11 (80–100 km per hour). Caught in this cabal were 306 yachts. Five yachts were sunk outright, with a further 100 suffering knock-downs (when the mast and sail are tipped over at 90° or more and the sail fills with sea water) and 75 were flipped over. Over the ensuing 24 h 15 sailors lost their lives and 125 required rescuing. Of those 306 yachts that began the race—only 78 were to complete it. The racing disaster forced a re-think of the industry's safety procedures.

Although the storm was fairly intense (979 mb) it was not extraordinary. However, the Fastnet storm, as it is now known had, quite literally a sting in its tail. Lurking a hundred or so kilometers behind the cold front was a comma-shaped cloud swirling around a largely cloud-free zone. Analysis soon showed that this cloud-free feature was created by a rapid downburst of air from the middle and upper levels of the troposphere. This downward-directed jet is known as a sting-jet and is a frequent cause of extreme winds in the mid-latitudes. These sting jets are not explained by the conventional Bergen model—and this is hardly surprising as the meteorologists of their day had no access to information from higher altitudes.

The effect of the sting jet is twofold. As we have seen it rapidly strengthens winds—in that case of the Fastnet storm, throughout a region around 150 km across. This is caused by the rapid descent of cold, dry stratospheric air. However, the jet also punches through the cold front and dislocates its southern portion from the rest of the storm. Rapidly descending cold air essentially drives the cold front so quickly that it moves independently of the rest of the storm. The warm front then wraps around the center of the low pressure forming a second front with the characteristics of an occlusion to the storm's rear. Over the space of a few hours, the cold front and the associated low plow through the storm's warm sector while the trailing occlusion wraps around the core of the low, its heart driven by the descending flanking column of cold air. It is along this curling frontal boundary that the strongest winds are found. The standard Bergen model fails to predict this kind of explosive behavior. However, a second model, developed some 70 years later does fit the bill. The Shapiro-Keyser model, as it is known, was developed in 1990, and adequately explains these explosively developing storms. Warm air is trapped in a pocket at the lows core, which is aggressively undercut by cold air that is surging in on its southern flank. Much like a tropical hurricane, within the warm core, air can ascend rapidly. This causes a very rapid deepening of the low's central pressure, which, in turn, intensifies the winds that wrap around it.

The New Zealand Bomb

In the Southern Hemisphere, a very similar set-up drove the explosive development of a storm over New Zealand in March 2012. This storm brought fairly widespread damage to the southern end of New Zealand's North Island. As the low, knick-named "the bomb", approached from the south west during the early morning of March 3rd, winds picked up to over 80 km per hour, with one gust measured at over 120 km per hour at around 5 am. After 6 am all wind speed readings ceased at Hawera station when the site suffered a power cut as a result of the storm. Meanwhile to the north of the storm center, warm air dragged over the mountains underwent the Foehn Effect. Warm moist air cools more slowly and warms more slowly than dry air (4–9 °C per kilometer versus 9.8 °C per kilometer, respectively). As the air moves over any mountains, the moisture is wrung out of it, leaving it bone dry. Condensation also releases latent heat further heating the air mass on its ascent. If the air cools by 10 °C when it ascends the windward side, it may then warm by 13–21 °C on the leeward side, depending on the amount moisture and its original temperature. Thus to the west (leeward side) of the central chain of mountains, stations recorded a sharp rise in temperature as the storm approached and tracked away to the north east.

The 1993 US Superstorm

Although the 2011 New Zealand storm was fairly severe, the 1993 "super-storm" that afflicted the eastern seaboard of the US was altogether more infamous and for good reason. This storm became a classic "Nor'easter": a storm that while positioned off the eastern seaboard, drags very cold but moisture laden air from the Sea of Labrador. Therefore, in addition to severe coastal gales, nor'easters also bring a dreaded harvest of snow during the winter and early spring months.

The 1993 superstorm began its life rather innocently over Cuba as a squall line on March 12th. Moving north eastwards from the Gulf of Mexico, and driven along by a strong jet stream, the storm rapidly intensified and brought devastation to many central and eastern States of the US. in variety of distinct ways.

The key to understanding this storm's explosive development is the geography of the eastern US. During the winter months the continental landmass cools rapidly and the bulk of it has a mean temperature well below zero. Meanwhile, drifting north along the eastern seaboard are the warm tropical waters of the Gulf Stream. When this particular storm struck, in March, the continental interior was still very cold and the contrast in temperatures between the land and sea, severe. As the storm developed and moved north east with the overlying jet stream, it dragged very warm, humid air with it. During the 13th of March this mass of warm, humid and unstable air was progressively undercut on its western flank by the cold continental air. Such a configuration allowed the development of a very active cold front and a rear-side sting jet that drove its northern portion rapidly to the east. Like the Fastnet and New Zealand storms, this set-up drove explosive development of the storm. By the time the low center passed Washington DC it was packing storm force winds, with gusts up to 180 km per hour.

At its peak the storm stretched from the Florida Everglades northwards to southern Canada. Entrained to its rear was such cold air that snow fell as far south as the Florida Panhandle: 10 cm were recorded in places. Further north, in Alabama, snow fall exceeded 40 cm. Indeed, Florida's snow was the most severe since 1899. Further north, the Superstorm dumped over a foot of snow over much of the north east of the continent. Snowfall totals exceeded 1 m over the Appalachians, with blizzards whipping depths to over a whopping 11 m in places. Indeed, calculations suggested that up to 27 billion tones of snow fell over the day it took the storm to pass.

As well as widespread hurricane-force winds, the active cold-front spawned several lethal tornadoes. The combination of tornadoes and storm surges across north-western Florida, dozens of people were killed in that region alone. From Texas north to Pennsylvania blizzard conditions were often mixed with thunder and lightning giving rise to the fairly exotic phenomenon of "thundersnow". Here, vigorous convection along the advancing cold front produced thunderclouds—cumulonimbus—while the undercutting cold air ensured that the precipitation was in the form of snow rather than the usual rain or hail.

In the storm's immediate aftermath ten million people went without power as storm force winds and heavy snow brought down power lines. In all, 318 people were killed with over 40 % of the country's population feeling the effects of the storm. Many thousands of homes remained without power for over 3 weeks. The diversity and spread of damage, the number of people affected and the severity of snowfall—in places exceeding an entire winter's average—ensured that the 1993 super-storm will be remembered as the storm of the century, for many.

Like the Fastnet and New Zealand storms, the systems power was enhanced by an upper level jet directing cold air rapidly downwards towards the ground to the rear of the cold front. As the storm tracked northeast the cold front accelerated eastwards. With rapid uplift of the warm air to its east, there was ample opportunity for the initially shallow low to explosively deepen. The track of the storm ensured that warm air was always available along the system's eastern flank as the upper level jet drove it north east. In the end this favorable track, coupled to the very constructive topography of the eastern United States ensured that the system had ample power to devastate the eastern States.

The Tropics: A Quick Guide

Let's now leave the chill of the mid-latitudes behind and examine the weather over the Earth's tropics. Arguably, weather in the Tropics is a simpler affair than more southerly and northerly latitudes. Weather is driven predominantly by convection—the rise and fall of air that is dependent on its temperature and relative density. In the tropics, air flows from the flanking high pressure areas, centered around 30° north or south of the equator. The core of the region is a belt of calm conditions and severe thunderstorms, famed as the Doldrums. However, this apparent simplicity is deceptive. Aside from the seasonal northward and southward migration of the storm belt that forms the system's core, the storms within the belt are not continuous. Rather they form distinct pockets that migrate from east to west, within the general airflow: some of these are the easterly waves.

However, that's not all. Lurking within the general easterly flow of air lies a pattern of variability akin to the Rossby waves embedded within the mid-latitude jet streams. Named after its discoverers, Roland Madden and Paul Julian, the Madden-Julian oscillation (MJO) is a slowly propagating wave that moves eastward against the prevailing wind. These waves begin their life in the Indian Ocean, before moving slowly towards the Pacific at 14–29 km per hour. As these waves move, they take with them a region of enhanced convection on their leading edge, much like easterly waves. Linked to them, but with a shorter wavelength and a faster easterly velocity are so-called Kelvin waves. These regions have enhanced convection associated with their leading edge as well as strong westerly winds lurking in their rear. These features often appear to kick-start MJO events, but with faster movement—up to 70 km per hour— pull away in front of MJO waves over the Pacific.

MJO waves manifest themselves first with enhanced rainfall, initially over the western Indian Ocean. This is followed by a period of increased drought. In most cases they die out over the cooler waters of the Eastern Pacific but on occasion they travel right around the globe, weakened in the Atlantic, before re-strengthening once again in the western Indian Ocean. Waves emerge and propagate roughly every 1–2 months, predominantly during the northern summer months. There is some hint that these waves (unlike their speedier Kelvin Wave cousins) may help initiate El Niño events (Chap. 2), but why some do and some don't is as yet unclear. Certainly, the west to east movement of MJOs would help encourage a reversal of the normal east to west pattern of wind and rainfall, generating the reversed pattern that characterizes the El Niño.

On extraterrestrial planets similar tropical waves are thought to play a crucial role in the development of a phenomenon known as super-rotation, where the equatorial atmosphere moves more rapidly from west to east than the ground (or deeper atmosphere) underneath (Chap. 10). Moreover, there are also some suggestions that pairings of equatorial Kelvin Waves and westward moving equatorial Rossby Waves[3] not only kick-start MJO events but help initiate tropical cyclones. They may also initiate trains of storms that pummel the northern, mid-latitude, Pacific coastline during winter months.

In the Pacific warm waters are driven towards the west by the prevailing winds. If the Earth lacked landmasses, this water would simply spread out evenly around the tropics near where heating was strongest. However, Australia, Indonesia and Asia get in the flow of this warm surface water. Therefore, it backs up in the Western Pacific, with a similar, though less extreme parallel development in the western Atlantic and Indian Oceans. Where the warmest waters are you find the greatest extent of convection. Warm, humid air rises most strongly here leading to the highest rainfall totals—in most years. Over the western Pacific you also find the greatest preponderance of tropical storms. These roughly circular areas of low pressure may start out as easterly waves over the Atlantic, before pushing westwards across the Pacific. In the summer months, when these easterly waves are located furthest from the equator they are influenced by the Coriolis Effect and can become organized into circular storm systems. Over the western Pacific where the summer monsoon rain belt intercepts these waves the stage can become set for the rapid development of storm systems. When these develop winds in excess of storm force 12 (110 km per hour) they take on the name "Typhoon"; in the Atlantic "Hurricane", "Cyclone" in the Indian Ocean and "Willy-Willy" to the north of Australia.

Such tropical storms afflict the western edges of the Pacific and Atlantic in particular, but as northern storms continue to move away from the equator they become entrained in the overall

---

[3] Equatorial Rossby Waves are indicated on surface pressure charts by pairs of high pressure and low pressure that alternate along the equator and move from east to west like their larger mid-latitude cousins. Each low or high to the south of the equator has a matching high or low (respectively) on the northern side.

northerly flow of air and enter the mid-latitudes. Here they deliver tropical warmth and moisture to countries as far north as Canada and the UK. A similar, but less extensive process happens with Australasian cyclones that enter the southern mid-latitudes north of New Zealand.

The tropical storms are an obvious deviation from the regular pattern of calm simplicity. However, they are not alone. Regional topography plays a crucial role in the development of other tropical storms. Over Sumatra, during the summer, easterly winds deliver moisture and vigorous storm systems develop over the highlands that run along its length. These storms become organized into waves as they move out over the warm waters of the Straits of Malacca, towards Singapore and south western Malaysia. These Sumatras typically develop during the evening before slamming into the coastline, bringing overnight storms. Sumatras bear a striking resemblance to a number of other terrestrial and otherworldly storms and we will hear more about these later (Chap. 7).

Finally, within the tropics and between the tropics and the poles are Rossby waves. These are evident as paired areas of low or high pressure that flank the equator. These waves not only organize convection along the equator but help transport momentum to and from the equator. They will come into their own in Chap. 10, when we look at the circulation of the atmospheres of hot Jupiters.

Interconnections

The tropical weather systems and those of the mid-latitudes and Poles talk to one another in a number of ways. High above the surface, at the level of the tropopause, air escapes the tropical Hadley cells and leaks northwards into the mid-latitude Ferrel Cells and the Polar Cells. At the surface, particularly during the summer and early autumn, winds blowing around the limbs of the Horse Latitude high pressure areas bring tropical warmth into the continental interiors of China and the US in the northern hemisphere, and across Australia and New Zealand during the southern summer. These airflows carry tropical disturbances, including hurricanes, across more polar latitudes. In the summer of 2014, the remains of

Hurricane Bertha swept across the Atlantic towards the UK. Along the way the decaying storm became embroiled with a mid-latitude frontal system and morphed into a rather aggressive frontal storm. Although it was greatly reduced in its severity, it still packed a lot of tropical energy. Fronts crossed the UK, from Southern Scotland to northern France. Having driven south in it, I can't emphasize enough the effect of all that tropical heat. Although winds were not severe, the rainfall was impressive. Leaving Glasgow and heading south over the Southern Uplands, the sky was nearly black, with waves of torrential rain that made driving a rather "interesting" experience. By the time we reached the Midlands, in the heart of the UK, the rain had cleared east and a very blustery, autumnal airflow sent clouds briskly across the sky.

Although Bertha did take a fairly southerly track, the passage of hurricane remnants to the north of the UK is a more common experience in September and October. Such storms are largely welcome across much of the UK as they usually redirect the prevailing winds, bringing considerable warmth from the south. As autumn begins to tighten its grip, these northerly-tracking extra-tropical storms bring a return to summer temperatures across Western Europe.

Other connections include the link between Africa and central Asia. As was mentioned earlier, air that began life over West Africa may also be driven north eastwards towards southern Russia during the summer months. This can bring excessive heat and in some instances violent summer storms.

During the autumn and winter months cold fronts often penetrate far to the south over Africa. Although these were previously mentioned, their role in transporting sand and dust all the way to the Americas cannot be understated. This delivery system affectively cycles air from the mid-latitudes to the tropics and back again over North America.

A more limited but sinister link connects South America and North America. During the spring, when parts of the Amazon Basin are set ablaze by farmers clearing forest for farming, smoke fills much of the basin then becomes entrained in air moving north across the Gulf of Mexico. Much of this activity is illegal but goes on nonetheless, because the world demand for cheap beef continues to rise with its growing population. Other forest areas are cleared for palm oil or other cash crops.

Now, if you are sitting back watching the television (or reading this book) and thinking that none of that matters to you, then think about this piece of research. Published recently in Geophysical Research Letters, Pablo Saide, an atmospheric scientist at the University of Iowa, and his colleagues, used NASA's Aqua satellite to monitor land clearance in the Amazon. Analysis showed that when fires were lit across the Amazon Basin, the smoke became entrained in air moving northwards across the Gulf of Mexico and into the heartland of North America. Rather surprisingly, the amount of smoke in the air traveling north over the mid-west clearly enhanced the severity of the April 2011 tornado outbreak: the deadliest on record. Modeling revealed that black smoke particles from the burning Amazon enhanced low-level cloud development over the Mid-West; and enhanced low level wind-shear both of which are known factors in stimulating the formation of tornadoes. During the spring months this warm, humid air was given an extra kick from the aerosols hidden within it. That cheap side of beef might just cost you more than a few bucks…

However, it must be stressed that aerosols have a variety of effects. While they appear to ramp up supercells (and thunderstorms in general) conversely, sand storms, over the Sahara appear to depress the severity of Atlantic hurricanes. Over the Indian Ocean aerosols decrease the intensity of the south west summer monsoon and the easterly jet that overlies it. Ironically, this allows for a greater intensity of tropical storms to form over the Arabian Sea before and after the summer monsoon. This is apparently the result of a weakening of the upper level easterly winds and the southwesterly monsoon winds as a result of the effects of smoke and other aerosols. The role of these aerosols is discussed more fully in Chap. 2.

From September onwards the Asian monsoon swings into reverse gear. As the vast bulk of Asia cools down during the autumn, the westerly winds re-establish far above it. Over Siberia a vast pool of frigid air develops that pours outwards, south east-wards across China, India and Indonesia. By December this air has flooded as far south as north Western Australia and the bulk of the Indian Ocean. Over Indonesia and the Philippines the winds bend from a northerly to a more easterly direction. As this air passes over the warm waters of the western Pacific the air becomes unstable and generates a lot of rainfall particularly across the Philippines, Borneo and eastern Indochina. Indeed, much more rain falls over this segment of easternmost Asia during these months than it does throughout the summer monsoon. This humid cool air mass then delivers rainfall along the northern territories of Australia. Some of the air is sufficiently unstable that it generates tropical storms, but the majority falls in smaller, so-called mesoscale systems, similar in size and severity as the easterly waves that drive west from Africa and into the Pacific.

Although the frigid air mass is considerably tempered by the passage across the warm Indian Ocean, it serves to bring air from the Arctic all the way south towards the Antarctic Ocean. Although less extensive, the same process happens over North America. Cold, Arctic air floods southwards behind storms, which are moving up the continent's eastern seaboard. This Arctic air, again modified by its journey, crosses the Caribbean towards the northern edge of South America.

In the Southern Hemisphere the lack of significant land masses spanning several degrees of latitude limits the movement of air in this manner. However, air still enters the tropical circulation from the polar westerlies to the east of New Zealand and the east of South America. In particular during the summer months a strong area of converging air is found stretching south east from Papua New Guinea to about 30° S, 120° W. This region, known grandly as the South Pacific Convergence Zone is distinctly tropical in nature at its western end.[4] However, as you track to the south east, westerly winds increasingly mix into the tropical circulation and instead of spawning tropical disturbances generate broad mesoscale storms in a manner analogous to the Mei-yu/Baiu front over Eastern Asia (Fig. 1.18).

There is one more connection worth mentioning from its influence on the US and the UK. The Madden-Julian Oscillation (described above) has a rather sinister connection to extreme winter rainfall patterns in these parts of the world. In non-El Niño years the MJO can generate large clusters of tropical thunderstorms

[4] There is an equivalent zone of convergence over South America during the southern hemispheres' summer. Converging air in this zone brought severe flooding during the 2015–2016 El Nino event.

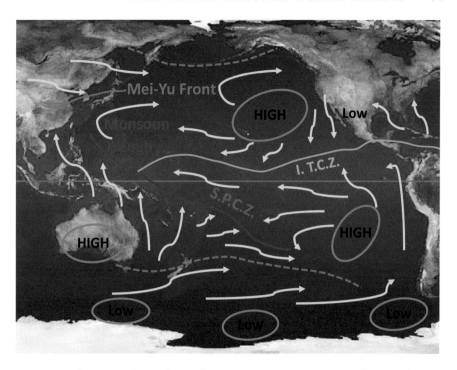

Fig. 1.18 The major boundaries between air movement in the Pacific. To the north west, over China, Korea and Japan lies the Mei-Yu front. South of this lies the monsoon trough where tropical cyclones regularly develop in the late northern summer and early autumn. South of the equator the monsoon trough extends towards South America as the south Pacific Convergence Zone (S.P.C.Z.). Like the Mei-Yu front this feature connects the tropical circulation with the westerlies that lie further towards the Poles. The major equatorial band of convergence is the Intertropical Convergence Zone (I.T.C.Z.). Arrows indicate averaged direction of air flow

(mesoscale disturbances) in the western Pacific as the oscillations propagate towards the east. Indeed, during these cooler years, MJO activity is enhanced. Where the La Niña is established (Chap. 2), or when otherwise a large blocking anticyclone develops over the northern Pacific, warm moist air can become entrained within the winter flow of air that runs eastward towards the Oregon and Washington State coastlines. The jet stream, flowing just to the north of the Hawaiian Islands picks up the moisture from the tropical MJO thunderstorm complex, then whisks it eastwards towards the Oregon coastline. Within the westerly flow, the Rossby wave that bears the blocking anticyclone drifts westwards towards eastern Russia, while the strong flow of air along its northern flank weakens.

Blocked to the north, the jet stream then breaks loose and drives along the block's southern edge towards the Oregon coastline. This super-charged air encourages the development of deep low pressure areas off the northwestern coast of the lower continental States. From here a vigorous train of low pressure areas can develop which then batter the western coastline. In particular, the winter of 1996–1997 brought $2–3 billion dollars worth of damage to the Pacific Northwest, in the form of extensive flood damage. These patterns, although not common, are far from rare and have earned the title the "Pineapple Express" after their origin near the Hawaiian Islands. The "Express" brings a succession of rain-filled storms that have earned the pattern its other name—an *Atmospheric River*—a stream of densely packed moisture that can bring catastrophic floods as far south as Central California.

The Madden-Julian Oscillation also influences the intensity of summer monsoon rains across southernmost North America. Eastward propagating waves intensify then weaken rainfall associated with the monsoon. Consequently, the MJO affects the moisture content of the air across the entire continent and the intensity of rainfall along the Polar Front jet stream further to the north.

Despite their origin in the equatorial Pacific, the flow of air from these mesoscale storms can influence the weather much further afield. A pattern, similar to the Pineapple Express developed during the winter of 2013–2014, with a persistent blocking anticyclone located over the northern Pacific. Once again clusters of tropical thunderstorms developed in the western Pacific associated with the MJO and injected a large amount of moisture into the atmosphere north of the Tropics. On this occasion, the block remained anchored in the northeast Pacific and the injection of energy and moisture drove the Polar Front jet stream into a persistent pattern, arcing over Alaska then down across the Mid-West then out into the North Atlantic. With the block locked in place the jet to follow this path for months on end. Thus the Pacific blocked pattern conspired with the intense tropical thunderstorms to sweep storm after storm across the heart of the UK throughout the three winter months. Flooding and storm damage were unprecedented, even coming as they were on the back of record breaking summer floods in 2007 and 2012. Much of south west

England lay under water for 3 months, while 50 m of the Victorian train line that ran along the Devonshire coast was left hanging mid-air, when storms washed away the embankment that it was cited on. Thus distant effects in the Pacific can play havoc with the weather half a world away.

Such connections are an essential means by which gases are mixed throughout the atmosphere as well as ensuring the efficient transport of heat from warm to cold regions. Much of the surface and lower atmospheric connections on the Earth are driven by topography that neatly diverts airflow—in particular diverts air between different latitudes. Without this the Earth's atmosphere would obviously still function; after all, you don't see Jupiter's "tropics" overheating while the poles become so cold that its gases collapse out. Instead of these overt north–south transfers of heat that defy Gustav Coriolis's effect, the atmosphere would display enhanced east-west transport of energy, while waves transferred large quantities of this energy to the north and south. The remainder of the heat would be dissipated by convection so that the atmosphere would retain its stability.

Conclusions

The Earth's weather and climate is a very complex beast, indeed. Only in recent years have the connections between different latitudes and altitudes become apparent. The exchange of energy and momentum between different climatic regions is perhaps the most important process that happens on the Earth and probably any planet. Such exchanges moderate the extremes of temperature and on our planet at least ensure a relatively steady delivery of moisture between regions.

Yet this pattern is prone to instability. The jet stream can become locked into waves and troughs that deliver an often unwelcome barrage of meteorological phenomena such as droughts or floods. The underlying reasons for this change in behavior are as yet unclear. However, the effects can be devastating.

On other planets, the same underlying principles will apply, therefore, the more we understand about conditions within the earth's turbulent atmosphere, the more we will understand about the myriad of new worlds that are being discovered, as well as the other planets of the Solar System. Yet, this is a two-way street: from our understanding of the greenhouse effect on Venus we've learnt about how we can modify our own world through our activities. Once again, precisely how our behaviors lead to the changes we observe requires further elucidation. It is to the extremes of terrestrial climate that we turn to next as we begin our journey outwards from the Earth to the rest of the universe.

References

1. Saide, P. E., Spak, S. N., Pierce, R. B., Otkin, J. A., Schaack, T. K., Heidinger, A. K., da Silva, A. M., Kacenelenbogen, M., Redemann, J. & Carmichael, G. R. (2015). Central American biomass burning smoke can increase tornado severity in the US. Geophysical Research Letters, Retrieved https://www.researchgate.net/publication/271384910_Central_American_biomass_burning_smoke_can_increase_tornado_severity_in_the_US. Available free on "ResearchGate.
2. Evan, A. T., Ramanathan, V., Kossin, J. P., & Chung, C. E. (2011). Arabian Sea tropical cyclones intensified by emissions of black carbon and other aerosols. *Nature, 479*, 94–97.
3. *Madden Julian oscillation.* Retrieved from http://www.cpc.ncep.noaa.gov/products/precip/CWlink/MJO/MJO_1page_factsheet.pdf.
4. Madden Julian oscillation impacts on the US. Retrieved from https://www.climate.gov/news-features/blogs/enso/what-mjo-and-why-do-we-care.
5. *Kelvin waves.* http://www.ess.uci.edu/~yu/class/ess228/lecture.6.adjustment.all.pdf.
6. Stevenson, D.S. (2016) An underlying predictability in winter weather patterns in the North Atlantic Basin. *Hypothesis,* 14(1): e3, doi:10.5779/hypothesis.v14i1.483.

2. Climate Oscillations in Space and Time

Introduction

The East Asian Monsoon is perhaps the most important climatic feature of the Earth. Across 60° of latitude and around 80° of longitude, winds reverse direction from summer to winter and back again, year after year. The cycle brings rains to two continents: Asia in the northern summer and northern Australia in the southern summer. The present monsoon has been active for tens of millions of years and undoubtedly, systems like it have been active on Earth since the first continents emerged from the dark blue waters of the early Earth billions of years ago. This chapter examines the broad nature of the monsoon and the broader context of global climate in which it exists. In Chap. 6, we will see how the Martian landscape also experiences its own ghost-like version of the terrestrial East Asian Monsoon. Finally, in Chap. 10 we will see how some planets may take a monsoon pattern of winds to extremes with circulations that run from pole to pole, year after year.

The East Asian Monsoon resides in the broader canvas of terrestrial climate and as such is subject to the influences these other cyclical patterns of climate exert upon it. Of these the El Niño Southern Oscillation, or ENSO for short, is perhaps the best known, but others exist in the Atlantic and Indian Ocean basins. The ENSO, and one like it in the Indian Ocean, has a global reach, but unlike the East Asian Monsoon, does not seem to follow a predictable pattern. While the Asian Monsoon is clearly driven by the Sun and the tilt of our planet, the ENSO and Indian Ocean Oscillations are more like the backwards and forwards sloshing of water in a disturbed bath. The interaction of the two oscillations and the monsoon influences the well-being of billions of our planet's citizens and similarly, one must expect that such patterns

© Springer International Publishing Switzerland 2016
D.S. Stevenson, *The Exo-Weather Report*, Astronomers' Universe,
DOI 10.1007/978-3-319-25679-5_2

will exist on other Earth-like planets that we will be discovering in years to come. As such, although this chapter may, on a universal scale seem rather parochial, it will undoubtedly relate to other, foreign worlds in ways we can only imagine at present.

The Present Monsoon

Over nearly a quarter of our planet's surface the pattern of wind reverses direction every 6 months. The driving force for this massive redistribution of air is the asymmetric distribution of land on the Earth's surface. For the last 200 million years or so, there has been a concentration of land to the north of the equator, and more importantly for much of this time the land has been mountainous, to varying degrees. It is the combination of land and mountains that really spurs on the monsoon pattern of airflow that typifies many of the continents today.

The underlying physics of the monsoon is simple: land heats up much faster than water. because land has a lower specific heat capacity than water, it takes less energy per unit of mass to warm land than it does water. Moreover, soil and rock are poor conductors of heat, so that when the Sun heats the land, most of that energy is initially trapped in the topmost layers. This excess heat is then available to warm the air above. Oceans, by contrast, not only are made of water on our planet- which has a rather high heat capacity (several times that of rock)—they also have this annoying habit of moving around. Such movement distributes what heat energy they absorb from the Sun both vertically to some extent, and horizontally to a much greater extent. Thus, although water can store an awful lot of heat energy, it tends not to warm up very much compared with rock exposed to the same solar battering.

This means that during the summer the land is much warmer than the water around it; and, in the winter, land cools down much faster than any neighboring water. As pressure varies with temperature, the higher the temperature the higher the pressure. This causes the air to expand, lower in density and then rise under gravity. Such warm, low density air, rising over the hot land lowers the pressure at the surface, but keeps it higher aloft.

With low surface pressure air is drawn in from the surroundings, in this case the neighboring oceans.

Over India, the air begins to warm in March and, by May the atmosphere at the surface regularly cooks to over 40 °C. Adding to this mêlée is the presence of a jet stream and a zone of convergence called the Inter-tropical Convergence Zone (or ITCZ, for short). High above the developing surface low pressure area an easterly jet stream develops and begins vacuuming away the rising air. This jet continues westwards, ultimately crossing West Africa, contributing to the same effect here (Chap. 1). Although a surface low pressure area develops as early as April, it doesn't begin drawing in much moisture until the easterly jet has developed over southern India in June. The ITCZ is the zone of maximum convergence, or bringing together of hot moist air on the planet. To re-cap, this band, which is experienced at the surface as the Doldrums, contains the zone of maximum precipitation in the tropics and moves north and south with the Sun. When the "Indian vacuum cleaner" is working at its strongest in the summer, this band is pulled northwards across the sub-continent and brings the most intense rains with it. Its arrival effectively marks the beginning of the summer monsoon in June across the south; and its end in September when it pulls back towards the equator, propelled in part by increasingly strong north easterly winds to its north.

This combination of hot surface and upper level vacuum cleaner ensures that air rises very efficiently: however, it is not quite enough in itself to drive the full summer monsoon that we see today. Instead it turns out that India's rather well-known range of mountains to its north does the final trick: the Himalayas. North of this lies a large expanse of land over 5000 m above sea level: Tibet. Both of these immense structures serve to funnel air upwards in a focused belt, intensifying further the surface low pressure area. Having a plateau 5000 m up, means that during the summer there is intense heating of the middle troposphere and it is this that finally drags the summer monsoon winds upwards across all of the Asian sub-continent. Together, all of this suction (the heating of India and Tibet; the Himalayan mountain front and the easterly jet above) that finally drags in cooler and much moister air from the Indian Ocean.

You can see the effect of the Himalayas and Tibet by comparing the northernmost extent of the summer monsoon rains. Although Iran and the Gulf States are heated as strongly as India, it is only India that experiences the full, wet monsoon. Heated by the land and driven upwards by various highlands and the Himalayas to the north, moisture-laden winds condense their cargo of moisture into a raging belt of thunderstorms. This delivers a couple of meters of rainfall over the summer months. This is the summer monsoon.

Weaker monsoons also occur in other parts of the world. Over the south western states of North America, strong heating in the summer also generates a low pressure area that sucks in air from the Gulf of California and the Gulf of Mexico. A low level southerly jet stream develops along the eastern flank of the Rockies which helps drag moisture northwards and develop sporadic but crucial rainfall in the hot summer months. Over Africa, the summer monsoon is also marked by the development of low pressure over the Sahara; an easterly jet (or rather two easterly jets) over the Sahel and the arrival of the ITCZ over western Africa. Ironically, as this moves north, a change in the circulation of the South Atlantic, caused by the onshore monsoon winds, brings cooler waters towards the equator and an end to rainfall across the West African coast. This effect isn't as apparent over southern India because the waters to its south are always warm.

As the position of the overhead Sun moves back south again in September, the zone of maximum heating moves south with it. Convection weakens over northern India and the easterly jet stream weakens, falters and ultimately dies away. The moist southwesterly airflow is replaced by a cool northeasterly flow and the rains retreat back south towards the Indian Ocean. Over northern India pressure rises and the polar front jet stream becomes established south of the Tibetan plateau: the winter season has begun.

This pattern is now fairly well understood. However, embedded within this greater understand there remain a number of issues, such as patterns of strengthening and weakening that vary on decadal timescales, along with a more generalized weakening of the summer monsoon over recent decades. Some of these variations are caused by changes to the pattern of ocean and atmospheric circulation over the Indian and Pacific Oceans, while others appear to relate to the effects of soot and other pollutants.

Clearly, if we also seek to understand the past Monsoon we can get a better handle on the factors that will affect the present one as climate change takes hold. This chapter looks squarely at the monsoon and how climatologists have begun to probe how this great weather system operates. Through an improved understanding of these changes scientists, economists and politicians can begin to plan more carefully for the future. For within the hands of the Asian monsoon, lies the fate of more than two billion people.

Shifting Continents, Shifting Climate

On the Earth, and presumably many other rocky planets, the surface of the planet is in a state of constant flux. Heat is generated within the mantle by a combination of radioactive decay and heat delivered by a cooling core. On top of this is a cold, dense, upper, rocky layer called the lithosphere, which includes the crust on which we live. Comfortable though we are this is an unstable situation, where the dense, cold lithosphere wants to sink under gravity towards the hot core.

While the oceanic lithosphere is easily dense enough to sink into the mantle, the continental crust is far too light and fluffy to descend far. Thus, as the lithosphere fractures and shuffles in response to these forces, continents drift around like blocks of polystyrene on a pan of boiling water. Drifting continents mean that occasionally they collide and when they do they generate chains of mountains, such as the Himalayas or the Alps. Elsewhere, a combination of strong pulling forces acting on the their edges and hot currents rising from below can split continents asunder, forming new ocean basins, much as you see in the Gulf of California or the Red Sea.

Shifting continents bring an inevitability of concurrently shifting climates. For the continents themselves, the process of plate tectonics, as this is known, can carry them north or south from one climatic belt to another. This is a fairly passive and relatively sluggish process that can take tens of millions of years. Alternatively, where a continental nugget moves across the path of the prevailing winds it might direct warm ocean currents and the winds themselves further towards the poles—or it might end up over one of the polar regions. This kind of movement can have more sudden and profound effects on the climate of the globe as a whole.

Alongside the slow dance of the continents are periodic larger and more rapid redistributions of land called True Polar Wander (TPW). Here, the Earth shuffles the land on its surface to better balance its spin and momentum. Where the continents have moved closer to the North or South Pole the spin of the planet (and the pull of the Moon and Sun) tend to want to pull this extra mass into alignment with the Earth's spin. This involves a wholesale shifting of the continents on mass until they are more suitably redistributed around the equator. Such movements probably take around 10–20 million years and will clearly have dramatic consequences for the Earth's climate and the life that populates its surface. Although not too much will be said about these TPW episodes, hold the thought in the back of your mind when we examine the positions of the continents in each of the ensuing geological periods in Chap. 3.

It is to these seismic shifts and to the impacts these have on planetary climate that this chapter turns its attention. The focus remains on the Monsoon but painted within a broader canvas of continental drift.

Probing the Past Asian Monsoon

How does science begin to establish the pattern and driving force of a climate system 50 million years ago? After all one can hardly get a snapshot of cloud patterns in the fossil record… Or can you? As it turns out the geological record is incredibly helpful, if you know where to look.

In general there are two very useful methods available to climate scientists (and geologists in general). The first is the most obvious: fossil sea shells. A cursory examination of fossils tells you what sorts of organisms were present and where they were. This can tell you about the general conditions, and on closer examination sea shells can tell you a lot more. Sea shells are mostly calcium carbonate (chalk/limestone). This compound contains three oxygen atoms and is a product of the reaction between carbon dioxide, water and dissolved calcium. The water contains one oxygen atom for every two hydrogen atoms, and it is this oxygen atom which is of critical importance to climatologists. Oxygen comes in three nuclear flavors: oxygen-16 (^{16}O); oxygen-17 (^{17}O) and oxygen-18 (^{18}O). Of these oxygen-16 is by far the most common and is present at roughly 495 times the abundance of oxygen-18. Oxygen-17 is even rarer (only 0.037 % of the total oxygen atoms on the Earth) and not generally used in any analysis. Oxygen-16 has two fewer neutrons in its nucleus than oxygen-18, which makes it lighter and easier to lift against the pull of gravity. Therefore, water that contains oxygen-16 is lighter and easier to evaporate than water which contains oxygen-18. This means that on average water that contains oxygen-16 will evaporate under cooler conditions than water containing oxygen-18.

When the weather is warmer more water containing oxygen-18 evaporates and when it is cooler less oxygen-18 evaporates. Moreover, when there is heavy rain the amount of light oxygen (oxygen-16) is greater in surface water, including seawater, simply because it evaporates easier and thus more of it ends up in clouds which then produce rain.

At other times, when evaporation is high, but precipitation low, the amount of heavier oxygen-18 rises in surface water. The abundance of the different oxygen isotopes that remains in the seawater will then alter depending on the temperature and on precipitation; and because the isotopes vary in the sea water, they will also vary in the shells of organisms that are living in it and busy assembling their shells. By analyzing the concentration of these different oxygen isotopes in sea shells climatologists can probe the temperature of the water in which the organisms grew and the amount of rainfall that was occurring. So, where do you look for these shells? The answer lies in layers of hardened mud that are laid down in the oceans every year. In the right, and fortunate, locations there are bands of sediment in the rock that clearly chart the summer monsoon, either as bands of sea shells with alternating "heavy" (^{18}O-rich) or "light" (^{16}O-rich) rock.

By observing cyclical patterns in the deposition of soil and isotopes of oxygen, evidence has accumulated that the modern monsoon was established over 50 million years ago. At this time, the Tethys Ocean, although shriveled markedly in stature, still extended along to where the Tarim Mountains now lie to the north of modern Afghanistan. The Himalayas were rising, but undoubtedly had lower elevation than they do now. Alexis Licht and colleagues examined the ratio of different isotopes of oxygen in fossilized invertebrate shells (gastropods) that lived in the Tethys Ocean. Through the examination of oxygen-isotopes preserved in their fossilized teeth and shells climatologists revealed a clear monsoon-like pattern of drought and deluge that is characteristic of modern Asia. Paralleling this were changes to the pattern of dust movement across China.

Over China strong, cold and dry winter winds currently deliver dust towards the south. Again, these winds only blow in the winter, thus a seasonal pattern will be evident in sedimentary rocks laid down at the time. Matching the pattern of isotopes seen in the gastropods, the Chinese records clearly show a monsoon pattern, with the annual pattern of deposition running in cycles driven by the onset and termination of each winter monsoon.

Thus, Litch and co-workers showed that during much of the Eocene, from 50 to 34 million years ago, a monsoon pattern of rainfall was apparent, despite a lower Himalaya and a limited Tibet. Although the Tarim basin was flooded by the Tethys, the relatively strong Eocene monsoon appears to have been driven by the higher carbon dioxide levels of the day, which trapped more of the Sun's radiation and propelled a stronger circulation. The lower elevation of the land, and the presence of the Tethys to the northwest, would naturally decrease the strength of the summer monsoon as these weaken the contrast in temperature and pressure needed to drive the summer monsoon. However, with a higher concentration of carbon dioxide in the atmosphere—roughly twice what is it now—there was sufficient heating of the rising land to the north of India to drive a monsoon circulation during the summer months. The stronger greenhouse effect more than compensated for the deficiency in the underlying geography.

However, things were about to change. Although Tibet was on the rise, empowering the summer monsoon with extra lift, the rise of Tibet had a less positive effect. As Tibet and the Himalaya rose ever higher, they naturally intercepted more wind and, with the incumbent rainfall, began to erode ever more fiercely. Enhanced precipitation and erosion drags carbon dioxide out of the air. With Antarctica also sliding into its modern position over the pole and declining levels of carbon dioxide, the planetary greenhouse began to fail. As the Eocene transitioned into the Oligocene, 34 million years ago, the Asian Monsoon began to lose its way. Work by Guillaume Dupont-Nivet and colleagues (who also worked with Litch on the work described above) charted the decline in the Monsoon. Their research indicates that the drying out, or aridification, of Tibet happened as Antarctica assumed its modern position and began to freeze over.

This was a bad time for the monsoon. For to the north of Tibet, the Tethys had shriveled away (Fig. 2.1) and the supply of moisture to central Asia declined. Drought intensified over Tibet and central China, while declining carbon dioxide, coupled to

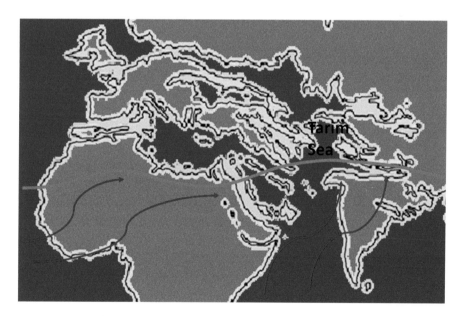

FIG. 2.1 An Oligocene (30–24 Mya) Paleomap indicating the position of the Inter-Tropical Convergence Zone (*orange*) and the likely direction of the prevailing winds during the northern summer. Moist winds blow directly across most of North Africa from the tropical Atlantic, keeping the current Sahara a verdant shade of green. Over Asia, moist winds blow from the Indian Ocean and from the Tarim Sea to the northwest of India. By the mid-Oligocene the Tarim basin has closed and Tibet is rising, while during the Miocene 20 million years later, the closure of the remnant Tethys ends North Africa's summer monsoon. Map by author

Antarctica's freeze-up ensured that the rise of air over Tibet and the Himalayas was weakened (Fig. 2.1), making the period between 34 and 20 million years ago India and Indochina's driest of the last 50 million years. In the end the ever-more lofty Tibet came to the monsoon's rescue. Up until this point the Himalayan range was providing much of the lift to the summer monsoon airflow. However, as Tibet assumed its modern height and dimensions, it boosted the overall lifting effect to the warm tropical air. Thus from 20 million years onwards, despite significant changes to the moisture supply—and declining global temperatures—the monsoon was reinvigorated and began its current pattern.

Milanković Cycles: Astronomical Influences on Terrestrial Climate

In 1930 Milutin Milanković published work that suggested that the ice ages were caused by cyclical changes in a number of astronomical factors: these were the shape of the Earth's orbit; its wobbling, or precession, on its axis; and the degree of overall tilt of the axis. The degree to which these different cycles coincided, summated or detracted from one another determined whether the world would warm or cool. Milanković had been working on these for the best part of a decade, motivated in part by close partnerships with climatologist Wladimir Köppen and geophysicist Alfred Wegener. Milanković's work built upon much earlier observations and theoretical work from Johannes Kepler and—and from even earlier—Greek philosopher Hipparchus. Kepler had derived three successful descriptions and mathematical formulations of the orbits of the planets around the Sun. These indicated that the elliptical orbit of the Earth altered over long periods of time (100,000 years as part of a 400,000 year-long "super-cycle"). In 127 B.C. Hipparchus had noted, from even earlier work, that the Earth's axis underwent precession over a period of around 23,000 years. This was evident by a gradual change in the position of the stars Regulus and Spica relative to the autumn equinox. Milanković triumph was to take these disparate observations and synthesize an effective, holistic model of the astronomical factors that govern the long-term evolution of the Earth's climate.

To get more of a flavor for these cycles you need to think of the Earth as a spinning top. Despite the Earth's massive size, it can still wobble about. In part this is caused by the pull of the Moon, Sun and Jupiter in particular, but it is also a consequence of having a surface that moves around through plate tectonics. The Earth's orbit around the Sun is an ellipse so that in January the Earth is nearly 4 million kilometers closer to the Sun than it is in July. The shape of this ellipse varies over a period of 100,000 years, primarily under the influence of Jupiter. At present the Earth receives around 6 % more energy from the Sun in January than it does in July. This 6 % inflation ensures that the northern winters are somewhat warmer than they would otherwise be. However, as this long cycle progresses, the northern winter will coincide with an even greater eccentricity that bequeaths the northern winter with up to 20 % more sunlight than it would get if the orbit was circular. Therefore, at the moment the northern winter is set up for a cooling trend, not the warming one we currently observe.

Superimposed on this cycle are two other ones involving the Earth's tilt. The overall tilt of the axis varies, from 21.5 to 24.5°, over the course of 41,000 years. When the Earth's tilt is less extreme, the planet experiences cooler summers but warmer winters than when the tilt is greater. This set up favors more snow-fall in the winter—as the warmer air holds more moisture—which then thaws less in the cooler summer. Imagine a long autumn, rather than a warm summer and cold winter as we have now. Finally, the axis precesses, or wobbles, so that currently when the Earth is furthest from the Sun the northern polar region is point-ing most towards the Sun. Meanwhile in our northern summer the North Pole is pointing at the Sun when it is furthest from it. That process will reverse in 10,500 years so that during the north-ern winter the Earth will be furthest from the Sun and thus be considerably colder for longer.

These cycles only matter to the Earth now, because most of the landmasses lie in the northern hemisphere. As land cools and warms much faster than water, the appropriate combination of trends will trigger cooling and glaciation across the northern land-masses: a slight axial tilt; a less extreme eccentricity in the Earth's orbit; and precession that leads to the North Pole pointing away from the Sun when the Earth is furthest from the Sun. While today we think of Milanković cycles in the context of cooling and ice ages, during earlier epochs there was a variety of impacts ranging from warming to cooling. In Chap. 3 we will see how Milanković cycles might explain three periods of extreme warmth of the Eocene, but here we will look to cooling and the formation of the Sahara desert.

The Death of the Tethys and the Birth of the Sahara

Between 20 and 7 million years ago the world assumed its present configuration of continents. Asia was already intact, with India welded firmly to its southern flank. However, to the west, the remnants of the Tethys and the modern Mediterranean Sea were still a battle ground of micro-continents. Africa was drifting lazily north as fragments of it shuffled around along the southern edge of what would become Europe. The Black and Caspian Seas were still part of a larger body of water called the Tethyan scraps while the Arabian Peninsula still had some water to close to its north. Although narrow, the Tethys was still able to supply a reasonable amount of moisture to the north of Africa.

Around 11 million years ago a pivotal event happened and the Sahara began to take shape. Zhongshi Zhang and colleagues showed that shortly after 11 million years ago, the Tethys had shrunk to such a point that the flow of warm moist air associated with the African summer monsoon effectively failed across most of the northern half of the continent. The process took a few million years to unfold but by 7 million years ago, most of the present Sahara had dried out and was filling up with wind-blown sand and other deposits.

As the Tethys shrank, the location of the summer doldrums— the Inter-tropical Convergence Zone (the I.T.C.Z.) began to retreat back towards the West African coast, to where it now resides during the summer months (Fig. 2.1). The ITCZ marks the zone of greatest convection on the planet and consequently the zone of heaviest tropical rainfall (Chap. 1). Before this time, the North African region was periodically afflicted by drought, but this process was cyclical, driven by the wobbling of the axis and changes to the orbit of the Earth around the Sun (Milanković Cycles). However, over most years the summer monsoon brought rainfall as far north as the southern Mediterranean coast. From 7 million years ago, this pattern changed so that drought was the norm and a verdant green the exception.

Prior to 7 million years ago there was no permanent drought. However, once the Tethys had narrowed to dimensions similar to those today, the flow of winds from the tropical Atlantic weakened and eventually assumed their present northern limit in what is now Chad: before this time the line separating the two Hadley cells extended into what is now northern Libya. What changed? The key moment appears to be the closure of the eastern end of the remnant Tethys—the Arabian Sea. When this shut, the drier air to the north of the modern Sahara was no longer able to retain enough heat and the African monsoon began to waver. This was caused by a reduction in the amount of moisture delivered from the shrinking Tethys. With its critical support gone, Milanković cycles began to dominate the climate and deserts began to wax and wane across the region. These began nearest to the Mediterranean coastline, but steadily advanced towards the south over the ensuing 5 million years. By 3 million years ago the modern Sahara was largely in place.

Within the last few thousand years the Sahara has periodically turned green when Milanković Cycles maximized Saharan warmth. However, this is only when the Earth has its greatest tilt (24.5°) and its northern hemisphere is tilted towards the Sun when it is closest to it. During this period of maximum summer heating—last experienced 6000–8000 years ago—the normally dry Sahara has blossomed, albeit briefly.

An intriguing additional factor in the Sahara's development must have been the closure of the Mediterranean Sea. As Africa has continued to move northwards, the seaway at Gibraltar has periodically been slammed shut, only to be breached by the waters of the Atlantic. When the Mediterranean-Atlantic seaway first closed around 6 million years ago it meant that the Mediterranean soon dried out. Although the Mediterranean contributes a lot less moisture than the ancestral Tethys did, its temporary loss (which may have happened several times around 5–7 million years ago), would further have limited the supply of both heat and moisture to the north of Africa. Sooner or later, the progressive northward march of Africa will close the gateway once more and lead to a renewed loss of the Mediterranean Sea.

This evaporation of the sea will be a prelude to the final act. As Africa continues its slow but steady march north, the Mediterranean will finally close, along with the final remnants of the Tethys: the Black and Caspian Seas. Their waters will drain as the land rises and replaces their pristine blue with snow and ice. Although these seas are small fry when it comes to supplying moisture to Africa and Asia, their loss will exacerbate drought in both continents over the next several tens of millions of years. The Mediterranean mountains will block the flow of winter westerly winds from the Atlantic into the Middle East, further reducing the region's rainfall.

Pangean Monsoons

The slow dance of the continents has ensured that some parts of the globe have had rain in abundance and drought in others. During the late Permian and continuing into the Triassic Pangaea undoubtedly experienced some form monsoon climate. The northern flank of the Tethys was always in turmoil. Every few tens of millions of years micro-continents were breaking away from what is now northern India and Saudi Arabia and colliding with what would eventually become southern Asia. Further west, the suture between Africa and North America was also mountainous (Fig. 2.2). This sort of set up would favor seasonal (summertime) ascent of air and the inflow of warm, moist air from the Tethys. Such monsoons were likely weak by today's standards, but they would still be a factor.

Indeed, the situation could have been rather complex. For to the south of the Tethys Ocean was Gondwanaland which undoubtedly experienced monsoon-like inflow of its own as it heated up and cooled down with the progressing seasons. Like the Sahara of today, it would seem likely that the ever-present underlying Milanković cycles would have driven periods of relative drier and wetter climatic conditions.

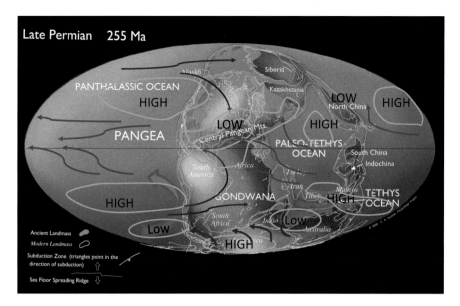

FIG. 2.2 Possible Late Permian northern summer monsoon pattern over Pangea. The central Pangean mountains (the ancestors of the modern Appalachians) act like the Himalayas and Tibet of today and focus warm prevailing winds blowing in from the warm Tethys ocean. During the summer, a strong area of low pressure develops and focuses rainfall on the windward, eastern slopes. To the west, a combination of higher pressure and prevailing winds blowing over the dry interior ensure that the future North American region is dry. Winds are shown by thin arrows of different colors, while a likely cold ocean current flowing up the west side of what is now South America maintains dry conditions to the south west of the Pangean Mountain range. Paleomap courtesy of Christopher Scotese

The location of the Pangean Mountains, only marginally north of the Equator would have meant that any Monsoon regime would have been rather limited in extent. By focusing the formation of seasonal low pressure areas, mountains are extremely effective at organizing the flow of air across continents. On a different scale the Rockies organize both a summer Monsoon across the south western United States but also organize the westerlies into a series of waves that ensure that the central continental U.S. is very cold in the winter (Chap. 1). On the Earth of today, Tibet and the Himalayas direct a strong north-south flow of air, but one that spans more than 50° of latitude (30°N to around 20°S over the southern Indian

Ocean). The Pangean Mountains, by virtue of their location, would have directed an air flow over, perhaps 10°–20° of latitude, that extended from around 5°–10°N of the equator to the northern half of Gondwanaland in the south.

The arrangement of the continents is critical in directing the movement of the air, but also the underlying ocean currents. The natural inclination for the air and water is to move broadly east-west or west-east under the direction of the Coriolis Effects and this is true of the oceans as well. Continents get in the way directing movement in more meridional (north-south) directions (as opposed to zonal or east-west directions), which in turn directs warmer or colder air in those directions. In general such masses of land encourage the warming of the polar regions by giving them access to warmer air from the tropics. Indeed, in today's atmosphere the warming planet displays more rigid and extreme buckles in the jet stream that are directing greater warming towards the Poles.

Superimposed upon this general trend are bumps and troughs in the pattern of temperature change. Most significant are the El Niño and La Niña patterns, which bring year-long alterations in the flow of air across the Pacific and beyond (Chap. 3). In any paleoclimate such patterns would also have been likely where the arrangement of the continents permitted it. We look, now, at these global phenomena and put them in the context of the broader pattern of terrestrial climate.

Problematic Children: El Niño and La Niña

Were the Earth a bland ball of water things would be so much simpler. Winds would blow according to the competing forces of pressure and the Coriolis Effect. On regional scales air would blow in a largely easterly direction towards the equator or in a westerly direction towards the poles (Chap. 1). Differences in pressure would ensure some north south movement, but primarily winds would be orientated 90° to this direction.

Of course, the Earth is unfortunately dotted with continents. Not only do these contain mountains that block and divert the movement of air, they also block the corresponding movement of water. This is no more apparent than in the Pacific, where the prevailing easterly winds (the north east and south east trade winds) that blow towards the equator drive a large mass of warm water towards Indonesia and the Philippines. Unable to advance easily through the myriad of islands that separate the Pacific and Indian Oceans, the water backs up into an enormous pool of particularly warm water. As long as the prevailing winds are easterly, the water remains pooled in the Western Pacific (Fig. 2.3). To the east South America intercepts the westerly winds and flow of cold water around Antarctica. Much of this water manages to flow around the southern tip of South America, but a large portion is forced to flow northwards up the South American coast towards the equator. Near the Galapagos the prevailing easterly winds divert this flow of cold water west, towards the Central Pacific, where it eventually sinks below the surface. In a normal year the Western Pacific is therefore warm, and the Eastern Pacific cold.

This prevailing pattern, with cold to the west of the basin and warm to the east is part of a much larger pattern known as the Walker Circulation. Thus, embedded within the prevailing easterly flow towards the doldrums, is a west to east flow in the Pacific Ocean. In physical terms the Walker Circulation is a bit like a very large standing wave: a fixed wave-like pattern that permeates the entire flow of air across the tropical regions.

In the Atlantic, the same pattern exists, but on a smaller scale, with cold currents flowing up the west coast of southern Africa towards the equator and a cold current moving south westwards past the Canary Islands to the north west of Africa. These cold currents descend under the warm equatorial waters off the west coast of central Africa leaving a warm pool of water west towards the Caribbean. In the Atlantic, the draw of cold water is enhanced in the northern summer when the African monsoon winds (Chap. 1) move inland towards the Sahara from the southern hemisphere. The Indian Ocean is forced into the opposite pattern with a warm east and a cold west. The forcing factor is the warm pool of warm water over Indonesia and the strong low pressure area that is associated with it. The Walker circulation in the Indian Ocean is strongly affected by the Indian monsoon which radically

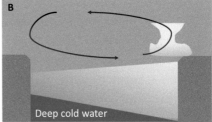

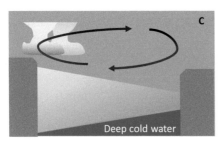

FIG. 2.3 ENSO. (a) Normal (non-ENSO) circulation. Winds blow along the equator towards the west—this is the Walker Circulation. Over the Pacific, winds cause warm water to back up against Indonesia giving rise to the greatest convection and heaviest rainfall. To the west of South America cool, dry air descends and blows west. Winds cause cold water to upwell which further enhances the sinking of air at the surface. (b) In an El Niño year the Walker Circulation weakens while the strength of the Hadley Cells supplying it strengthens. Warm water sloshes back eastwards towards the west coast of South America, bringing the maximum convection and rainfall with it. The cold upwelling along the coast stops. On occasion the sloshing stops mid-Pacific leaving both Indonesia and western South America relatively dry. (c) In a La Niña year the Walker Circulation strengthens and there is greater than normal upwelling of cold water along the west coast of South America. More warm water than normal backs up north of Australia enhancing rainfall along the east coast of Australia and over Indonesia. The pattern of circulation over the equatorial Pacific, in turn, alters the tropical circulation across the Atlantic and Indian Oceans with knock-on effects to the climate there

reverses wind flow across the entire ocean basin between winter and spring. This makes the overall pattern of circulation very disjointed with strong seasonal variation across the Indian Ocean that is not strongly reflected in the neighboring ocean basins.

Every now and then, the prevailing easterly winds weaken across the Pacific and the large pool of warm water near Indonesia sloshes back east, over the top of the cold waters moving north from Antarctica. Now, the waters lying in the Eastern Pacific become very warm, while the water in the normally warm west becomes cooler (though hardly cold). When the waters in the Central Pacific become warm a weak El Niño is declared (Fig. 2.4). The more warm water that sloshes to the east, the stronger the El Niño is. NOAA will declare an El Niño only when sea surface temperatures are at least 0.5 °C above normal for 3 or more months to the east of the 120°W longitude. Strong El Niño conditions are coined for El Niños that last 7–9 months. Anything longer and the term El Niño episode is used. These reversals in the temperature of the ocean surface are reflected in changes in the air above. On occasion the ocean sloshes warm water to the east but, as

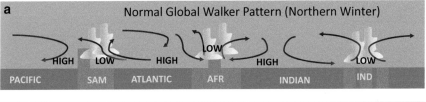

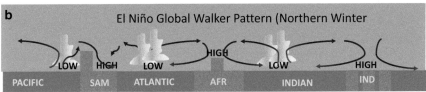

FIG. 2.4 The effect of an El Niño on the pattern of airflow in the, tropics. A comparison of normal and El Niño weather patterns. The Walker circulation is the patterns of airflow along the equator and tends to produce high pressure over the eastern pacific and low pressure over Indonesia (a). This pattern broadly reverses during an El Niño event (b). Although the most severe impacts are in the Pacific basin, the El Niño causes effects further afield. The Amazon basin dries out with pressure generally rising. This is also true of Indonesia, the Sahel and equatorial Africa. The Indian Ocean has its own equivalent pattern to the ENSO (El Niño Southern Oscillation) and may or may not run concurrently with the Pacific pattern. When it does, the Indian sub-continent tends to get weaker rains in the summer

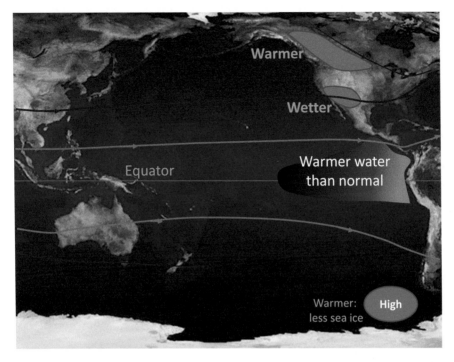

FIG. 2.5 Changes to the jet streams during El Niño events during the northern winter. The largest change, outside the tropics, affects the northern hemisphere's polar front jet stream (maroon). This strengthens along its southerly branch while the northern branch weakens. Frequent low pressure areas move along this into Western United States bringing large amounts of rainfall, while the south eastern states dry out. Across the eastern seaboard milder than normal conditions prevail. Meanwhile, the sub-tropical jet (orange) strengthens over Australia bringing drier conditions. Further south, near Antarctica, high pressure areas become more frequent west of the peninsula bringing warmer conditions. East of the Peninsula, cooler conditions prevail

in 2014, the atmosphere above does not fully respond. In these instances some of the weather phenomena associated with El Niño occurs but not all.

In a full blown El Niño the easterly trade winds weaken and die away and storm systems migrate towards South America. The Peruvian coast gets a soaking, while northern Australia and Indonesia suffer drought. The El Niño usually strikes in the late autumn through to the early spring, causing a significant realignment of the jet streams over the Pacific; however, in some years such as 1997 or 2015 it begins in the early summer. During the winter

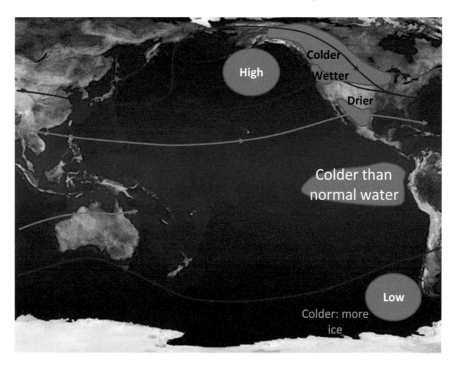

FIG. 2.6 Changes to the jet streams during La Niña events during the northern winter. Over the Northern Pacific the polar front jet stream (maroon) breaks into two branches with a prominent blocking anticyclone present south of the Aleutians. During La Niña episodes the northern branch has more energy than the southern branch. The southern branch wobbles more and brings heavier rainfall to the Pacific Northwest, while California and Mexico experiences drought. This set-up can become prolonged, establishing mega droughts in the South Western States. Over the southern hemisphere, the Polar Front jet stream bulges northwards affecting more of South America while a cooling trend affects Antarctica west of the peninsula. East of the Peninsula the situation is reversed with warmer conditions

months the northern sub-tropical jet—which normally lies across southern California is diverted southwards, while to the north, the polar front jet, with its attendant winter storms, also moves south and pummels the western states (Figs. 2.5 and 2.6). California often experiences flooding as a succession of low pressure storms swing south eastwards from the central north Pacific. The Indian monsoon is usually weaker during El Niño years and much of the Amazon experiences lower than normal rainfall. These changes are associated with changes to the Walker Circulation (Fig. 2.4).

Further afield, these oscillations have a more subtle effect on climate, with milder winters in Western Europe more typically associated with La Niñas. Figure 2.7 shows how the El Niño can establish broader changes to the circulation of the atmosphere over the northern hemisphere through the formation of far reaching Rossby wave. Even within that pattern is an interesting sub-pattern. La Niñas often bring blocking high pressure areas over Western Europe from late November through December. In many instances these bring relatively cold but dry conditions to the U.K.

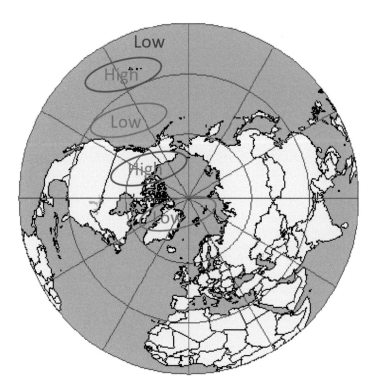

FIG. 2.7 Rossby Waves associated with the El Niño. Low pressure (*red*) over the central Pacific (paired on either side of the equator) is linked through a series of south-west to north-east trending high and low pressure areas that extend across the North Pole. These trans-latitudinal waves interact with and organize the longitudinal Rossby Waves that are moving around the North Pole within the polar front jet stream (Fig. 1.14). Similar patterns extend to the south east of the equator over the southern hemisphere. Map courtesy of Paul Anderson (http://www.csiss. org/map-projections/index.html)

When the Arctic Oscillation—an episodic pattern afflicting the circulation of air around the Arctic basin—is in its negative, high pressure dominated phase, a La Niña can work with it to produce extreme cold. The La Niña of 2010, for example, brought the coldest December on record to the UK. Blocking developed to the north of the UK in late November which directed a succession of winter storms across the UK with abundant snowfall. At the close of December, normality resumed and the La Niña helped maintain a very mild pattern, with February ultimately becoming one of the mildest on record. What appears to matter is how many waves separate the equatorial Pacific from the Atlantic. This can vary from three to seven and ultimately it may well come down to which wave number is present when the El Niño or La Niña kicks in. It would seem the latter tend to lock the system down in whatever mode it is in. So for example, during the 2015 El Niño there was already a 3 year long pattern of blocking over the northern Pacific and Alaska—something more normally associated with La Niña conditions (Fig. 2.6).

Although December 2010 was extreme, blocking patterns are common in December in La Niña years. The only difference is the position of the block, which more often than not lies over the UK or just to its east. December 2000 also saw a very cold pattern in the UK following a weak La Niña, and again this was associated with a negative Arctic Oscillation. Conversely, the winter of 2007 had a La Niña blocking pattern over the UK but this time the Arctic was positive and December, although fairly cold, was far from remarkable, with a very mild winter ensuing. Thus, in Western Europe La Niñas have effects, but they are embedded in much broader patterns, dictated by what is going on over the Arctic—and that is only partly linked to patterns further south.

Most significantly, El Niños stir up the Pacific waters, releasing heat from deeper layers as well as redistributing heat that are otherwise locked to the west. For this reason El Niño years tend to be the warmest ones in any one decade. The grand-daddy of the lot was the 1997–1998 El Niño which set records that were only matched if marginally broken in 2014, but shattered in 2015.

What is clear is that the ENSO seems to run to its own beat and has (thus far) shown little if any influence from the rising temperatures caused by anthropogenic global warming. However, within this end of the story is a growing uncertainty regarding the influence of rising CO_2 and (over decadal timescales) rising temperatures on the strength of the Hadley circulation. Models predict and observations suggest that as temperatures rise the Hadley cells expand towards the poles. Although the movement is relatively slow on human timescales it will influence the patterns of rainfall across the globe. Expectations would include enhanced rainfall across the southern Sahara and northern India where abundance wouldn't cause concern. However, an expansion of the Hadley Cells would reduce rainfall to the north of the Sahara, in the Mediterranean; reduced rainfall across the mid-West and also throughout parts of southern Australia where conditions are already marginal. Superimpose the effects of El Niño and La Niña and some parts of the globe could really suffer in terms of food production or the availability of fresh water.

Whether these expectations are met, is something we will have to wait and see, as we continue our experiment with the Earth's climate.

The triggers for El Niño events are far from understood. There are clear connections between Madden-Julian Oscillation (Chap. 1) and the El Niño pattern. Both are westward propagating systems, with the MJO driving westerly winds in its wake, which in turn drive a pool of warm water eastwards to a depth of 100 m. The intensity of MJOs is enhanced 6–12 months ahead of El Niño events, but once the El Niño is established MJO events tend to melt away, becoming almost non-existent. The 1982 El Niño, one of the strongest in decades, was driven by a strong MJO event that began in May 1982. By July of that year, it had initiated the El Niño which persisted into the following spring.

While MJO activity is lower during El Niño events, it is heightened during La Niña events and observations suggest that prior to El Niños the westerly winds running in the wake of MJOs are stronger, while the easterly winds running in their path are stronger ahead of La Niñas. Together, this would tend to suggest MJO activity plays a significant role in determining the broader pattern of weather in the tropics by influencing the movement of surface and upper level winds around the storm complexes. However, the precise nature of the connection remains to be determined.

The Future East Asian Monsoon Under the Cloud of Global Dimming

The far-reaching effect of seemingly innocuous presence of airplane contrails was revealed when terrorists attacked the United States on September 11th 2001. Then, in the attacks aftermath, planes were grounded and temperatures across the entire 48 States changed by more than a full degree Celsius. This is an astonishing observation. If you were to add together the entire area corresponding to a day's worth of flights, it would be a tiny fraction of the total area of the United States. Yet these 50 m long aluminum tubes generate trails of water vapor that can be kilometers long. As there are several thousand flights a day the high icy clouds that the combined flights generate is substantial. The authors, David J Travis (University of Wisconsin-Whitewater) and colleagues concluded that the 1.8 °C change in the region's diurnal temperature profile was the result of less cooling at night and more warming in the daytime.

Therefore, the effect of contrails depends on what time of day they appear. During the daytime, contrails primarily reflect incoming visible radiation and have a cooling effect. At night time, the reverse is true with the contrails serving to trap outgoing infrared radiation giving rise to a warming effect. The net effect (the balance between the day and nighttime influences) is to cause warming as the trapping of infrared radiation is greater than the cooling, reflective effect. This explains the observations made after the 9/11 attacks. There was a greater diurnal range in temperatures when there were no contrails: less cooling in the day *and* less warming at night giving rise to greater temperature maxima and lower temperature minima. This observation has a bearing on what will happen to the Earth in a billion or so years. If warming produces more cirrus clouds then the Earth might overheat faster than if it produces more reflective cumulus clouds. This is looked at again in Chap. 3; and again in Chap. 10, when we look at extra-solar planets.

Beyond, the confirmed influence of airplane contrails, there are other chemical pollutants that have had a varying influence on global climate over the last hundred years. During the first half of the Twentieth Century industries in Europe and the US belched copious amounts of black carbon soot into the air. We loved our fossil fuels, but not enough to burn them completely. Currently, Russia, China and the progressive incineration of tropical rainforests are the biggest sources of such particulate pollutants (Fig. 2.8).

Aside from the negative effects of soot on our health, such particulates absorb incoming radiation and cause very diverse effects on temperature that depend critically on what sort of material is present and at what height in the atmosphere it is found. Even one type of pollutant, black carbon which is a mixture of unpleasant carbon-rich compounds, has diverse effects. Black carbon absorbs radiation and causes warming within the layers in which it is most concentrated. This is simply because it absorbs infrared radiation and visible radiation efficiently—after all, that's why it's black. Beneath that layer, radiation that would have been received on the ground is blocked, resulting in it being cooler than if the soot was absent.

Particulates, as these particles are known, may also stimulate cloud development, which can further reduce temperatures on the ground. Black soot resulted in extensive smog across many industrial areas in the West in the 1950s and now in the East. The same chemicals we in the West pumped into the air six decades ago can be seen poisoning cities in India and China today. Sadly, this is a lesson in public health that has not been learnt. The drive for wealth apparently supersedes the necessity of preserving human health. In China, alone, there are 1.3 million deaths per year associated with pollution from burning fossil fuels.

Black soot has considerable effects on rainfall. Through their warming effect on the middle and upper atmosphere particulates will raise the air pressure in this layer relative to less polluted surrounding air. By raising the temperature in the middle layers of the atmosphere, the air column in which the soot particles are suspended in becomes also more stable and less prone to convection. Areas of warm air at middle and high levels in

Atmospheric Heating

Surface Cooling

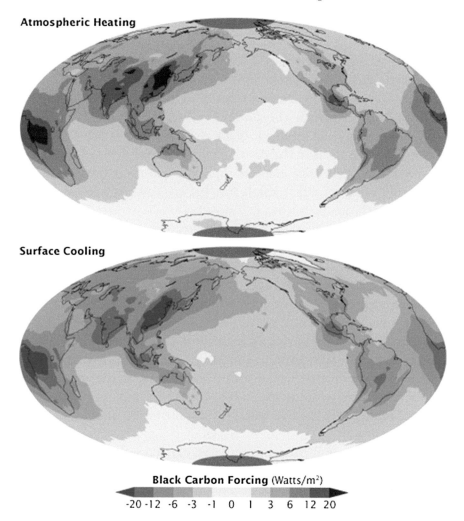

Black Carbon Forcing (Watts/m²)

-20 -12 -6 -3 -1 0 1 3 6 12 20

Fig. 2.8 Computer models showing the heating and cooling effects of black carbon soot. The *upper panel* shows the heating effects of these compounds in the middle and lower troposphere, caused by the absorption of energy from the Sun. The *lower panel* shows the cooling effect at the surface caused by the same materials. These effects alter wind flow across many regions of the globe as well as directly impacting human health. Not shown is the effect of soot on snow cover, which has an additional heating effect. Figures adapted from: http://earthobservatory.nasa.gov/Features/Aerosols/page3.php

the atmosphere can also have a blocking effect on the jet stream which can pin it into particular patterns which, in turn, can bring an over-abundance of precipitation in some areas but a deficit in others.

The clearest impacts of black soot pollution are seen in areas of snow and ice, and in Asia, where complex changes to the regional weather and climate have unfolded in recent years. While in Asia the greatest impacts appear to be on the Asian Monsoon, further north and south black carbon soot has very negative effects on the temperatures of the Polar Regions. Measurements show that by darkening the surfaces, black carbon soot has increased the amount of snow melt in the Arctic (and presumably the Antarctic), as well as on mountain glaciers. Increases in melt have reduced the albedo of the Earth, which has led to a net warming of the planet as a whole.

Significant changes have been observed the strength of the East Asian Monsoon which parallels the growth of cities and industry. Modeling work around the middle of the first decade of the millennium, in principle, suggested that the East Asian Monsoon should strengthen because soot blowing onto the snow cover on Tibet causes it to melt faster. This in turn should make Tibet warm up faster in the spring and summer. In turn, this should lead to an earlier and stronger monsoon over India and southern China. However, observations actually show a decrease in the strength of the Asian Monsoon since the 1970s. Unlike in previous geological periods, such as the Eocene warmer conditions are not strengthening the monsoon.

The weakening monsoon is particularly evident over northern China where rains often fail altogether or arrive much later than normal. This failure, or delay, is a significant contributor to the decrease in air quality over China's northern and eastern cities, primarily because the monsoon airflow pattern does not become established until later in the summer. This leaves a stagnant pond of heavily polluted air over the north of the country. When the summer pattern does emerge, the south westerly winds blow the pollutants out into the Pacific.

Interestingly, where modeling did show a failure of the monsoon, it was over eastern China. These models indicated once again that a warmer Tibet should bias any trends and lead to a stronger monsoon. In this instance the increasing warming power of Tibet sucks air more evenly in from the south and south east so that less moisture blew northwards over mainland China. This left parts of the Yangtze basin dry. The researchers concluded that more soot should give a warmer plateau and stronger summer monsoon, yet, that this was likely offset by the effect of a Pacific ocean warmed by the enhanced concentration of greenhouse gases. Thus even though Tibet was warming, the Eastern Pacific was warming faster, which reduced the temperature gradient between land and ocean and thus weakened the Asian monsoon.

The waters were muddied further by research carried out by Veerabhadran Ramanathan (Scripps Institution of Oceanography, University of California) and colleagues and published in Proceedings of the National Academy of Sciences. Their research indicated that even though soot should intensify temperature differences between land and sea, the effect was drowned out by the effect of particles on the region as a whole. These particles caused more overall cooling at the surface that drowned out the effect of the soot lying on Tibet. This effect dominated the impact of soot and thus, overall, also weakened the monsoon.

Thus, the initially simple picture of more soot means more warming and a stronger Monsoon is clearly insufficient to explain the changes to the climate of southern and eastern Asia. This is no trivial observation. The climate of the region critically determines the supply of food for more than two billion people. Moreover, the climate also can enhance or mitigate the health effects of the soot and other noxious gases liberated by the burning of fossil fuels. Therefore, it is vitally important that we understand the discrepancy between observation and theory. Not simply for the prosaic reason that we the need to understand our climate better, but also because the well-being of our species depends upon it. Moreover, the impact of improved modeling will clearly extend our understanding of the climate of planets elsewhere.

Finally, another critical issue is the effect of dimming on the transpiration of plants. As well as the obvious negative consequences of destroying forests directly, reducing the intensity of sunlight reduces the rate of transpiration. The fewer trees that there are transpiring over Indonesia the less moisture is returned to the atmosphere upwind of India. Indeed, experiments show that fully 80–90 % of the water vapor entering the atmosphere over land comes from plants—not from simple evaporation from the soil. Although the "plants give us oxygen" story may not quite hold up to scrutiny (because most oxygen comes from bacteria and algae in the oceans), the contribution forests make to the planetary hydrological cycle is undeniable. Fewer trees will mean lower levels of atmospheric moisture and hence less cloud and precipitation downwind. Burn them and the risk rises for extra tornadoes in vulnerable regions (Chap. 1). Humans chop the forests down at the peril of climatic balance.

Thus while extra carbon dioxide should cause a stronger airflow across Asia, the effect of carbon dioxide is swamped out by the negative effects of soot and other particulates, especially sulfates—and there may be other effects caused by changes in the regional hydrological cycle. Sulfates tend to reside in the atmosphere for longer periods particularly, when they are high in the atmosphere (stratosphere and upper troposphere) and above the level at which rain can wash it out. Sulfates cause cooling by reflecting incoming radiation from the Sun. In the past, sulfates in the high troposphere and stratosphere, primarily originated from volcanic activity and from the oceans. However, since the industrial revolution sulfates also come from combustion of fossil fuels, in particular coal. Around 150 million metric tons of sulfur, mostly in the form of sulfate, is mobilized by human activity each year; and around two thirds of the sulfate in the air is from man-made activity. Although most of this contamination is low level—and thus is washed out fairly readily as acid rain—some does work its way higher and contributes to global dimming. On the positive front, overall in the West, most of this sulfate is now removed from the flue gases so that it has less impact now than it did 30 years ago. Despite this the problem is not solved and sulfates continue to plague many developing nations where less investment has been put in place to remove them from the effluent from coal-fired and oil-fired power stations.

The West cannot claim to be out of danger either. Notwithstanding nearly three or four decades of cleaner air, the lower levels of soot and other particulates produced by cars and other vehicles is still an issue. One only has to look across the skyline of Los Angeles, New York or cities in the UK to see a hazy layer of soot and dust which we all breathe in. This material ultimately blows around the globe affecting the formation of clouds and consequently precipitation. Combustion may be a profitable business but it is also a very messy and damaging one. The best way to deal with it is to replace it as the longer-term impacts of humanity's love of combustion becomes clear.

To put things in a broader historical perspective, an unusual and damaging event around 1500 years ago can help illustrate current risks.

How Global Dimming Caused Geopolitical Chaos in the Dark Ages

Can global dimming bring down an empire? That was the question raised by David Keys at the close of the last millennium. Keys presented a very interesting and rather persuasive thesis that suggested something rather catastrophic had set about the end of the Roman world—and the birth of our modern one. Although much of Key's 1999 book *Catastrophe* is clearly speculative, it makes for an utterly fascinating ride. Moreover, Keys is very careful throughout to underpin his hypothesis with some very real and ultimately quite disturbing science.

To summarize, in 535 AD there were reports of loud detonations to the south east of China. In 536 AD, thick yellow dust began to fall from the skies. Although such precipitations are not uncommon in the winter months (wind-blown loess from western China) these precipitations were followed by snow in July and frost in August: hardly something normal in sub-tropical southern China. Meanwhile, in Europe the Sun was reported to be dim, often taking on a bluish tint. This effect is something typical of airborne dust, which scatters the blue light. The winters in the following years were extremely harsh, while the summers were uncommonly cool.

Ice cores in Greenland and Antarctica reveal the presence of a spike in the levels of sulfuric acid (hydrogen sulfate) at this time. As we have seen, this key chemical is produced by the burning of fossil fuels and by volcanic activity. Given that Middle Age humans were not accustomed to burning fossil fuels, its presence in both polar locations in such abundance indicates that a large volcanic eruption occurred near to the equator around 535 or 536 AD. The location of the eruption was important, for only an equatorial eruption could scatter large quantities of sulfate into the atmosphere of both hemispheres, and as such only an equatorial eruption could affect the climate and geopolitics of both hemispheres.

An equatorial location narrowed down the likely culprits. The eruption had to be large and explosive, which further narrowed the search. Keys suggests that Krakatau—famous for its 1883 eruption—was the likeliest candidate. Working with Keys, Ken Wohletz (Los Alamos National Laboratory) produced some likely scenarios for an eruption that would produce the desired effects. Geographically, what is interesting about Krakatau is that the island that was obliterated in the massive 1883 eruption itself lies within a much larger caldera—a volcanic structure produced by the collapse of a much greater volcanic edifice. This arrangement, flanked by enormous eruption deposits suggests that prior to the 1883 eruption Krakatau had at least one other, even more powerful eruption that had obliterated the earlier, and much larger, volcano. The famous 1883 eruption was a mere blip that occurred after this. Volcanologist, Harald Sigurdsson, took part in the production of the TV series[1] that accompanied the book and although he was unable to definitively date the deposits, he did narrow the eruption to a (broad) window that encompassed 536 AD using radiocarbon dating.

Bathymetry reveals that the original volcano that gave rise to this massive outer caldera was at least 50 km across and perhaps, given the scale of the later volcano, 1000 m or more high. To produce the observed caldera, the eruption would have had to release over 200 cubic kilometers of rock—or 200 times that of Mt St Helens in 1980. Although dwarfed by much larger eruptions, such as the Siberian Traps (Chap. 3), this is still a substantial eruption that would have devastatingly global impacts today.

Wohletz's models show an eruption column that is perhaps 50 km in height, discharging rock and gases at a rate of over 1 billion kilograms (1 million metric tons) per second. Given the amount of rock, Wohletz suggested that the eruption was sporadic and took place over several weeks or months—similar to the 1883 eruption. After an initial violent phase, the volcano would have begun to collapse, generating a lower, broader eruption column and more violent explosions. Perhaps it was these that were recorded by the

---

[1] Secrets of the Dead; Channel 4 UK, 1999.

Southern Han Dynasty. Up to 150 cubic kilometers of seawater would have been vaporized either through contact with hot flows of volcanic debris or through direct contact with the underlying magma when the volcano began to collapse. This would drive a column of very hot steam high into the stratosphere, where it would combine with sulfurous gases to produce even more sulfuric acid.

An eruption of this magnitude could produce a cloud layer over 20 m thick in the troposphere, with a finer but thicker spray of sulfates and ash higher up in the stratosphere. This would have lasted for months, dimming the sunlight reaching the ground underneath. Therefore, beneath this translucent layer, global temperatures would have plummeted for several years leading to the cooling evident in tree ring data.

A more recent and precise analysis of ice core data paints and even grimmer picture to the one initially envisaged by Keys and Wohletz. Work by Michael Sigl (Desert Research Institute, Nevada) as part of a larger collaboration, indicates that there was more than one large eruption. Around 536 AD, the ice cores show a signature of possibly three large eruptions that probably went off in North America—most likely Alaska or British Columbia. These left a signature in Greenland's ice cores, but not Antarctica's and lowered temperatures by 1.6–2.5 °C. Devastating though these eruptions would have been to life in the Northern Hemisphere, they barely touched the Southern Hemisphere because the Hadley cells of the north and south barely communicate. However, these eruptions were followed 4 years later by an eruption in the tropics, possibly in Indonesia at the site of Krakatau. All four of these eruptions were very large—dwarfing Tambora and adding a significant baggage of dust and sulfate to the atmosphere. This final assault on the climate lowered temperatures by a further 1.4–2.7 °C. This cooling then lasted until 550 AD and making the newly arrived Dark Ages truly dark.

Quite aside from the obvious effects of prolonged cooling and general drying out of the atmosphere caused by more limited evaporation, cooling could have a more insidious effect. Keys pointed the finger of cooling as a causative agent of an outbreak of the Black Death which decimated the Justinian Roman Empire. He identified an effect of prolonged drought followed by abundant rain as a mechanism that would have allowed plague bacteria to jump from their resistant endemically infected rodent hosts to other rodents, including Black Rats. A drought would have led to famine and a drop in the resident population of rodents. The prolonged period of cooling would have caused drought because less water was evaporating from the oceans. Several years after the global catastrophe, when the planet warmed up once more, the rains would have returned and allowed the rodent population to boom. This larger population of rodents would then have come into contact with black plague and then humans. The Justinian trade routes then became the unwilling vectors that carried their lethal bacterial passengers around the Roman Empire.

There is another, more recently discovered means by which climate change could facilitate the spread of disease. Many pathogenic bacteria, such as plague (*Yersinia pestis*) or cholera (*Vibrio cholerae*) derive their pathogenicity from resident viruses. These viruses bequeath their bacterial hosts with additional traits that allow the bacteria to cause untold misery. For example the cholera bacterium contains circular pieces of DNA called plasmids which carry the gene encoding the poison that causes the potentially fatal diarrhea. This plasmid is clearly derived from a type of virus called a bacteriophage that infects these cells. Strains of the cholera bacterium that lack this plasmid are effectively harmless to humans. Similarly, plague bacteria harbor a set of related plasmids that encode features allowing the bacteria to invade tissues and survive in the bloodstream. The original *Yersinia* bacterium was more like the largely harmless *E. coli* bacterium that is a resident of the guts of every human on the planet. However, with a handful of additional genes, the *Yersinia pestis* bacterium became an aggressive invader of rodent then human tissues, which ultimately leads to the characteristic infections of Bubonic, Pneumonic or Septicemic Plague. The interesting thing about these bacteria is that temperature appears to play a big part in which of these disease-causing genes function. Some gene appear to work best in the flea with a body temperature of around 25 °C, while others work best at mammalian blood temperatures of 37 °C. Now, although the key driver of these genes will be the temperature of the host organism, playing with environmental temperatures will undoubtedly also impact on the working of the bacteria, as well as the behavior of the host insect and mammals.

Of greater concern is related research by an international team led by Martha Clokie (University of Leicester). Their research focused on the behavior of a bacteriophage that infects pathogenic bacteria in the tropics. Both the bacterium and accompanying viruses are present in soil throughout the region. When these viruses infect cells that can either go through what is known as a lytic cycle and bursts open (lyse) the cells; or they can go through what is referred to as a lysogenic cycle, where the virus remains dormant and only replicates with the cell. At 25 °C the viruses tend to replicate with the cell in a lysogenic cycle, while they preferentially use the lytic cycle at 37 °C.

You might think that having a virus kill the cells at higher temperature would be a good thing for us and that global warming would tend to favor the virus lifecycle where the pathogenic bacteria is killed. However, this is too narrow a view of evolution—and of climate. For one, global warming might not make the area in which the virus and bacteria live warmer. Indeed, if the area was tropical, at altitude, it might become wetter. This would make it cooler favoring the persistence of the virus within the bacterium. If that virus made the bacterium more pathogenic then humans could be in trouble.

Alternatively, at higher temperatures there is a clear evolutionary benefit to the infected bacterial pathogen if it can avoid death by lysis. This can happen in a number of ways that lead to the persistence of the virus and work against the virus killing its host cell. This change can happen through mutations to the virus or changes to the inner workings of the bacterial cell. If that virus happens to produce any toxins, or can otherwise affect the immunity of the human or animal the bacterium infects, that infected pathogen can become a lot more deadly. Indeed, it is clear that many human pathogens, cholera included (and perhaps others like the plague) have become modified by the addition of viruses that then became adapted to live harmlessly in the bacterium at higher temperatures.

The discovery that certain pathogenic bacteria might play host to their own pathogens certainly complicates out view of the natural world and should make us wary of experimenting with its climate. This is a new and active area of research and one worth keeping an eye on. At present the precise outcome of each of the complex evolutionary interplays needs further scrutiny.

No organism exists in isolation and the laboratory investigations will not reveal the full repertoire of behaviors these organisms show in the wild. These investigations do reveal a new possible impact of changing global temperatures. Bear this in mind in Chap. 3's exploration of the planetary turmoil we and other species have coped with over past millennia.

Conclusions

This chapter and the next set weather and climate within a broader terrestrial framework to examine the broader context of climate before moving on to discuss the climate of specific planets. Climate is the greatest natural phenomenon on the Earth and its impact on us, via natural catastrophes and more routine patterns, is as large as humanity's own influence on climate has become today. By considering these issues together a broader canvas of planetary climate is created, one that we can apply to the many new worlds we are exploring and discovering.

References

1. *Modelling of the possible 536 AD Krakatau eruption.* Retrieved from http://www.ees.lanl.gov/geodynamics/Wohletz/Krakatau_6th_Century.pdf.
2. Knirel, Y. A., Lindner, B., Vinogradov, E. V., Kocharova, N. A., Senchenkova, S. N., Shaikhutdinova, R. Z. et al. (2005). Temperature-dependent variations and intraspecies diversity of the structure of the lipopolysaccharide of *Yersinia pestis. Biochemistry, 44* (5), 1731–1743.
3. Ramanathan, V., Ramanathan, V., Chung, C., Kim, D., Bettge, T., Buja, L. et al. (2005). Atmospheric brown clouds: Impacts on South Asian climate and hydrological cycle. *PNAS,102,* 5326–5333.
4. McKitrick, R. R., & Michaels, P. J (2007). Quantifying the influence of anthropogenic surface processes and in homogeneities on gridded global climate data. *Journal of Geophysical Research, 112*, D24S09.
5. Zhang, Z., Ramstein, G., Schuster, M., Li, C., Contoux, C., & Yan, Q. (2014). Aridification of the Sahara desert caused by Tethys Sea shrinkage during the Late Miocene. *Nature 513*, 401–404.
6. Licht, A., van Cappelle, M., Abels, H. A., Ladant, J. -B., Trabucho-Alexandre, J., France-Lanord, C., et al. (2014).Asian monsoons in a late Eocene greenhouse world. *Nature, 513*, 501–506.
7. Dupont-Nivet, G., Krijgsman, W., Langereis, C. G., Abels, H. A., Dai, S., & Fang, X. (2007). Tibetan plateau aridification linked to global cooling at the Eocene–Oligocene transition. *Nature 445.*
8. Shan, J., Korbsrisate, S., Withatanung, P., Adler, N. L., Clokie, M. R. J., & Galyov, E. E. (2014). Temperature dependent bacteriophages of a tropical bacterial pathogen. *Frontiers in Microbiology, 5*, Article 599, 1–7.
9. Sigl, M., Winstrup, M., McConnell, J. R., Welten, K. C., Plunkett, G., Ludlow, F. et al. (2015). Timing and climate forcing of volcanic eruptions for the past 2,500 years. *Nature 523*, 543–549.

3. Tales of Mass Destruction

Introduction

This chapter illustrates the fluid nature of planetary weather and climate. Climate is a dynamic creature that changes with the activity of the planet's central star, the Sun, and with other dynamic forces that operate on and within planets. The climate of the Earth has repeatedly changed dramatically over the eons since it has formed.

Moreover, humans have become particularly adept at altering their environment. These activities have taken the planet down a new and at present unpredictable path that will impact all other life forms. Yet, dramatic though these changes are, they will be transient in nature with Earth driving the final outcomes. At the end of course the planet will meet an inevitable death inside the dying Sun—but there will be lots of excitement on the way.

Global Greenhouses: Eocene, Permian and Anthropocene

Utter the words "greenhouse effect" and suddenly attitudes polarize in the United States and United Kingdom, though not as much elsewhere. Conversations change from idle chatter to politically divisive personal opinion. In order to get to the bottom of all this social and political angst we will look at global greenhouses in three different contexts: the Permian Mass Extinction; the Eocene Hyperthermals and most recently anthropogenic warming. As the underlying mechanism is the same in each case, we should and can expect similar outcomes, at least in terms of process if not (hopefully) the degree of change.

© Springer International Publishing Switzerland 2016   91
D.S. Stevenson, *The Exo-Weather Report*, Astronomers' Universe,
DOI 10.1007/978-3-319-25679-5_3

The Eocene Climate Maximum

Fifty five million years ago the world's climate experienced a rather puzzling incident. Starting rather abruptly, a period of around 86,000 years ensued when global temperatures rose between 6 and 8 °C above their current value. The warming, known as a hyperthermal, was truly global in its complexion. Occurring at the end of the Paleocene and start of the Eocene, the Paleocene-Eocene Temperature Maximum or PETM was followed two million years later by a second somewhat smaller temperature surge called the Eocene Thermal Maximum 2 (ETM2). Finally, 1.6 million years later at 52.6 million years ago a third hyperthermal (ETM3) occurred. The origin of these warming periods initially appeared to be utterly mysterious. Each period is associated with an apparent surge in the levels of carbon dioxide, without a clear reason in the biological record. So what happened?

Two largely competing, but successful, models have been proposed. Initial hypothesis suggested a geological signature in the hyperthermal, while the most recent proposal looks further afield to the cosmos. In 2004 Henrik Svensen and colleagues (University of Oslo) suggested that the event, which is superimposed on a longer period of global warming, was linked to the opening of the North Atlantic. Around 55 million years ago, large continental rifts were extending northwards, severing Europe from North America and Greenland. The climax of this process was the eruption of large volumes of basaltic magma from volcanoes in the area. The remnants of this outburst stretch from Northern Ireland, through western Scotland and northwards to Greenland and western Norway. Here, a magnificent series of basaltic plateau and extinct volcanoes play testament to the arrival of a particularly large batch of magma that would ultimately give rise to Iceland.

As this magma torched its way through the crust, Svensen and colleagues proposed that it interacted with large amounts of buried organic material in the form of sedimentary rocks. As these were progressively cooked by the intruding magma, around 300–3,000 billion metric tons of methane was released into the atmosphere. Methane is a potent greenhouse gas in its own right and this surge of methane would readily explain the sudden rise in global temperatures. The methane would then oxidize to carbon dioxide, which although a weaker greenhouse agent than methane, would still perpetrate a period of more prolonged warming, lasting tens of thousands of years. Together with volcanic carbon dioxide, the total input of CO_2 was probably double this (Table 3.1).

The alternative model is less dramatic but perhaps ultimately more compelling. In 2012 Robert M DeConto (University of Massachusetts) and coauthors published an alternative model that owed its origin to the work of the Serbian astronomer, engineer, mathematician and geophysicist Milutin Milanković. In Chap. 2 how the Milanković cycles influence the climate of North Africa in the last few million years was discussed, but their effects can be traced even further afield. In the Eocene their cyclical effect on warming in the Arctic is apparent in reconstructed temperature records.

In DeConto and his colleagues' work, it is suggested that Milanković cycles triggered the hyperthermals by themselves. Rather than an extreme bake-off involving volcanism, DeConto suggested that the alignment of tilt, precession and orbit tipped the Earth across a climatic divide where the northern permafrost catastrophically melted. They quote the apparently rapid rise in temperature over a short 10,000 year window, as well as the cyclical pattern of the events that appear to repeat with a two million year period. This period, corresponds to a longer cycle over which there are changes to the shape of the Earth's orbit and its axial tilt. The Milanković cycles, mentioned above, which are relevant to today's climate, are embedded within these (Fig. 3.1).

In the world of 55 million years ago, carbon dioxide concentrations were higher than at present to begin with. With 700–900 parts per million (ppm), the level then is more than twice the contemporary value of 400 ppm. More importantly, the arrangement of the continents allowed warm waters to circulate much more efficiently from the tropics to the poles. The combination of higher greenhouse gas concentrations and an underlying climate that was biased towards warmth ensured that the planet was ice free at the poles. Temperatures in the Arctic may have been as high as 24 °C during the summer months. Antarctica, although far from frigid, was at least cold enough to have permafrost underlying its forest and tundra. This was also true of northernmost North America and Eurasia, which were in the process of fissuring.

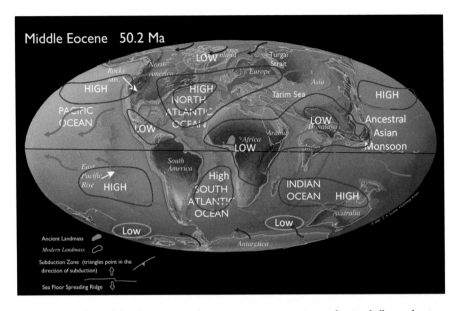

FIG 3.1 A plausible, but speculative representation of wind flow during the northern summer in the early Eocene. The continents were assuming their modern positions and the rising Himalaya were beginning to generate something akin to the modern Asian Monsoon. Most of the moisture for this came from the south and from the western Tarim Sea. Tropical conditions extended 50° north and south of the equator. Presumably, westerly (zonal) winds were weaker than present to allow much stronger transport of heat to the Poles from the tropics. Paleomap courtesy of Christopher Scotese

As the Eocene began, the planet crossed a threshold where the right combination of Milanković cycles and underlying atmospheric carbon dioxide would thaw the permafrost at both Poles. When this first happened, 55.3 million years ago, around 3000 trillion grams (or 3000 billion metric tons or gigatonnes) of methane would have been liberated over a period of ten millennia. This was enough to raise global temperatures for a few hundred thousand years, until the permafrost was exhausted and the methane converted first to carbon dioxide then inorganic carbonate rocks on the ocean floor.

With the ebb and flow of the Milanković cycles, the permafrost would partly reform before thawing out once more. Yet each refreeze was less than the time before because the underlying climate remained warmer than it had been previously, thanks to the carbon dioxide that remained from past thaws. Eventually, the permafrost was largely eliminated and the process ended at around 52.6 million years ago. For the remainder of the Eocene, temperatures remained several degrees warmer than today, but gradually lowering levels of carbon dioxide and continued reshuffling of the continents would eventually bring the hothouse Earth to an end.

By 34 million years ago, India was fully engaged with Eurasia and the Tibetan plateau began to rise in earnest (as seen in Chap. 2). Around the same time Australia uncoupled from Antarctica, opening up the southern ocean and allowing the circum-Antarctic currents to become established. This largely cut Antarctica off from the rest of the planet and allowed the air over it to chill even more starkly. Antarctica then swung round so that it sat directly over the South Pole. As Tibet rose, carbon dioxide was flushed out of the atmosphere as Tibet's rocks were eroded. A chilling Polar Antarctica reflected more of the Earths radiation in the winter and released heat more efficiently than a Polar Ocean ever could. Over the next 20 million years the continent would slide into a near eternal deep freeze. As it disappeared under highly reflective ice, and carbon dioxide continued to fall, the stage was set for the next wave of ice ages—the underlying level of carbon dioxide is the driver of the change. The Milancović cycles are superimposed on top of these and provide the icing on the meteorological cake.

Were something akin to the Eocene Hyperthermals to occur today, the outcome would be disastrous. Such a rapid rise in global temperatures would cause significant melting of the ice caps and the flooding of much of the planet's most habitable regions for humans. Sea levels were a good 200 m or more higher in the Eocene than at present. Although part of that rise is to do with changes to the depth of the ocean floor, much of it is clearly linked to the melting of the polar ice caps. In this regard, it is important to remember that we are increasing carbon dioxide at an unprecedented rate. From 1800 until present we have added nearly a third extra mass of this greenhouse gas to the atmosphere. At 400 ppm (and with a rise of nearly 15 ppm per decade) we are well on our way to recreating the pre-ice age hothouses of the last 60 million years. While it seems highly unlikely that we will see a temperature rise of over 5 °C in the next 85 years, a still sizeable rise of at least 2 °C is likely, given our predilection for generating this greenhouse gas. Since 1900 sea levels have risen by 20 cm and this rate is accelerating. The most recent measurements from NASA suggest that sea levels are rising faster than predicted and will rise a further minimum of 30 cm by the end of the Twenty first Century. A rise of a meter is now increasingly likely.

Critical though such an event would be for modern man today, 55 million years ago, it was also crucial in an altogether more profound way. For around this time, the Eocene warmth converted much of continental Eurasia into a lush tropical swamp, despite it lying at a latitude similar to that of today. Somewhere within this swamp the first primate evolved. Thus, while much of the planet drowned under warm tropical waters, our immediate ancestor was taking its first steps along the branches of some ancient, greenhouse gas-driven tropical paradise.

If we inadvertently create a similar climate in today's world, then would it be so bad? Well, the rise time for carbon dioxide was around 10,000 years in the Eocene. It's happening in around one quarter to one eighth that time-span now. Evolution through natural selection is a remarkable thing, but its capacity for delivering rapid change is not boundless, and while the resulting ecology could be a fruitful one for new species, it might prove ruinous for humans and many other existing life forms.

A Bad Day in the Permian

Although the Eocene saw a rise of around 7 °C over the space of around ten millennia—very much a blink in the cosmic eye—the rise appears to have been embedded in a much more gradual rise in temperature that dates to the end of the dinosaurs, ten million years earlier. Indeed, during most of the time the dinosaurs were around, the same combination of favorable plate tectonics, high sea levels and higher than present carbon dioxide ensured the planet was a few degrees warmer than it is now. Thus the Eocene warmth was not catastrophic in its outcome. However, 200 million years earlier things were a little different.

251 million years ago the planet's continents were arranged in one large supercontinent, encircled by a vast ocean, Panthalassa. A wide tongue of water extended inwards broadly along the equator, known as the Paleo-Tethys Ocean. This ocean was the immediate precursor of the Tethys Ocean but had broadly the same area and

shape. The Paleo-Tethys was partly separated from Panthalassa along its eastern flank by a string of micro-continents and island arcs that would ultimately become China, Indochina and Indonesia. Over much of what is now Asia, warm waters connected the Arctic basin with the Tethys.

At the heart of the continent lay a string of recently formed mountains that separated the future Africa from North America. These would have intercepted the prevailing tropical easterly trade winds and delivered abundant rainfall to an area roughly the size of the modern Amazon basin. North of the mountains, was a large rain-shadow, covering most of what is now Western Europe and continental U.S.A. Further south, under the belt of Horse Latitude high pressures would lie another arid region, which would grade through semi-arid steppe to taiga and tundra far to the south. Over the southernmost portion an ice cap remained—the remnant of a much larger continental ice sheet that had extended across much of Gondwanaland 50 million years earlier.

The region that is now Siberia was partially covered in shallow seas, with much of the remainder forested, much as it is now. Figure 3.2 shows a speculative, but plausible arrangement of high and low pressure areas associated with the climate of the day.

The micro-continents that slice across the eastern end of the Paleo-Tethys are important as they will isolate this basin from the wider deep circulation that filled Panthalassa. Despite a slightly dimmer Sun (96 % its current luminosity), higher levels of carbon dioxide and a favorable arrangement of continents, which directed warm currents north and south around their edges, ensured that the Permian was a largely warm epoch.

Four million years after the planetary snapshot shown above, something dramatic was to afflict the planet. For reasons, largely unknown, that may link to the formation of the supercontinent 50 million years earlier, a hot plume of rock rose from the base of the Earth's mantle and began to whittle away at the crust under Siberia. Perhaps after doming the crust somewhat, Siberia split open and, as the late, great volcanologist Maurice Kraft would have said, the blood of the Earth poured out. An area roughly bounded 2000 km from north to south and 4000 km from south west to north east was covered in hundreds of meters of molten rock and ash. The eruptions began around 251 million years ago and lasted

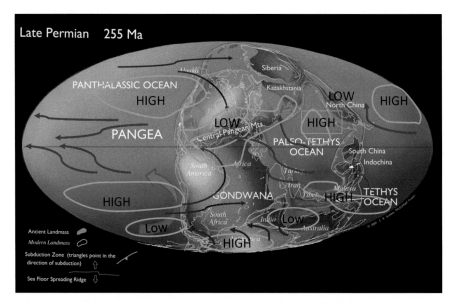

FIG 3.2 A re-iteration of the late Permian paleomap to illustrate the diversity of Pangean landscapes. Rumors of a vast uninterrupted continent dominated by deserts are vastly exaggerated. Equatorial regions were dominated by the Appalachians, formed when Africa and North America (Laurasia) collided. On the western flank of these mountains would be warm tropical rainforest, grading into savannah then desert to the south. A monsoon climate similar to India, would be likely. To the northwest of these mountains, central North America and Europe would have been hot desert, grading into steppe and tundra to the north. The Eurasian portion was largely submerged under fairly warm seas, giving it a climate similar to western Europe. The large southern ice cap was largely gone with steppe, taiga and tundra surrounding the pole. Underlying paleomap courtesy of Christopher Scotese

for a few hundred thousand years. Amongst the litany of disasters that ensued, lavas poured into the Arctic Ocean vaporizing methane clathrates on the ocean floor; they disintegrated and incinerated developing coal beds and also likely explosively decimated the permafrost along Siberia's northern edge.

Vast in their own right, such eruptions could not directly do much damage outside Siberia. However, at the end of the Permian almost all major land animals and much of those in the oceans were extinct. The planet was cooking in its own juices and life—aside from microbial life—was nearly obliterated in its entirety. How then do we go from the eruption of lavas in one corner of the globe to near obliteration of the Earth as a habitable planet?

The key to understanding the near annihilation of life was the temperature of the globe and the arrangement of the continents: both of these factors appear to have conspired with the greenhouse conditions established by the eruptions to change the composition of the oceans and the atmosphere.

Let's picture the scene. At the end of the Permian, dinosaur-like animals dominated the land along with many reptiles and the humblest beginnings of the mammals. The newly assembled Pangaea has a diversity of landscapes from a sizable ice cap in the south, through deserts, tropical swamps and open grassland. The levels of carbon dioxide are roughly three times that at present, giving rise to a very temperate world. Into this world 251 million years ago, poured over 1,700,000 cubic kilometers of basalt and pyroclastic material[1]—in places over 2 km thick. Such a volume of basalt would cover all of China to a depth of around 300 m, or the punier UK to a depth of more than 12 km. The eruption would have been accompanied by the steady release of various gases, including gigatonnes of sulfur dioxide, hydrochloric acid, hydrofluoric acid and much more substantially 1200–1800 gigatonnes of carbon dioxide—although the final amount of this may have been far greater (Table 3.1).

Table 3.1 Comparative changes in carbon dioxide levels, temperatures and the time over which these happened at two natural and one man-made greenhouse

Period	Temperature change (°C)	Period for temperature change (years)	Total mass of CO_2 (Gigatonnes/Gt)	Averaged rate for CO_2 addition (Gt/year)
Permian (PT) boundary	8–10 over 60,000 years	60,000	5400	0.09
Ecocene	5–8 over 10,000 years	20,000	2–7000	0.1–0.35
Anthropocene	0.9 over 200 years	200	~340 since 1901	3.4

Changes in the Permian were gradual, occurring over many millennia. Those in the Eocene, although less extreme were faster (perhaps 100 centuries). Currently, humans are setting the pace with the fastest rate of carbon dioxide rise and the fastest rate of temperature change (approximately 1° per century versus 0.05–0.08° per century for the Eocene and 0.016 °C per century for the close of the Permian). The figure for the anthropocene is an averaged one for the last 116 years. It hides the fact that most of the input of carbon dioxide has occurred since 1940

[1] This is an absolute minimum with the likely volume exceeding three million cubic kilometers, or more than twice the Deccan Traps in India.

Seth Burgess (M.I.T.) and colleagues narrowed down the Permian extinction to a window only 60,000 years wide sandwiched between 251 million 942 thousand years ago and 251 million 880 thousand years in the past. Through the examination of carbon isotopes, the authors identify a major change in the way carbon was being cycled through the ecosystems of the planet immediately before the extinction event. It takes another 500,000 years before the system appears to return to normal. In that 60,000 year window, the biosphere contracted rapidly as species after species became extinct. The 500,000 year period that followed marked the recovery of the planet, where the surviving species begin to radiate and fill the vast bank of free habitats that were now available.

The method of annihilation appears to have been anoxia: a lack of oxygen. Almost certainly this was accompanied by an altogether smellier killer: euxinia, a lethal combination of lack of oxygen and a wealth of hydrogen sulfide. The arrangement of the continents ensured that there was very limited cold, deep water circulating from the poles. In the modern Earth, this method drives cold, oxygen-rich waters from the Polar Regions towards the equator through the ocean depths. This guarantees a supply of oxygen to the planet's deepest waters maintaining an oxygen-rich ocean. Whenever the waters get too warm, ice will obviously be limited and this deep, cold flow of oxygen-rich water will cease. The Paleo-Tethys, in particular, was vulnerable to becoming anoxic as it was partly cut-off from the deep circulation along its eastern flank.

Instead of temperature differences driving circulation, in the relatively isolated Paleo-Tethys circulation would have been driven by evaporation from the surface. In this system a dense, warm, brine-rich current develops that takes warm, oxygen-poor water down into the ocean depths. Once the deep ocean has been filled with this warm, oxygen poor water, anaerobic (oxygen-hating) bacteria thrive. These generate noxious hydrogen sulfide gas as well as methane and carbon dioxide. Hydrogen sulfide is directly toxic to most oxygen-breathing organisms and it reacts with dissolved oxygen forming sulfates that rain out onto the ocean floor. Thus the oxygen levels are kept low by those organisms that hate it.

Investigations by Paul Wignell (University of Leeds) and Richard Twitchett (University of Plymouth) revealed that as the extinction got underway, oxygen levels, even at depths as shallow as

1 m, fell below those necessary for the survival of marine animals that lived there. Except at the very shallowest levels, the entire volume of ocean water on the planet appears to have become anoxic and there are many indications that it was probably flooded with hydrogen sulfide. The nail in the coffin of the Permian was the release of thousands of gigatonnes of carbon dioxide. Although carbon dioxide dissolves in rain water and ultimately precipitates out into the oceans, it has a fairly long residency time in the atmosphere. This ensures that with a sufficiently large and sustained supply, the amount in the atmosphere will build up. The release of carbon dioxide appears to have compounded pre-existing levels which were already high: perhaps three or four times the current level. Adding even more carbon dioxide was a critical event. Such a large amount of greenhouse gas ensured the ocean circulation would shut down and allow bacteria to ferment the gases that would eliminate life.

Although sulfate aerosols from the eruptions, as well as soot from incinerated forests, would have provided a temporary cooling blanket, soon thereafter the excess carbon dioxide would have taken charge of global affairs and driven the temperatures sky-high. Fossil evidence indicates that global temperatures rose sharply at the end of the Permian and this almost certainly relates to the massive release of carbon dioxide by the Siberian eruptions.

With temperatures likely 10 °C higher than they are at present, the oceans became so hot at the surface that they ceased to circulate in all but the feeblest of manners. Deep waters became effectively oxygen free and the marine extinctions that ensued were as inevitable as they were catastrophic. These conditions would favor the generation of more hydrogen sulfide, which is produced under anaerobic conditions. Warm water also holds less dissolved gas, so warmer waters would favor the release of the gas to the atmosphere. Further compounding this effect would be the addition of large quantities of sulfate from the volcanic eruptions, which can be chemically reduced to hydrogen sulfide by bacteria living in the oceans. Therefore, soon after the cooling blanket of sulfate was washed out, even more hydrogen sulfide would have been available from the oceans to further poison the atmosphere.

Meanwhile, on land animals were cooked, while what remained of the polar caps melted and flooded many other habitats—not with clear ocean waters, but with a noxious brew containing sulfurous gases. That complex life survived at all is a testament to the strength and ingenuity of evolution. Were such a catastrophe to unfold today it would be far from clear if humans would survive. Adding to the potential woes, halides such as hydrochloric and hydrofluoric acid, which were released by the extensive volcanic eruptions, would deplete atmospheric ozone. Potentially, this would have exposed surviving species on land to lethal amounts of ultraviolet radiation.

At the time, mammals were far from the top species. Most large animals were reptilian in nature. The problem these animals have is their metabolic activities, the level at which their cells tick over is dependent on the external temperature. Such cold-blooded, or poikilothermic, animals cannot maintain a constant internal temperature and merely respond to that of their environment. This spells double-trouble for the reptile. If the environmental temperature exceeds its safe level and it cannot do enough to cool down, the chemical machinery in its cells will cease to function. The high temperatures of the Permian would have been bad enough in this regard, but the high temperatures would also have sped up the metabolism of the animal. Metabolism in all complex animals depends on respiration and in such organisms this requires oxygen. While oxygen would likely have been available in broadly the same quantities as it is now, the gas hydrogen sulfide competes with it. High levels of hydrogen sulfide would have prevented the cells from generating the energy they needed to keep the organism alive in such high temperatures: many organisms would simply have asphyxiated under these competing demands.

Yet, much like the fallout from the milder Eocene greenhouse, as far as evolution was concerned, catastrophe was the mother of invention. While many of the dinosaur-like large animals went extinct, the gaps that opened up were soon filled by the ancestors of the Dinosauria and most importantly for humans, by the modern mammals. Were it not for this short but violent period of annihilation, we would not be here today.

What is most interesting is the suggested rate at which temperatures rose. You might imagine that such a catastrophe was associated with that 10 °C rise over a couple of hundred years, but apparently not. The vast release of carbon dioxide happened over a fairly protracted period of time. Instead of rapid change, the increase was 1 °C every 6000 years, on average. Hold that figure in your head as we move on to look at the current rise in global temperatures. If you have a head for another figure it is worth comparing the outpouring of carbon dioxide in today's world with that in the Permian mass extinction: burning fossil fuels is currently releasing around 7 gigatonnes (Gt) per annum. Compare this with 0.018 Gt released on average by the Siberian Trapp eruptions every year of the 300,000 years or so that the event took place. The numbers for our contemporary period are rather frightening in comparison.

Perhaps, as a group of M.I.T. and Chinese researchers have suggested, alongside the global spectacle of extreme volcanism, there was an altogether more pervasive yet unseen killer. The researchers suggested that a lone group of bacteria may have helped annihilate much of the contemporaneous life on Earth. Although this idea may seem "out there" in terms of its impact, it is far from ridiculous. These *Methanosarcina* species have the capacity to consume abundant organic chemicals called acetates that are found in decomposing organic matter and convert it into methane—our very potent greenhouse gas. Although widespread, the activities of this bacterial species are largely held in check by a lack of a crucial metal cofactor: nickel. Without this element the enzyme that converts acetate to methane is inactive. Unfortunately, so the story goes, the basaltic lavas prevalent in the Siberian Traps are loaded with this element. The researchers propose that as the eruptions unfolded, nickel that was abundantly present in the ash fertilized the ocean depths. As this element rained down, the *Methanosarcina* set to work, growing rapidly in number and releasing vast amounts of methane. In concert with the carbon dioxide from the eruptions, the methane added greatly to the greenhouse effect—turning the planet into the hellish oven that led to the wipe-out of plant and animal species that demarcated the end of the Permian.

In the end the Permian mass extinction probably had many immediate causes, but with very little doubt the final squeeze of the trigger was the release of the Siberian Trap lavas. This set about initiating fatal anoxia in the oceans; the emission of gigatonnes of carbon dioxide; the release of copious hydrogen sulfide; and possibly runaway growth of a particular strain of oceanic bacteria that vented methane. A planet that was already close to its temperature maximum was shoved over the edge and 95 % of ocean species and greater than 70 % of those on land were wiped off the planet. That the Permian mass extinction is the only one known to have drastically affected insects is also likely to be significant and a testament to the event's severity. Insects are very sensitive to oxygen concentrations in the atmosphere (and while larvae, in water). This could be another hint that oxygen levels fell, while hydrogen sulfide rose. Could such an event happen again? Perhaps, given the manner we are fiddling with our climate and taking into account the vagaries and unpredictable nature of our planet's geology. It must also be a consideration when we look further afield at life elsewhere in the universe.

Finally, we turn our attention next to the anthropocene global greenhouse: the era where humans are affecting the planet at a rate far in excess of any natural cause. How does our global greenhouse measure up to the two that we have already discussed?

The Human Factor

It's an interesting and perhaps unavoidable fact that most people's belief systems are based on what those around them subscribe to, rather than being derived from their own personal experiences and learning. After all, why bother going to the trouble of deriving something from first principles when you can draw on the accumulated knowledge of your peers. Such a system of derivation works fine for pretty much everything we do in everyday life, whether it is choosing the best brand of pizza or finding the cheapest deals online. However, when it comes to systems that require more than daily experience, or ones which require derivation over time intervals longer than a human life-span, the collective body of current peer knowledge is often a poor guide.

Instead of consulting peers for opinion, when it comes to most of the subjects that actually impact on our longer term survival we substitute our opinion with those of experts in whatever field is relevant. Therefore, if we have chronic pain we consult a physician. Although we might make some initial diagnosis ourselves, ultimately if the pain persists we out-source our opinion for that of an expert who has been formally trained to diagnose and treat our malady. Similarly, if the lights don't work in our home, we hire an electrician. In each case the person we consult is formally trained in the relevant field.

Now, aside from matters of life and death, most of us like to think we are experts in popular science and don't look beyond the information that often validates our personal view point. Most people simply don't have access to the original data or peer-reviewed work that is published. Instead we access the canned views that are available in popular media outlets or the Internet. This is hardly surprising when the language of primary research is fairly impenetrable to the armchair scientist, and more importantly may cost an arm-and-a-leg to access in the first place. This limited access is a serious problem, for it is the peer-reviewed work that is most likely to be valid and reliable—two keywords in science. After all anyone can say what they like on the Internet: there is absolutely nothing to guarantee that it is even vaguely true. Thus many inaccurate, untruthful and often downright bizarre interpretations of climate science appear, which will be summarized in the next section.

Anthropogenic Global Warming

There are four different viewpoints that people of different persuasions adopt when looking at the climate data. By far the most commonly adopted in the scientific establishment is that the Earth's atmosphere and oceans are warming mostly as a consequence of rising levels of carbon dioxide and, to a lesser extent, methane and other greenhouse gases produced by human activities. The Sun and other factors, such as the tilt of the Earth (discussed in Chap. 2), contribute to, but are no longer driving the observed changes.

The second viewpoint is that the Earth is warming but that it's part of a natural cycle, most probably driven by the Sun.

The third viewpoint, which partly overlaps the second, is that the Earth is no longer warming and that this "climate pause" demonstrates that rising carbon dioxide plays no part in global warming (or in extreme views) the global greenhouse at all. Given that temperatures have now risen above the 1998 peak for 2 years in a row, this view has perforce somewhat fallen by the wayside.

The fourth point of view is probably the most radical, and posits that the Earth is in fact cooling, not warming, and that we

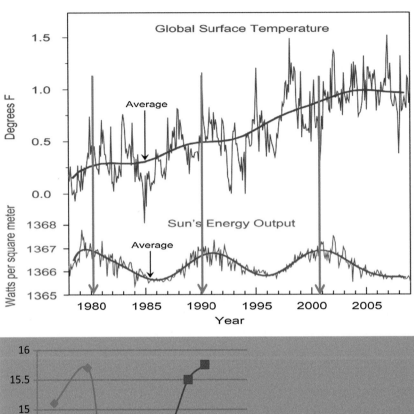

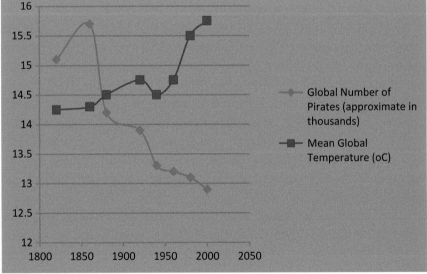

Fig. 3.3 Myth 1—Global surface temperatures are driven by sunspot number/changes in the output of the Sun. (Solar output is linked to the number of sunspots through complex changes in surface and atmosphere of the Sun driven by the sun's magnetic field.) The data clearly show that there is no correlation between the two—however, from the matching pattern of waves in both curves, the Sun appears to add or detract from

are heading for a mini ice-age. The cooling Earth, in this view, is driven by lower solar output. Although this point of view is rarely heard now with the current El Niño in full swing, it was a dominant skeptic viewpoint throughout the last 15 years. We can expect this view to become more pronounced from 2017 onwards if the current El Niño is followed by a generally cooler La Niña, as is often the case.

Dissension

Let's tackle the second viewpoint: that there is warming but that it is part of a "natural cycle" and linked (presumably) to the Sun and also to a lack of volcanic eruptions that would otherwise cause cooling. The problem with this thesis is that the solar output declined over the last four decades in terms of sunspot numbers, yet temperatures continued to climb with each decade being warmer than the one preceding it (Fig. 3.3). The warming trend goes back at least until the 1960s. Before this point, the trend continues but it is more complex (Fig. 3.4). When one compares the solar output and the temperature record there is an apparent match between the two when one looks at low amplitude changes in temperature. The Sun may well be modifying the temperature variation by a small fraction of the total (roughly 0.01–0.02 °C) but it is clearly not a driving influence in dictating temperature.

What about volcanic eruptions? We looked at the potentially serious effects of these in Chap. 2, since the Dark Ages were truly dark as a result of up to four catastrophic eruptions around the 530s–540s AD. However, to claim that a lack of eruptions would produce a continuous upward trend is clearly untrue. Volcanic eruptions produce dips that are superimposed on the background trend. They do not drive long term trends. Even massive historical

FIG. 3.3 (continued) the overall pattern in temperatures by a fraction of a degree (*the three arrows*). Below left—a well known negative correlation between the number of pirates (only an approximate value) and global temperatures. Maybe pirates hold back global warming? For more on this idea visit: http://sparrowism.soc.srcf.net/home/pirates.html Many more spurious correlations can be found at: http://www.tylervigen.com/

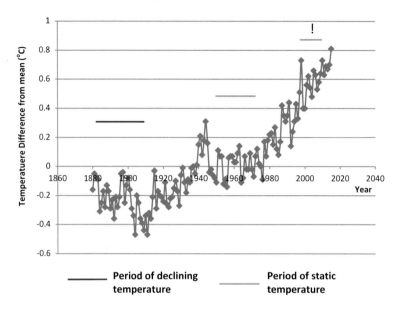

Fig. 3.4 Myth 2—There is (or was) a standstill in global temperatures. Human perception is a fickle thing. While most of the media prevaricated over the so-called "climate slowdown of 1998–2013" and its alleged significance, a brief sortie into the past shows that the alleged flat "trend", shown by a *green bar* and *exclamation mark* is nothing of the sort. The slow down, even when restricted to 1998–2013, is virtually unrecognizable in the trend. From 1885 to 1915 temperatures fell relative to the mean for the period shown (1880–2014). Temperatures were static (with variation) from the early 1940s until the mid-1970s—twice the length of the recent "slow-down". However, there is absolutely no doubting that over the period, where most records are available (over one billion readings), temperatures have increased significantly by 0.96 °C, relative to the mean for the period

eruptions such as those of 535–550 AD only lowered temperatures for a couple of decades. They did not suppress it for over a century. Therefore, one really cannot claim the current upward trend in temperature is down to a lack of eruptions.

The last climate skeptic position—and it is perhaps the most widespread and popular—is that there is a cooling trend. A very popular British Tabloid newspaper and one of its columnists strongly advocate this position, particularly when there is a cold winter. After all, if you start your record at 1998, between then and 2013 there does indeed appear to be a slight but statistically significant downward trend in the data. There are many websites dedicated to this pattern and each has its own slant. The problem with this interpretation is twofold. Firstly, 1998 was an anomalously warm year, powered by the strongest El Niño on record at that time. The 1997–1998 El Niño brought temperatures as high as 29 °C across the central and eastern Pacific, with a rise of 4 °C above normal along the South American coast. Most of North America enjoyed its warmest winter on record in that year—particularly the Northeast of the country: this is a common El Niño effect (see earlier in this chapter). Temperatures were on average 5 °C above normal for the December to February period. Similarly, the UK basked in a very mild and wet winter, which was followed by (at the time) record breaking wet March. The El Niño lasted into the following summer and ensured that 1998 had record warmth, globally.

Clearly, if you start from an anomalously high position—and there can be no contesting that 1998 was anomalous—then subsequent years may fail to peak above this. Indeed, with a lack of significant El Niños since then (2006 and to a lesser extent 2014, excluded) it is hardly surprising that global temperatures struggled to exceed this large rise. However, in 2014, without an El Niño, global temperatures surpassed those of the extreme 1997–1998 El Niño. That should be an eye-opener. Many skeptics disputed the numbers for 2014 when they were first published as there was a small, but significant possibility they were simply on a par with 1998, but later analysis confirmed the numbers. 2015 went on to be broadly equivalent to the temperatures of 1997–1998, with a moderate to strong El Niño. Before the year was even out, and the El Niño at full strength, the 1998 record had been smashed, along with that set in 2014.

As Fig. 3.4 shows, there is clearly no slow-down: the trend continues upwards as predicted by Anthropogenic Greenhouse Warming (AGW) models. If you want to argue that inclusion of 2015 isn't fair, as it's an anomalous El Niño year, then you must also remove 1998. Do this and the trend clearly persists: El Niño or no El Niño there is a consistent upward trend in global temperatures in defiance of any *natural* background factor. Reflecting this, in 2015 the NOAA released comprehensive analysis which thoroughly debunked the cooling hypothesis.

If still unconvinced by this argument, house prices are a useful analogy. From 2008 until 2013 or 2014 house prices declined in most parts of the western world. If you take the climate skeptic approach and apply it to house prices then the trend in house prices is downwards. While that might be true over a period of 5 or so years, no one would argue the trend in house prices has been down for the last 130 years. Nor would they argue that the trend in house prices will continue downwards for the foreseeable future. The recession of the last few years, now largely over, was simply the latest blip in a series of blips, which have punctuated the upward trend in house prices over the last few hundred years.

The cooling trend interpretation is perhaps personified by US meteorologist Joe Bastardi who has argued for "astronomically driven cooling" since at least 2011. Bastardi also claims that the southern ice cap has increased to record levels, which in his and many other climate skeptic's minds indicates that there is a "natural cycle" in the amount of ice between the north and south polar regions.[2] However, the data (Fig. 3.5) clearly tells a different tale.

This brings us to another argument used in favor of "no global warming", that the amount of sea ice around Antarctica has increased. Therefore, the argument goes there can be no global warming as this would surely decrease the amount of sea ice. Yet a casual examination of the full data on ice caps reveals something rather different (Fig. 3.5). There is no doubting that the trend in Arctic sea ice cover is down and has been for at least the last four decades. Antarctica paints a similar, if slightly more complex picture. The volume and hence mass of ice on West Antarctica is in serious decline—broadly matching the Arctic sea ice. The mass and volume of ice in East Antarctica did show a smaller increase, but this too may now be in reverse. Overall, the change in Antarctic ice is downward and has been for at least 10 years (Fig. 3.5). Therefore, far from increasing the amount of ice on Antarctica is decreasing and this is clearly in line with Arctic sea ice. Thus the skeptics' claims that there is a see–saw cycle of ice between North and South is clearly at odds with the observations of declining ice at both Poles.

If Antarctic ice is in decline, then why is the amount of sea ice increasing? Take a lot of ice perhaps 1 km thick, melt it and dump the slush and fresh water in the ocean and there will be two

---

[2] http://www.forbes.com/sites/larrybell/2013/05/26/meteorologist-joe-bastardi-blaming-turbulent-weather-on-global-warming-is-extreme-nonsense/2/

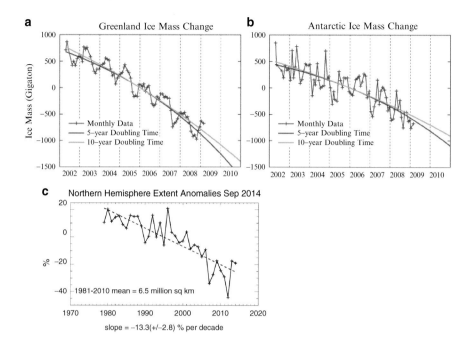

FIG. 3.5 Myth 3—Ice cover is decreasing in the Arctic but increasing in the Antarctic. These three graphs show the change in ice since 2000 recorded by satellite. Graph (**a**) shows the change in mass for Greenland; (**b**) the change in mass for the Antarctic as a whole, and (**c**) the extent of Arctic sea ice cover. In no instance is the mass (or extent, in the case of the Arctic) increasing. East Antarctica does show an increase in mass (not shown) but the Antarctic, as a whole, shows a decrease in mass. West Antarctica is losing mass at a rate roughly four times the gain in the east. Skeptics pointed to the increase in Arctic cover from 2012 to 2013 and suggested the ice cover was recovering. Do you agree? Arctic data is from the National Snow and Ice Data Center website (http://www.arctic.noaa.gov/detect/ice-seaice.shtml) and Antarctic and Greenland data from GRACE

effects. Firstly, the slush can cover a much larger area than the intact ice on land and secondly, the influx of fresh water, which naturally floats on top of the denser salt water, raises the freezing point of the water. This allows for more extensive freezing come the winter. Thus sea ice can increase in *total area* even as the total mass of ice declines. Moreover, the spread of fresh, icy water over the surface of the ocean alters precipitation and evaporation leading to greater snowfall, which in turn increases sea ice. However, the gain in thin sea ice in no way compensates for the extensive loss of the thicker ice sheets on and around the continent.

Another very common misconception is that carbon dioxide and methane are unimportant greenhouse gases because the wavelengths over which they absorb infrared radiation overlaps water, which is far more abundant. Therefore, the argument goes, even increasing carbon dioxide and methane levels has no effect because there is no more energy left to absorb because water vapor has already absorbed it. But if the current concentration of water vapor was all that was needed to set the global temperature, then changing its concentrations would have no effect on temperature. This would be regardless of the presence or absence of any other greenhouse gas. This is blatantly untrue and it can be shown in the laboratory that changing the concentration of water vapor will alter the absorption of energy. This is through line broadening and other effects (Fig. 3.7). If this argument were true nor would we have to worry about the Earth overheating in a billion years or so (Chap. 5): increasing the concentration of water vapor would have no effect. Furthermore, that "water vapor has already absorbed all of the energy" makes a very testable prediction. If this is true then the amount of energy escaping the Earth will be the same now as it was in 1970 or 1990. We can test this as we have satellite data of our planet's outgoing radiation. Unfortunately for the skeptics the amount of outgoing long wave radiation has decreased at precisely at those wavelengths over which both carbon dioxide and methane absorb. This is in agreement with the idea of greenhouse gases contributing to warming but contradicts the nonsense that water vapor has absorbed all the radiation that there is to absorb (Fig. 3.6). If you are unhappy with that argument then the First Law of Thermodynamics is not in your favor. Energy is conserved: less energy escaping the planet means that it must be retaining that difference in energy and thus be warming up. Stratospheric temperatures are also on a decline. Again, this is perfectly reasonable if more energy is being trapped in the troposphere and is consistent with the First Law of Thermodynamics.

The problem with water vapor is that under terrestrial conditions it is very close to its triple point where it will either freeze or condense. As such water vapor responds to temperature rather than drives it. Lower the temperature and water condenses out.

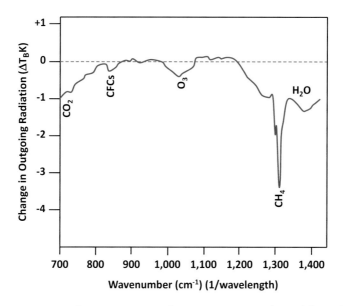

Fɪɢ. **3.6** Testing predictions. Most skeptic sites say that although methane and carbon dioxide are greenhouse gases, their effect is insignificant compared to water; or that all of the available radiation has already been absorbed by water so these gases can have no effect. This is testable. If this is true then increasing the amount of these gases will not affect how much radiation escapes the planet. How could it, if it had already been absorbed by water vapor? However, by comparing data from satellites over the last few decades it is abundantly clear greenhouse gases are absorbing more long wave (infrared) radiation (making the line dip into the negative portion of the graph). Methane has a notable effect in the portion of the spectrum shown here, despite absorbing radiation at wavelengths that completely overlap with water vapor

Raise it and water evaporates. Consequently, water vapor has a short residency time in our atmosphere—around 9 days on average. This is how long a water molecule has in flight before it condenses with others and precipitates out. As such water vapor responds to temperature rather than drives it. Indeed, water vapor only really becomes a driving greenhouse gas at high temperatures—above 70 °C—when conditions become difficult for liquid water to condense at terrestrial atmospheric pressures.

On planets with temperatures above 100 °C, liquid water is impossible unless the atmospheric pressure exceeds ours. Here, the vapor pressure will depend solely on the temperature and water will become a strong greenhouse agent, just as carbon dioxide does at terrestrial temperatures. It is likely that such a moist greenhouse was critically important just after the Earth formed; and, unfortunately, will come into play once more on the Earth in the distant future. The grisly fate of the Earth is examined in more detail later in this chapter. For now, it is carbon dioxide that is well above the temperature at which it can condense (or rather re-sublimate). Therefore, it is this gas, with its long residency time in our atmosphere, which has the greatest role to play in modulating climate.

If unconvinced by this argument then we could apply the same "logic" to Venus. If the argument for water vapor having absorbed all the available radiation was applied to Venus, then if we were to suddenly remove 99 % of this planet's atmosphere's carbon dioxide the temperature wouldn't go down. This would be because the carbon dioxide left over will be "absorbing all of the available energy". Indeed, if the skeptic argument is correct then the trivial amount of water vapor in the Venusian atmosphere (Chap. 5) should override the carbon dioxide and poor old carbon dioxide won't do anything to Venus's heat. Or imagine you've got 10,000 dollars in earnings and you earn another $500. Do you tell the IRS that the $500 is irrelevant because the $10,000 is much bigger? No, you add the figures together and reluctantly fill in your tax return as appropriate. Similarly, the effect of carbon dioxide and methane is additive to that of water. It's not that water overrides absorption, for as was said before if the current water had absorbed all that was available then the planet's temperature would not respond to the overall concentration of greenhouse gases: it would simply reflect a baseline concentration of water, which it does not.

Carbon dioxide therefore *is* a greenhouse gas, which has not been in question since John Tyndall experimented with various pure gases in 1859. Tyndall observed that carbon dioxide, ozone and water vapor all strongly absorbed radiant heat, yet oxygen, nitrogen and hydrogen did not. Various measurements since then have refined the contribution carbon dioxide makes to absorption of heat (Fig. 3.7 below).

Figure 3.7 also illustrates the effect of changing the concentration of a greenhouse gas. Even if we could guarantee that every available photon of light at the appropriate wavelength that could be absorbed is being absorbed, increasing the concentration of greenhouse gases will still elevate temperature. At higher pres-

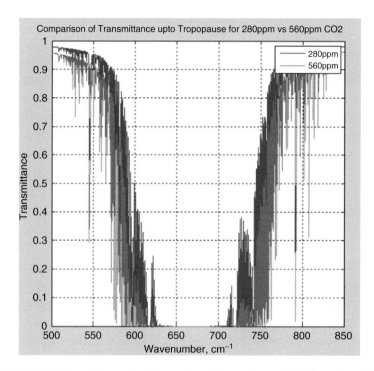

FIG. 3.7 Collisional broadening—the absorption of infrared radiation at two different concentrations of carbon dioxide: 280 and 560 ppm (0.028 % and 0.056 %, respectively). At the higher concentration carbon dioxide absorbs energy across a broader range of wavelengths, increasing its potential as a greenhouse gas (*green curve*). *Graph sources:* http://scienceofdoom.com/2011/04/30/understanding-atmospheric-radiation-and-the-%E2%80%9Cgreenhouse%E2%80%9D-effect-%E2%80%93-part-twelve-curve-of-growth/ and https://www.cfa.harvard.edu/hitran/ for the source data

sures there are more collisions between greenhouse gas molecules (of any type). This has the effect of altering the energy of the electrons that link the atoms on these molecules together. The net effect is that this causes the molecules to absorb energy across a broader range of wavelengths. Effectively, at higher concentrations the gases behave more efficiently as greenhouse agents—and this effect on greenhouse gases can result from an increase in the concentration of any gas that these greenhouse gases are mixed with. So, if we were to crank up air pressure on the Earth, but not change the concentration of greenhouse gases, the strength of the global greenhouse would also increase. This process is called collisional broadening and will be returned to in Chap. 5 when we look at how Venus and the Earth will "twin-up" once more, in the future.

Finally, many skeptics take to doubting the temperature record effectively in its entirety. The reason for this endemic view, cited since the 1960s, is the urban heat island effect. Essentially, the argument boils down to the location of many thermometers or other devices used to record temperature. The argument is that if you site these devices in built-up areas, the extra heat generated within the city by the combination of solar heating, respiration, air con units, vehicles and planes will naturally raise the temperature. Moreover, as cities have grown in size the effect as naturally increased over the last century or so. This is a good argument, but with one very obvious flaw (quite aside from all those rural and latterly satellite measurements): global temperatures did not rise significantly from 1998 to 2013—and we've seen many skeptics argue jhmtf that they declined slightly. Hopefully, you can see the problem: the human population, and the size of cities, has clearly not declined since 1998 (there are around 1.3 billion more people since 1999). Therefore, wouldn't one expect a rapid increase in the heat-island effect, rather than a leveling off? What's worse for proponents of the heat island effect is that when measurements are made of cities 42 % of them show a lowering of averaged temperatures, not a rise, relative to their surroundings. This is because temperature measuring readings are increasingly taken from park locations rather than actual concreted areas.

There are over one billion measurements in the global temperature record, and they can't all be wrong. Neither can you pick and choose data points in order to make a case for or against man-made global warming. If you accept the validity of a climate hiatus in the last decade, despite the many studies disputing it requires accepting a generic rise in temperature since 1885. The record cannot be right only in the bit you choose to validate your argument. More importantly, in most analyses that are done, readings from those stations that are most prone to the heat-island effect are discounted in favor of any nearby rural measurements, thus the effect would only ever have limited impact, if any at all.

There is one peculiar historical perspective underlying these arguments: the birth of the industrial era, which most historians would put at around 1800. However, many skeptic websites erroneously suggest that the industrial era began as late as the 1940s. If you subscribe to this view, then there can be no link between historical rises in carbon dioxide and global temperatures since carbon dioxide levels began rising around 1800.

Another dubious argument by climate skeptics is a rather odd mathematical one. Some skeptics will say that natural variability is around 0.1 °C per decade, which is reasonable. They will also say that temperatures have only risen by 0.1–0.2 °C or so over the last two decades. Therefore, the current rise is supposedly within the bounds of natural variability. But the temperature record supports a rise of roughly 1.0 °C (0.96 °C if you want to be precise) over the last 130 years or so: this is roughly ten times faster than the rate of rise during the Eocene Hyperthermals, the nearest comparison to the period of current warming that we know of in the geological record. The rise is erratic but is utterly inconsistent with any natural cycle, solar or otherwise. The only factor that varies in line with this rise is the level of greenhouse gases—notably, but not exclusively, carbon dioxide. Table 3.1 puts things into a somewhat grim perspective.

Given all of this, despite likely upward and downward bumps along the road, the general path of the global temperature will be an inevitable rise unless carbon dioxide levels begin to fall sharply soon. Assuming a worst case scenario society will continue to burn all fossil fuels until these are exhausted, or at least become uneconomical to harvest. In this case, assuming we don't turn our attention to methane hydrates, or experience some form of runaway where there is a catastrophic release of methane, the temperature peak will happen in around 300 years. Carbon dioxide levels will be back at where they were in the Precambrian, but with a more luminous Sun, the temperatures will be rather more torrid. Bjorn Lomborg has even argued in "The Skeptical Environmentalist" that it may well be more cost productive to ride out the effects of global warming rather than stop it happening in the first place. Although Lomborg's opinions are controversial, riding out the storm seems a more likely political and social outcome than cutting production or investing in new means of generating energy. So, if we decide not to tackle warming by changing energy production what will the likely costs be? The principle cost to the West will be increased insurance costs tackling the output of climate change: increased rainfall in some places causing flooding and increased drought in others, which limits food production or the availability of water. There is obviously cost associated with loss of coastlines through sea levels that are rising faster than predicted. Rising seas also bring enhanced coastal flooding, even when the land itself isn't permanently flooded. To illustrate this trend, there is now a data set for coastal flooding in the UK which covers the last 100 years. This was produced by David Haig and colleagues and is referenced at the end of this chapter. There is an obvious upward trend in the numbers of floods as well as their severity during this period. The problem with these costs is that you can make guesstimates upfront, but the precise regional impacts on damage to infrastructure, loss of food production, etc. are hard to determine and clearly prone to a lot of debate. This doubt affords people the flexibility to do as little as possible. However, just to underscore the point that climate is changing and we are the likeliest culprits, we'll look at a couple more pieces of evidence.

Drifting Hurricanes Amid a Changing Climate

The Earth's climate is certainly changing, but it is true that the specifics of this change are still being established. Within the mêlée of data a few trends are becoming clearer. Take temperature extremes, for example. Analysis of the temperature records for Europe reveal that on any given day temperatures are far more likely to break record highs than they are record lows. Martin Beniston (University of Geneva) has analyzed the temperature records of the last 60 years from 30 different climate stations across the continent. The ratio of days per year with record high temperatures to record low temperatures increased from 1 to 1 in the 1950s to 1 to 5.1 in 2013. Indeed, despite the alleged climate slowdown over the last 15 years, the rise in the ratio of record highs to record lows has accelerated. It appears a slight rise in average temperatures has been masking a much more marked rise in (high) temperature extremes.

That's not all. Of rather more concern is a slow but steady shift in the distribution of tropical storms. Although much was erroneously made about changes in the frequency of such storm systems, something which did not hold up to scrutiny of longer term records, there has been a shift in the location of tropical storms over the last 31 years and in particular the latitudes at which they reach their maximum intensity. James Kossin (NOAA) and co-workers analyzed the hurricane track data for the northern and southern hemispheres.

Analysis reveals that there has been a gradual migration north and south of the equator amounting to approximately 1° of latitude (53 km north and 62 km south) per decade. This movement is consistent with the gradual expansion of the Hadley cells that deliver hot, humid air to the Doldrums and drier, cooler air back to the Horse Latitudes (described in Chap. 1). The expansion of these large cells, that dominate the central 60° flanking the equator, is an observation that matches predictions for global warming caused by changes in the levels of carbon dioxide and other greenhouse gases. As these cells expand the amount of wind shear reduces further north and south, which in turn enhances the conditions favoring the formation of tropical storms. Most of the migration is apparent within the Atlantic and Pacific basins, which host the greatest number of tropical storms. The northern Indian Ocean shows the least migration of storm intensity—and interestingly this follows the opposite pattern, with storms becoming more intense towards the equator. However, given the relatively small number of storms the Indian Ocean spawns, the effect of this pattern is drowned out by that in the Atlantic and Pacific basins. Storms, meanwhile to the south of the equator in the Indian Ocean, follow the same pattern as the Atlantic and Pacific Oceans. Quite why the northern Indian Ocean bucks the trend for the rest of the planet isn't clear, although changes to the strength of the Asian summer monsoon, which has generally weakened, may be a factor (Chap. 2).

Endemic Misunderstandings

Within the skepticism fraternity—and much more widely within the scientific community—lie a number of underlying misconceptions that unfortunately help to crystallize the ideas of natural variability as a cause of the current temperature trends. Most notably within the literature are many comments relating to the historically recent climate swings that are perceived erroneously as being global events. In particular the "Medieval Warm Period" and the subsequent "Little Ice Age" bring a lot of attention to apparent global trends. If you subscribe to these ideas then the current upward trend in temperature is clearly not going to be anything new.

However, when these climate swings are examined in more detail, rather large holes appear in the natural variation picture and the local and regional reality of these swings emerges. The Medieval Warm Period is endemically quoted in most climate skeptic literature, but perhaps more significant is the so-called Little Ice Age. For it is from this point onwards that anthropogenic global warming ensues. Let's examine each of these very interesting climatic periods in more detail and unpick the myth from the climatic narrative.

The Medieval Tepid Period and the Luke-Warm Little Ice Age

Around 1000 years ago the Viking Empire was extending from northern Europe across to Iceland, Greenland and eventually North America. As the true European discoverers of the western continent, they took advantage of relative warmth that brought a largely ice-free North Atlantic and a climate suitable for the growth of some crops and livestock, such as cows and pigs, in southern Greenland. The climate over the southern and eastern side of Greenland was more typical of western Scotland with temperatures up to 2 °C higher than they are at present in that region. Although by no means "warm", conditions were benign enough to allow the formation of permanent Viking settlements for more than 100 years.

In all the Medieval Warm Period, or MWP for short, as this strange climatic blip is known, lasted approximately 300 years from 950 to 1250 A.D. its presence leading many to conclude that the current warm period was simply another upward blip in temperature equivalent to the MWP. However, like so many things, appearances are deceptive.

Initially, the climate used to determine the extent of the period was obtained from Europe and the periphery of the North Atlantic basin. This data revealed that the warmest region, which rivaled or even exceeded the temperature maxima of the 1960–2000 period.

However, with further, and more extensive, analysis of global temperature data it soon became clear that the MWP was restricted to Southern Greenland and the neighboring Atlantic basin. Elsewhere, globally there was clear evidence of a significant cooling event.

For example, the so-called Medieval Warm Period did not affect the Anasazi tribes of south western North America—or the Maya in Central America. Indeed, the same pattern of North Atlantic warmth coincides with an extreme mega-drought in the region encompassing southern United States and Mexico. Prior to the MWP, the Anasazi had developed a wonderful cultural system, with complex cities, many carved into and constructed within cliff faces, such as the famous Mesa Verde Complex in New Mexico. As the Vikings made their way across the Atlantic, the Anasazi civilization collapsed: the underlying reason was persistent drought associated with regional cooling. Temperature records from tree rings reveal that while the Vikings enjoyed warmth, the Anasazi endured temperatures that were consistently lower than they are now.

Similarly, but somewhat earlier, the Mayan civilization collapsed in the period 850–950 AD and there is at least circumstantial evidence that they too were reduced in power by increasing drought in Central America. This begs the question, what meteorological phenomenon united the demise of the Anasazi and the success of the Vikings? The answer was a prolonged La Niña (Chap. 2).

For much of the period from 950 to 1250 AD the central and eastern Pacific basin was afflicted by La Niña conditions. This brought cooler, drier air across southern North America and much of Central America. La Niñas bring very predictable changes in global temperatures, with the coolest conditions around the shores of the eastern Pacific (Fig. 2.5). However, these conditions also redirect the polar front jet stream so that after dipping southwards across the Western States region, it arcs it back northwards towards Greenland. This brings considerable warmth to the north western Atlantic basin and Western Europe. Thus, while the Anasazi lands cooled down and dried out, the Europeans enjoyed warmth (and Viking raids).

Across the rest of the globe it is apparent that much of central Eurasia was colder than average, while the La Niña brought warmer and much wetter conditions to the western Pacific and

New Zealand, with temperatures in the latter region up to 0.75 °C higher than the average for 1960–1990. There is circumstantial evidence that Australia was also affected by prolonged La Niña conditions, but climate records for this region during this period are scarce. However, it is clear that what was meat for some was murder for others. The Medieval Warm Period was simply a regional event, not a global one: the current warm period *is* global and has no natural counterpart.

Similarly, a lot is made in Europe and North America of the *Little Ice Age* (LIA, for short)—the period lasting roughly 2–300 years between 1600 and 1900 where severe winters and cool wet summers afflicted Europe and the Atlantic seaboard of the U.S. However, once again, more extensive analysis of global temperature records (and we are primarily looking at tree-ring evidence and in some instances isotopic evidence) the timing and extent of the cooling clearly varies dramatically from region to region. China had its coldest period of the last 1000 years around 1200 A.D., while the mainland US was coldest in the nineteenth century. 1200 A.D. is, as you may notice, when Western Europe had the Medieval Warm Period... Scandinavia's lowest temperatures of the last millennium were around 1600 A.D. while the UK suffered the most extreme cold of the period in the 1700s. Indeed, climate data indicates that while central England shivered, parts of North America and Labrador were as warm as the average for the period 1960–1990. The Little Ice Age, it seems, was more a scattering of ice cubes, rather than a global freeze.

The sporadic nature of this period unfortunately disposes of the supposed link between global cold and the lack of sunspots, known and the Maunder Minimum that took place during the 1600s. Much was and still is made of the lack of sunspots and the cold period. Since there was no global chill, but instead a period of enhanced climatic instability that spanned well beyond the sunspot minimum, there can be no causative link. The current warming trend is then even less likely to be associated with a change in sunspot numbers as it clearly wasn't in the past, as the sunspot numbers would not affect Scandinavian temperatures alone. Currently, there are a few skeptic websites that suggest we are (or will be) heading into a mini ice age (one even claims that we are already there). The driver, it is suggested, is a weak solar cycle. Yet, as historical records show there is no link between the two.

There is something of a link, albeit a weak one between warm summers in the northern hemisphere and solar maxima, but there are plenty of summers that really don't fit the bill, at least in Europe. The moral of these two tales is to beware of cherry-picked data.

Closer analysis of the temperature record also reveals some interesting regional biases that seem to cause similar misperceptions. The temperatures across continental US (and this was a surprise to me) fell so far in the 1930s that it took nearly 60 years for them to reclaim the ground. This was not a pattern that was repeated elsewhere on the planet. Quite why the continental US dipped more strongly, then warmed more slowly than the rest of the planet is unclear but it is likely to relate to the overall pattern of regional climate and airflow, or perhaps to the release of additional pollutants that dimmed the surface. For example we've already seen in Chap. 2 how the tragedy of the terrorist attacks on New York and the Pentagon in 2001 affected regional temperatures. To reiterate, immediately after the events of 9/11 planes were grounded across the entire continent. Over the following 3 days, trans-continental temperatures showed an increase in the diurnal (day–night) range of 1.8 °C compared to the days running up to the attacks. Clearly, something as seemingly insignificant as airplane contrails could drastically affect temperatures.

All of this evidence shows that the idea that temperatures will not be affected by increasing amounts of methane is clearly flawed. By making a straightforward comparison of the amounts of carbon dioxide and the rate of temperature change with the Eocene and Permian "Hyperthermals", hopefully the current changes can be put into perspective. With the three tales of climate warming behind us, it's time to look at how drifting continents led to changes in the strength of Monsoons across Asia and North Africa. These illustrate the volatile nature of the global climate and the underlying role for carbon dioxide and plate tectonics in regulating the terrestrial climate system.

The Effect of Global Climate Catastrophes on Life

Each of the climatic catastrophes had radically different effects on life on Earth. The snowball, or slush-ball, episode coincides with the emergence of complex life. Although this may be just that, coincidental, a global glaciation event would eliminate many forms of life, clear niches of competition and create new environments for life's re-emergence and diversification. Although one would certainly regard the connection as circumstantial at present, there is a clear case for the glaciations having led to the development of complex, multi-cellular life. As such, we may owe our existence to this period.

Similarly, the Permian-Triassic boundary marks a near wipe-out of complex life on Earth and is clearly contemporaneous with the eruption of the Siberian Traps. Although it appears clear, the eruption of vast amounts of carbon dioxide appears to have conspired with an increasingly anoxic ocean to drive the global mass extinction. However, despite the severity of "The Great Dying" once again, although somewhat protracted, within ten million years of the event life had regenerated, firmly establishing both the mammals and the dinosaur lineages and henceforth the birds.

The first Eocene event is perhaps the closest analogy with the current phase of warming. While the end-Permian event may have taken 60,000 years, much of the PETM warming was concentrated within 10,000 years (although 20,000 years probably elapsed for the full warming to unfold). The Eocene saw the extinction of some marine species, particularly those cold-loving species found at greatest depth. Like the PT extinction, this was probably as a result of the deeper oceans becoming anoxic as a result of poor circulation. However, the Eocene also saw the evolution of many large species of mammal and the first primate. Undoubtedly, at least from our perspective, the Eocene Hyperthermals were critical for our evolution. Without the Eocene warmth, there would be no modern humans or the large mammalian fauna we preyed upon at the close of the last ice age.

Strange Tales from the Frontiers of Political Climate Skepticism

The "UK Independence Party Climate Policy 2012—Keeping the lights on" had some wonderfully silly material in it. The document contained a graph that purported to show the temperature records for the last ten thousand years. This indicated that temperatures now are no higher than they were in the Medieval Warm Period around 1000 years ago. Sadly, the data was shown to be flawed 3 years before UKIP used it. The Medieval Warm Period has been shown to only have been warm in the north west Atlantic. It was in fact cooler elsewhere. The 2014 document states that the current 0.8 °C rise is part of a "natural cycle"—and they favor the Sun—even though the pattern of solar radiation and temperatures do not match.

UKIP also describe carbon dioxide as a trace gas essential for plant growth (a common theme). Yes, but they don't seem to understand simply whacking up the concentration will not increase crop yield. Plants do not respond in this manner to changes in carbon dioxide: other factors are at play, which limit the beneficial effects of extra carbon dioxide. More importantly, cereal crops, upon which the world depends for the bulk of its calorie intake respond poorly to increasing carbon dioxide, whereas competing broad-leaved weeds do rather better. Worse still, geological evidence (PETM and ETM2) suggests that increasing the level of atmospheric carbon dioxide results in a stunting in the size of mammals—which now, of course, includes us.

As the science above reliably shows, increasing carbon dioxide increases its potential to absorb infrared radiation. UKIP, a potential party of government, do not seem to understand this, grossly underrating the effects of carbon dioxide and other greenhouse gases and openly seek to expand the consumption of coal.

Worse still was the bizarre claim by Maurice Newman, who chairs the Australian prime minister's Business Advisory Council. Newman had adopted the "world is heading for a mini-ice age scenario", also touted by Piers Corbyn, of "Weather Action" in the UK[1] and several others on the Internet. Newman claimed that there was evidence that the world is set for a period of cooling, rather than warming. This would leave the nation ill-prepared for the coming climate catastrophe. There are several pundits in the cooling camp that even suggest that we should burn more coal to avert this coming disaster. That begs the question (particularly in a country blessed with a (relatively) benign climate like Australia), are you burning the coal to make carbon dioxide to avert cooling because you realize CO_2 is a greenhouse gas (and causes warming), or are you expecting such a downturn in temperatures that the average Australian will have to invest in a coal fire to keep warm? Strange indeed.

[1]Weather Action pdf: WANews13No5.pdf ((2013) On some publications, Piers Corbyn has claimed that the mini-ice age is upon us.

Yet, this apparently rosy tale for mammals comes with an interesting caveat; one which impacts on the most currently fashionable skeptic position that extra carbon dioxide is a boon for plants. While the Paleocene (the immediately preceding geological phase) saw a spurt in the size of mammals, as did the late Eocene and Oligocene, the warmest period of the Eocene saw mammalian sizes decrease. Indeed, this was not an isolated event limited to a few species: this shrinkage in mass was across the mammalian board and occurred not once but twice, first 55 then 53.5 million years ago. Philip Gingerich of the University of Michigan has attributed this temporary shrinkage to a loss in the nutritional value of the abundant vegetation that the hyperthermals stimulated. While many originally attributed the observed reduction in the mass of mammals to the increase in heat, clearly mammals evolved during the Mesozoic when the planet was warm and survive today in places such as Africa. Heat, per se, is not an issue for mammals. However, when one looks at the effect of increasing the amount of carbon dioxide on plant growth, yes plants increase in mass, but their nutritional value does not increase in turn, meaning any animal feeding on plants will need to eat more in order to prosper. Thus a herbivore will succumb to the contrary need to eat more to get nutrients, but be unable to do so because the sheer mass needed for survival will exceed what it can usefully gather and process. Thus, mammals shrank in size even though there was more food available. Bear this in mind when you see the latest skeptic argument: more carbon dioxide means bigger plants. Yet beyond the immediate concerns that global warming will undoubtedly have for us humans, more carbon dioxide won't do the planet (as a whole) any harm. If we look beyond the next 50 years and assume that we will do at best too little to avert a temperature rise of a few degrees for the planet, such a rise in temperatures might be beneficial over time to the future evolutionary history of life on the planet as a whole, though not likely to humans.

One must bear in mind that a projected three to four degree rise (if nothing is done to limit carbon dioxide emission) will be sufficient to melt the West Antarctic ice sheet. We know this because during the Pliocene, when carbon dioxide concentrations were comparable to those of today (400 parts per million, or 0.04%), temperatures were 3 °C above those of 1800. Two papers published by Tim Naish (Antarctic Research Centre, Victoria University of Wellington), David Pollard (Pennsylvania State University) and Robert DeConto (University of Massachusetts) and colleagues demonstrate that during the Pliocene, between three and five million years ago, the West Antarctic ice sheet periodically collapsed and reformed over intervals around 41,000 years long. Each collapse took, at most, a few thousand years to complete. The cyclical nature was driven by, you guessed it, Milanković Cycles, that continued to drive the longer term cyclical glaciations of the last Ice Age a couple of a million years later. The 41,000 year cycle is linked to the angle of tilt of the Earth's axis and is embedded within the longer 100,000 year-long cycles that were evident during the Eocene that were described earlier in this chapter. Interestingly, after 800,000 years ago the dominant 41,000 year cycle gave way to the current 100,000 year-dominated cycle. The origin of this change—the Mid-Pleistocene Transition—is currently unknown but may well relate to (what was overall) declining carbon dioxide levels in the atmosphere.

Remember that in the Pliocene changes to the amount of heating in the Arctic underpinned the warming: there was not a global increase in the amount of radiation the planet received, so the 2–3 °C rise that caused the polar melt was not a global phenomenon in the sense that it is today. Thus we should be wary of driving a global increasing in insolation and making a direct comparison with something that primarily impacted the North and South Poles. The effects of our activities are likely to be much broader in their impact.

The cyclical rise and fall in sea levels during the Pliocene amounted to roughly 10 m. Seven meters of that was from the collapsed West Antarctic Sheet, with a further three meters from the melting of ice on East Antarctica. Ten meters is rather a lot. Imagine all of our coastal cities progressively drowning—a process that is already underway as sea levels rise at their accelerating rate. Already somewhat alarming are measurements made by Fernando Paolo (Scripps Institution of Oceanography) of the floating portions of Antarctica's ice sheets. Overall, the rate of loss increased from 25 cubic kilometers per year from 1994–2003 to 300 cubic kilometers per year in the period since then. Meanwhile in East Antarctica the process of increasing ice thickness ceased. Now, for the first time since the Pliocene, Antarctica's ice is undergoing a profound net melt. For some floating shelves, in West Antarctica the rate of melting is now so high that it means they will have melted in their entirety by the year 2100. Now, remember, fortunately, these floating shelves do not contribute much to sea level rise as they are already part of the ocean. However, their loss will facilitate enhanced movement of the ice shelves that lie behind them, increasing their rate of flow into the ocean and thus their subsequent melting. This is because these ice shelves are partly grounded on the ocean floor, below and buttressed against islands and mountains in the neighboring ocean.

So while the entire West Antarctic Ice sheet won't have melted in the next 50 years, the effects of melting Antarctic ice will be progressive, with positive feedbacks accelerating the melting process. Even London is vulnerable, not only because of accelerating sea level rises, but because ever since the northern ice cap began melting 15,000 years ago, the whole of the UK has been tilting, upwards in the north and downwards in the south. Undoubtedly, a capital city such as London will be defended for some considerable time, but smaller towns may be less fortunate.

If we limit temperature rises to 2 °C—a worthy but a likely unachievable target—then the West Antarctic sheet will probably fray around the edges but otherwise persist. The world will be warmer, the climate, perhaps, but not definitely more fractious, but nothing dramatic will have happened. Several billion dollars extra on coastal and river defenses might well do the trick. For example, in the UK we currently spend £1.1 billion (around $1.7 billion US) per annum on flood damage costs. This amounts to roughly 1 penny per pound on income tax in the UK. The Parliamentary committee overseeing the spending projects an extra £1 billion cost per annum over the next 20 years.[3] This is regarded as financially acceptable by The Market and thus will continue. You can easily see that, globally, the cost of flood defenses will exceed $500 billion by 2030: this figure is from the World Bank. That's a lot of extra taxation—as well as increased insurance costs—merely to defend coastline many of us don't live on.

The long-term cost of defending the coastlines and river margins is likely to be prohibitive in the longer-term and when we decide that we are unwilling to keep subsidizing those directly affected by global warming we will surrender those lands. Areas such as peninsular Florida are already vulnerable to sea level rise. In the case of this low-lying state it's not so much flooding that is the problem (although it should be of concern); it's more the effect on the supply of fresh water. Here, the State's supply of drinking water is effectively confined to a narrow band sitting atop the saltier waters that permeate the deeper ground. As sea levels rise at a few millimeters per decade this layer is being progressively displaced by the rising tide of salty water. Residents of the State are already aware of the growing problem caused by sink-holes. As water levels rise—and this is increasingly acidic sea water—the limestone foundations of the State are dissolving away, opening up sink-holes that threaten property or cost increasing amounts of money to shore-up.

An Icy Future?

Here's where the current debate on global warming gets interesting. Imagine we burn most or all of the current available supply of carbon-rich fossil fuels. What will be the effects on the planet in the immediate and longer term? We can look at two scenarios: one where climate skepticism is overwhelmed and we restrict

---

[3] Source: http://www.parliament.uk/business/publications/research/briefing-papers/SN05755/flood-defence-spending-in-england

further excessive input of carbon dioxide to the atmosphere. In this scenario, energy production switches from predominantly fossil fuel incineration to greener technologies, so that we only release 1000 Gigatons (see Table 3.1 for a comparison with the Eocene and Permian-Triassic events). Carbon dioxide levels are restricted to 500–600 ppm for which the peak level occurs around the year 2100–2300. There is a delay in reaching the peak even if we turn off the supply around the year 2050 because of feedbacks within the Earth's climate system that will cause the release of more carbon dioxide from buried stores, including the oceans and thawing permafrost. Meanwhile, global temperatures continue to climb for another hundred or so years after carbon dioxide peaks, due to feedbacks involving the oceans.

With a cut-off of the supply by 2100 temperatures max out around 3–4 °C above those of present in around the year 2300 AD. Since carbon dioxide has a residency time running into thousands of years, the subsequent decline in global temperatures lasts for tens of thousands of years. Indeed, it could take over 50,000 years for the amount of carbon dioxide and global temperatures to reach pre-industrial (pre-1800) levels. During this time, humans successfully hold off the next glaciation of the northern hemisphere, which is due in 50,000 years (Fig. 3.8), until at least 130,000 years from now.

In the more extreme scenario we profligately consume all the fossil fuels we can and make no effort to switch to cleaner technologies until we absolutely have to. In this scenario we obliterate enough fossil fuels (around 1300 Gt) to release 5000 Gt of carbon dioxide. Consequently, carbon dioxide levels keep rising until at least the year 2150 to 1700–1900 ppm or higher. The final level depends sensitively on the nature and extent of the inevitable impacts that such profligacy will have on other carbon stores: these include clathrates on the ocean floor; permafrost carbon stores; and the Amazonian forest (whatever remains of this in 200 years). This scenario is altogether more unpleasant. Carbon dioxide levels max out 500–1500 years in the future, then take more than 300,000–400,000 years to fall to levels comparable with those of today. Indeed, one million years into the future they will still be above pre-industrial levels. The very slow decline is simply a consequence of

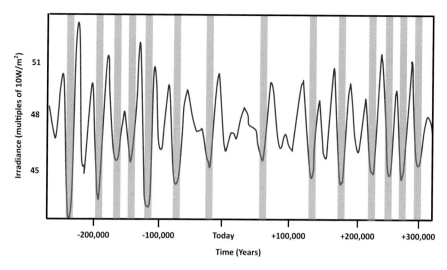

FIG. 3.8 Changes in the amount of radiation received at the Earth's surface at (the insolation) a latitude of 65° N, 1 month after the summer solstice. The value varies in a complex, yet utterly predictable way determined by the overlapping effects of the angle of precession; the axial tilt and the amount of eccentricity in the orbit of the Earth around the Sun. These Milanković cycles accurately describe the patterns of cold and warmth back at least as far as the Eocene and can be used to predict when the Earth should return to glaciation in the next few million years at least. Cold periods are indicated by *blue, translucent bars overlapping the curve.* Data used to construct this figure came from NOAA paleoclimate databases

the inability of the natural sponges for carbon dioxide—the oceans in the short term and weathering of continental rocks in the longer term—to cope with such a massive rise in the levels of this gas. Consequently, the planet would be looking at a rise in carbon dioxide not seen since the end of the Permian—and over a much shorter interval of time than the Siberian Trap event. Warming would be so severe that the planet would lose almost all of its ice. Even the mighty sheets of East Antarctica could not hope to cope with such a prolonged thaw.

What then would be the immediate impacts for humans? Well, at the moment a sea level rise of 7–10 m over the next few hundred years is probable with a modest 2–3 °C rise in temperature, if West Antarctica does what it has done under similar conditions in the past few million years. The precise timing, as was said earlier, is unclear but has a very definite geological precedent. Sea levels rise because of three factors: extra melt-water; extra expansion from higher temperatures; and finally a gravitational boost from the redistribution of mass. The Antarctic ice sheets are so massive that their combined gravity holds water pinned up against the continent. Melt this and the gravitational pull of Antarctica declines and the waters slosh outwards to cover more of the globe. As was said previously, such a melt will have significant financial implications for all of the nations of the planet—at least those with coastline.

Take a more severe scenario with a 2000 ppm carbon dioxide level and you look at a 5–9 °C rise in global temperatures lasting centuries, which will be followed at 3000–5,000 AD by a peak rise in sea level of around 70 m (approximately 230 ft). Such a rise would inundate much of the low-lying areas of the continents and necessitate a massive reorganization of human infrastructures, including roads, cities and the food supply. Now, that's a long way off, but once again, the geological record is clear: increase carbon dioxide and you raise temperatures. Raise temperatures and you melt ice and raise sea levels. The amount of carbon dioxide that is released would vastly exceed the capacity of the natural systems to absorb it, therefore, temperatures would take nigh on five millennia to fall by a single degree Celsius.

At least once we consume whatever all available fossil fuels, the inevitable (eventual) trend in carbon dioxide levels is downwards. Carbon dioxide is vulnerable on any planet with water, as we've seen. If it is not continually supplied it will be taken up by the planet's rocks and oceans and slip silently out of the atmosphere. Thus, regardless of whatever point we choose to stop burning fossil fuels the supply of the gas in the atmosphere will decline.

The driver of this change is the Sun. Since the planet and Sun formed, 4.56 billion years ago, the Sun has gradually become more luminous. As this occurs, the amount of radiation received by our planet has naturally increased. As it has done so the planet's meteorological systems have slowly pepped up. As they've done this, the rate of weathering and erosion has increased, in turn. More heat means more evaporation and more evaporation means a stormier and wetter planet. This in turn accelerates erosion, which removes more carbon dioxide from the atmosphere. The two forces, declining carbon dioxide and increasing solar luminosity, have worked in unison to keep the planet's temperature roughly constant—and well within the range of habitability. In summary: more Sun; less CO_2; stable conditions. Even in the extreme scenario where we incinerate all of our coal reserves and cause a massive release of methane from clathrates on the ocean floor; from the permafrost and from the tropics, we will still face declining carbon dioxide levels from 400,000–500,000 years in the future.

Regardless of whether we decide to maintain a global greenhouse or not, our ancestors will face the challenge of dwindling carbon dioxide and rising temperatures. The long-standing cooling–warming, Milanković cycles will be overwhelmed by the effects of the brightening Sun. A steadily brightening Sun will raise temperatures faster than a declining level of carbon dioxide can balance it: the planet will then steadily warm. This process will begin somewhere in the next couple of hundred million years.

For other examples of how the planet can really (and ultimately) irrevocably change in response to changes in the concentration of greenhouse gases and the intensity of radiation the Earth receives from the Sun, let's look further forwards and backwards in time.

Visions of Hell: Terrestrial Snowballs and Fireballs

As continental drift continues for the next several hundred million years at least, the current pattern of drought and deluge will continue until the oceans finally boil away. But before we contemplate the rather torrid death of planet Earth, we must turn the clock back 750 million years to when planet Earth was disappearing under a sea of ice. This intriguing phase of terrestrial climate saw the planet utterly transformed in its appearance, but also saw the greatest wave of evolutionary innovation that would ultimately lead to us. The snowball Earth marked both the planet's greatest (identified) freeze and a critical role for a carbon dioxide as the driver of its eventual release.

Snowballs and Slushballs

During the period from 800–700 million years ago the planet appears to have done something unpleasant to its resident life forms. There were significant changes to the abundance of the two stable isotopes of carbon: carbon-12 and 13 during this period which imply photosynthesis took something of a knock. Plants prefer carbon-12-carbon dioxide over the rarer but still relatively common carbon-13 (about 1 % of the abundance of carbon-12). During the periods of apparent freeze, the level of carbon-13 fell by 14 parts per thousand—not a large figure in itself, but one which is extremely significant for life on Earth. Such a change implies one of two things: either the planet suddenly released a very large amount of carbon-12 from hidden stores, as in the Eocene hyperthermal (earlier in this chapter) or photosynthesis largely ground to a halt and little carbon-12 was being taken out of the atmosphere. These excursions in the abundance of carbon-13 were the largest in the preceding 1.2 billion years: this event, whatever it was, had to be severe.

Moreover, during this interval, 750 million years ago, the planet began to form deposits called banded-iron formations. These iron-rich rocks form when oxygen concentrations in the oceans are low. Iron sulfide becomes more common and the iron remains in a less oxidized state. When oxygen periodically rises, these iron compounds effectively rust, becoming more oxidized. Such rust is less soluble in water than the less oxidized forms of the metal and they rain out on the ocean floor, forming alternating bands of rust. On the Earth, most banded iron formations were produced around 2.45 billion years ago when the planet's oceans and atmosphere began to fill with oxygen. In the intervening years, oxygen concentrations were too high to allow much iron to build up in the oceans. Yet, 750 million years ago, something happened that made the oceans more anoxic, or oxygen-poor. Life, it seemed, was in trouble and the evidence was in the rocks. The only way by which photosynthesis could so extensively fail would be if the oceans froze, temperatures became too low for the process to happen, and the availability of light to the surface ocean was restricted. An extensive glaciation, in which much of the Earth's oceans were frozen tight, would do the job.

The idea that the entire planet could freeze over might seem rather fanciful, however a substantial, if still controversial body of evidence suggests that this happened several times during the late Precambrian, and possibly much earlier, as well. The idea that the planet could freeze over developed over the last half century from a mixture of diverse observations that had implied glacial conditions existed near to the equator.

The original proposal goes back to the work of Sir Douglas Mawson in the first half of the twentieth century. Mawson had examined glacial deposits in southern Australia and assumed, without knowledge of plate tectonics, that Australia had been glaciated at its current latitude. It is now understood that the geological formations Mawson described were formed when Australia lay further to the south than it does at present. Later, in 1964, the idea of a global glaciation was re-examined by W. Brian Harland. Harland identified Proterozoic deposits in Svalbard and Greenland that also implied that glaciation had occurred while these regions lay within the equatorial region. Harland identified interwoven glacial and carbonate rock formations that implied alternating periods of deep freeze and deep heat. However, Harland lacked a model to explain how these alternating periods of climate could have occurred. It then fell to the Russian climatologist Mikhail Budyko to develop a climate model whereby ice sheets would grow until the planet was largely covered. His positive feedback loop sent ice over enough of the planet that the reflectivity, or albedo, of the planet rose to the point at which it reflected enough energy to cause further cooling.

Budyko concluded that although the planet could theoretically cool by such a mechanism, it couldn't possibly have happened. This was because Budyko had no means of envisioning a reversal: given his premise, the planet would cool, freeze over and remain so indefinitely.

In the 1990s Joseph Kirschvink resurrected and extended the idea further, coining the term "Snowball Earth". But how did planet earth freeze over if it happened? Kirschvink proposed that around 750 million years ago the supercontinent Rodina lay close to the equator, but extend Pole-ward along its southern flank (Fig. 3.9). Such a large area of land would cool effectively during the southern winter and serve as a repository for snow and ice.

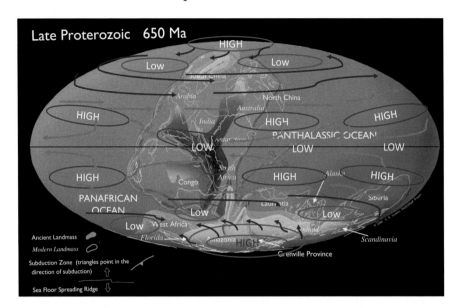

FIG. 3.9 Possible global wind patterns for the latest Precambrian period around the supercontinent Rodina. Like today westerlies dominate the mid to high latitudes with easterly trade winds dominating the tropics. The presence of a mountain chain down the middle of this belt would encourage the formation of a warm tropical low pressure over the middle of this supercontinent. With a faster spin (21.5 h day) the largely unimpeded westerlies flowing around the northern Polar Region would be stronger than those seen around Antarctica today (the Roaring Forties). An ice cap dominates the southernmost portion of the continent much as it does today. Underlying maps courtesy of: C. Scotese

Around 750 million years ago, Rodina began to fragment as pieces that would ultimately form Asia, Europe and North America broke away on their journey towards the north. As the break up ensued, these continental fragments began subducting increasing amounts of ocean crust. As mountains rose around their flanks, enhanced weathering set about lowering the concentration of carbon dioxide. Meanwhile, the relatively weak Sun (around 94 % its current luminosity) conspired with the lowering atmospheric concentrations of carbon dioxide to set the global temperatures tumbling.

Temperatures were on the fall and increasing snowfall meant that the land became even more reflective. With less heat absorbed by the Earth, the planet cooled further and the process accelerated in a spiraling period of positive feedback: enhanced snowfall meant more ice; more ice meant lowering temperatures and lowering temperatures meant more snowfall at increasingly low latitudes.

This was the planetary freeze of Mikhail Budyko. Estimates suggest that these periods of glaciation lasted between 4 and 30 million years and were repeated.

So if the planet froze, why did the cooler climate not persist? The key to solving this puzzle is our favorite greenhouse gas, carbon dioxide. When the ocean was sealed off and the land frozen, carbon dioxide was not taken up by weathering—or by photosynthesis. Moreover, carbon dioxide was being released by volcanic activity. Consequently, the level of the gas built up in the atmosphere. When it reached 10–100 times its current value, the warming effect of carbon dioxide overwhelmed the cooling caused by the presence of the global ice sheet. Within a few hundred years (at most), the ice melted. With melting, the planetary albedo decreased, which further accelerated warming. This time positive feedback drove melting and warming. Estimates suggest that in the immediate aftermath of the glaciations, carbon dioxide might have raised the temperatures to an average of 32 °C—something that will not be seen again for another billion years.

Evidence for the surge in carbon dioxide levels is found in the form of the carbonate rocks (cap carbonates) that were found in Svalbard. These carbonate deposits are formed in contemporary tropical conditions, thus their presence immediately above the glacial deposits implies that temperatures rapidly soared from frigid to torrid over a very short interval of time. In particular, the deposits in the Neoproterozoic are thick. This implies that a particularly large supply of carbon dioxide gas was present.

Although the evidence for a global freeze is strong it is not infallible. For one the presence of carbonates implies that the oceans were fairly alkaline. For if the concentration of carbon dioxide was high, the increased acidity it would cause would prevent carbonate rocks from forming. One way around this is if during periods of deep freeze, the oceans filled with calcium and magnesium ions—possibly from underwater volcanic activity. These would then react with the atmospheric carbon dioxide once the ice melted and the environments came back into contact. However, this remains a problem, particularly as a rather different mechanism—the oxidation of methane that was released abundantly—would not only fit the observed changes in the ratio of carbon-isotopes but could also produce the observed carbonate rocks.

Moreover, analysis of the carbon record implies a cyclical event—much like the modern day (Pleistocene) ice ages where periods of glaciation alternated with periods of comparable warmth. Indeed, perhaps four or five periods are identifiable where the ratio of carbon isotopes varies in a pattern suggestive of repeated climatological catastrophe and recovery. This pattern implies that the entire snowball period was driven and controlled by changes to the planet's tilt and orbital shape (a la Milanković Cycles). Although there is nothing wrong with this idea in principle, the problem is whether the planet can drive a massive spike in carbon dioxide levels only to rapidly removing the excess over a period of a couple of hundred thousand years. As carbon dioxide has a fairly long residency time in the atmosphere, this process seems less tenable.

Finally, the evidence for a global glaciation is piecemeal. Glacial deposits are found in some places but not others. The presence of a global freeze depends on where those pieces of continent lay at the time of the glaciation. Resolving this depends on a good geological record and unfortunately, nature is not kind: plate tectonics has scrambled and destroyed much of the rock record. Moreover, much of the geological record can be interpreted in different ways. For example, rocks called dropstones are usually formed when glaciers carry rocks from mountain to sea (or lake) and when the ice melts, the rocks fall to the bottom of the water. However, these can also form in so-called debris flows: massive currents of rock and mud that flow down-slope, underwater. Other deposits, which can be interpreted as glacial, can also be formed in this environment. This introduces some degree of ambiguity in the rock record.

All of these caveats accepted, the prevailing view driven by the greatest body of evidence still favors some form of global freeze, but that the Earth was completely frozen is certainly in doubt. Instead of a snowball, evidence favors more of a slush-ball. For one, life was not completely erased: photosynthesis was maintained, albeit at a lower level than was seen before the glaciations occurred. Computer models certainly struggle to produce a completely frigid Earth. Instead they cover roughly two-thirds to three quarters of the surface in a solid sheet of ice, with the remaining land more like Alaska or the Antarctic peninsula: chilly but not rigidly frozen. There is evidence of material being rafted over open water, probably on icebergs. This further suggests a planet that is largely frozen, but one retaining substantial areas of open-water—that is open-water dotted with icebergs. In these oases life persisted, somewhat chilled but still robustly intact.

Raindrops Keep Falling on My Head: The Controversial Climate of the Early Earth

Let's keep journeying back in time. If we go back to the end of the Archean era we can begin to answer some deeper questions about the nature of our planet's atmosphere. How can we find out what the Earth's climate was like 3 or ever 4 billion years ago? What we do know is that by 3.5 billion years ago photosynthesizing bacteria known as Cyanobacteria were busy in the planet's shallow seas and lagoons constructing structures called stromatolites. These living buildings of colonies of bacteria are still found today in places like northern coastal Australia. Their structures are so unique that they are easily recognizable when preserved in rocks. The presence of these macroscopic fossil structures tells you that there was abundant liquid water and that, globally, temperatures must have been moderate enough for liquid water to persist. So far, all seems well: except that the Sun was only around 75 % its current luminosity. This means that the Earth should have been rather chilly—so chilly in fact that the later Proterozoic snowball would seem positively balmy by comparison: picture present day Mars to get some idea.

Faced with this conundrum scientists will first turn to our favorite greenhouse gas, carbon dioxide. This isn't unreasonable as it should have been abundant—indeed it could have been a hundred times as abundant as it is now, given its source in what should have been a sea of volcanic activity. The problem with this scenario is that such a high level of carbon dioxide will make its presence known through the formation of the cap carbonates described above. Huge amounts of carbon dioxide will dissolve in the oceans and form carbonate rocks. Such abundance is not seen in the geological record. Next scientists will turn to line broadening. This was mentioned previously in the context of the effect of raising carbon dioxide concentrations and its capacity to act as a greenhouse gas. The higher the concentration the broader the range of wavelengths carbon dioxide can absorb infrared, because the molecules bash into one another and alter the properties of the molecule (specifically, the energy of its electrons).

Now, you can broaden the absorption of carbon dioxide without increasing its concentration if you have a higher density of atmospheric gas, overall. The additional, non-greenhouse gas will boost the absorption power of whatever greenhouse gas is present. Nitrogen can serve this purpose and it is clearly very common within the Earth's atmosphere. Therefore, if nitrogen or other gases were more common and the overall pressure of the Earth's atmosphere higher, carbon dioxide could act more effectively as a greenhouse gas than it does now, because it will be able to absorb heat energy over a wider range of wavelengths. This would be a simple solution to the conundrum of the missing CO_2, except that it isn't true.

Sanjoy Som (University of Washington) and co-workers used an ingenious method of determining ancient atmospheric density: fossilized raindrop impact craters. Raindrops falling on fresh volcanic ash—in this case laid down by very hot flows of material—turn into concrete-like material which can be preserved. The researchers examined a group of South African volcanic rocks called the Ventersdorp Supergroup. Within these rocks the impact craters produced by raindrops were preserved and indicated how quickly the drops fell to the ground. The faster the drop falls, the larger the crater, for any given size of raindrop. The speed of descent is proportional to the density of the air. The denser the air is, the lower the raindrop's terminal velocity will be. Analysis soon showed that the size of the craters was the essentially the same as modern-day ones—unless of course Archean raindrops were much larger, and hence more massive, than modern day ones. This meant that the density of the Archean atmosphere was effectively the same as todays: bang goes that idea. The result also confirmed that carbon dioxide levels could not have been much higher than present as this, too, would clearly have raised the density of the atmosphere.[4]

So, what are we left with is a warmer planet and, apparently, no way to make it any warmer. Methane produced by anaerobic bacteria could be the answer, but ultimately this methane would

[4] While going to press, Som has published another research article in the journal Science that suggests that the density of the atmosphere was actually a lot lower than it is now.

have to be at a fairly high concentration and there is, at least, little direct evidence to support its presence. Indeed, the presence of
some minerals such as magnetite and banded iron indicates that if
there was methane present it couldn't have been very abundant. If
it were more common in the atmosphere it would have prevented
these materials from forming.

Enter the most prosaic and likely correct hypothesis that the
ancient Earth was less reflective than it is at present. If it was
darker in color than it is now then, much like black card heating faster in the Sun than silver card, then the early Earth would
absorb more energy and warm more strongly than it does at present. To understand this idea further we need to look a little more
closely at the history of the planet.

At some undetermined time, several tens of millions of years
after the planet formed the temperature of the atmosphere fell
below the condensation point of water. A deluge would then have
begun that soon, thereafter, covered the bulk of the Earth in liquid
water. At this point the only visible land, if there was any, would
have been a scattering of volcanic islands. Modern plate tectonics was likely non-existent as the super-hot mantle would need
to cool further before an organized system of convection currents
could begin. The Earth's crust and topmost solid mantle, the lithosphere, was likely too hot and buoyant to subduct as it does now.
And, although a different system of plate tectonics wasn't far off
in the earliest and hottest Earth this wasn't quite able to operate.

Run the clock forward a couple of hundred million years and
things in the cooling mantle begin to work more like they do now.
Hot currents rising from the core generate massive waves of (likely)
underwater volcanism. These eruptions produce large undersea plateau that serve as the starting points for continents. Even so, at 4.3–
4.2 billion years ago most of the land is still under the dark oceans.
Now, take a step back and look at the modern Earth. The lightest
areas, those that reflect the most light, are clearly the continents—
particularly those areas free from vegetation. The deserts and ice
caps are very reflective in comparison to the dark, blue oceans. 4.3
billion years ago there were no land plants to the land was barren
and thus reflective. There was so little of it that the planet as a
whole was dark in color in comparison to that of today. A darker
surface absorbs more light and hence can warm more thoroughly. A
dark early Earth, as Minik Rosing from the Nordic Center for Earth

Evolution proposed, would neatly circumvent the problem of not having sufficient greenhouse gases to warm the planet. The lack of living organisms is also of benefit early on. Life, in particular plants release chemicals that form a photochemical smog—clearly visible as a haze over forests during the day. In today's atmosphere this haze serves as a source of particles that can condense water vapor to form clouds. Without life such a haze would not have been present and the planet likely less cloudy. Fewer clouds would also mean that the planet would appear darker than it does now and would be more able to absorb energy from the Sun. A difficult problem appears to have found a simple solution. Though one should always bear in mind H.R. Mencken's adage, "A complex problem has a simple solution. And it's wrong," it does appear as though the issue of the "faint early Sun paradox" is probably solved in a very prosaic manner.

Parallel Lives: The Formation of the Earth and Venus

To conclude this chapter we bring the Earth's climate into close parallel with that of our sister world and continue backwards in time to the era in which they formed.

The Earth and Venus began their lives in very similar ways: neither was at all pleasant. Both planets were likely bequeathed with approximately the same abundance of light elements and consequently would have had approximately the same amount of greenhouse gases at birth. What of the earliest atmosphere? That would have been interesting. Within the first one to two thousand years (and before the Moon formed) the atmosphere of both the Earth and Venus would have been supercharged by frequent large impacts. At such high energies transient atmospheres of vaporized rock and metal would have cloaked both planets. However, by 2000 years or so of their formation the most frequent bombardment stopped and the atmosphere would have cooled. From this point the hellish brew of materials that formed the early atmosphere would have rained out into the magma ocean underneath. Over the next two million years the surface would have solidified. At this point the surface temperature would still have been around 500 K (227 °C)— about half that seen on modern day Venus. An increasingly dense atmosphere of water vapor, carbon dioxide and nitrogen would have maintained hellishly high temperatures while the surface cooled slowly underneath. The atmospheric pressure would also have been similar to that on present day Venus: around 100 times the pressure of the Earth's current atmosphere at the Earth's surface.

At some point, perhaps 60 million years after the Earth took shape a Mars-Sized object, side-swiped the Earth. Likely ejecting the earliest atmosphere, this giant impact blasted most of the Earth's surface off into space along with a sizable amount of mantle. In the collision the Earth's greater gravity pulled the denser portions of the impactor—its metallic core in particular—back to the Earth's tormented surface. A thick soup of rocky debris formed a cloud then ring around the Earth. Within 1000 years of the catastrophe the Moon began to take shape. During the time the Earth would have re-gained its atmosphere of gassified rock, at least transiently. The temperature of this atmosphere would have once again hovered somewhere above 1800 K (1523 °C). The Earth also received a kick to its spin and a new tilt on the universe. The stage was set for the formation of today's climate systems. Hovering only 25,000 km over the Earth, the rapidly orbiting Moon generated tides up to 30 m high.[5]

Modeling by Kevin Zahnle (NASA Ames Research Center) indicates that once again—within a few hundred years—this rocky atmosphere would have cooled and rained out to the surface. Left behind was a very dense, hot atmosphere dominated by water and carbon dioxide. The surface pressure would have dwarfed the contemporary Venus and returned to the pre-impact value of more than 100 bars of pressure from carbon dioxide and several hundred bars of pressure from steam. This Venusian atmosphere was, once again, kept hot by heat radiated upwards from the magma ocean underneath. Additional heat was also being pumped into it (and the rocky sea beneath) by the infant Moon's immense tides. Such tidal heating still pumps a whopping 10 trillion, trillion Watts of power into the Earth's oceans today and drives the tides. Tidal heating would have been orders of magnitude greater when the Moon first formed and hovered over the Earth's atmosphere like some vast

---

[5] Finding any definitive value for the height of tides seems to be more trouble than it's worth! Values range from 1 to several tens of meters with few references giving anything conclusive. Take 30 m as speculative…

bloody orb. It is interesting to note, that at this time Venus would have likely been the more habitable of the two worlds, with a cooling atmosphere and a solid surface.

This primordial vision of hell was unstable. The Sun couldn't supply sufficient energy to keep the atmosphere this warm, and by 20 million years into the Earth's history, the surface cooled and solidified. This cut off the supply of heat energy that was required to sustain the global greenhouse. With the energy input per second reduced from over 140 W per square meter to a value more similar to that of today (0.5 W per square meter) the water vapor cooled to the point at which it could condense as rain. Now, don't go getting all dreamy-eyed about the Earth suddenly becoming habitable. The loss of the water from the air still left over 100 bars of carbon dioxide which kept the temperature high. Thus, despite an increasingly deep ocean, the atmospheric temperature was still around 500 K (227 °C). This is far above that at which water boils at the Earth's surface today. This was possible because the atmospheric pressure was that much higher, and with it the boiling point of water.

This final Hadean greenhouse was also unstable. Without a constant input of energy from the surface (or the Sun) heat was gradually lost to outer space. All the while, carbon dioxide was dissolving in the oceans and precipitating out as carbonate-rich rocks on the ocean floor. This appears to have been true for Venus, although as Chap. 5 describes, this is still controversial. If there was no subduction only around 10 bars worth of carbon dioxide could be removed from the Earth's atmosphere and dumped into the interior at any one time. However, once the interior was cool enough to allow some form of subduction, greater masses of carbon dioxide could be dumped into the mantle. The main issue is how long this took. Current estimates put the loss of the carbon dioxide-rich atmosphere at some point between 10 and 100 million years after the Moon formed. The Earth would still have had an atmosphere far richer in carbon dioxide than today (a few percent of its total mass rather than the current fraction of one). Yet, even this relatively low level of carbon dioxide couldn't be maintained and as the atmospheric temperature continued to fall carbon dioxide levels fell to levels comparable with that of today. With this final decline in carbon dioxide levels, down went the temperature from more than 200 °C to around 10–15 °C. Thus, by 4.4 billion years ago the atmosphere was rather similar to that found around us—except that it was effectively oxygen-free. The dark ocean compensated for low levels of greenhouse gases and ensured that the planet remained largely ice-free.

Venus being closer to the Sun would have been warmer but there is a good chance that it, too, had extensive oceans and relatively benign temperatures at 4.4 billion years ago. Current views of this world suggest that it lay somewhere between having a hot (50–100 °C) ocean or one that was cooler and more similar to the Earth's. Were the ocean hot, it probably didn't last more than a few tens of millions of years before evaporating. However, a cool ocean could have been stable for hundreds of millions or perhaps even a couple of billion years. The precise figure depends on how much cloud cover Venus had and what kind of clouds were most prevalent. A lot of cumuliform clouds would have kept it cooler for longer by reflecting solar radiation back to space.

Returning to the Earth, things weren't quite as pleasant as the vision of a cool, planet wide ocean first suggests. Punctuating this broad sweep of time, through to the beginning of the Archean, 600 million years after the planet formed, were a succession of massive bombardments. Although there appears to have been a spike in the rate of such collisions around 3·9–4.2 billion years ago (the Late Heavy Bombardment), the entire period is likely to have been rather violent. Impactors—mostly asteroids initially, but later massive comets, would have brought objects up to a few hundred kilometers across to the Earth at tens of thousands of kilometers per hour. Although such monsters were likely rare, compared to hundreds of objects tens of kilometers in size, these would still have packed enough energy to boil substantial portions of the ocean—and penetrate the crust through to the mantle. A single 300 km-wide projectile would vaporize a significant depth of ocean directly and generate a thick water-filled greenhouse above what remained. Anything above 500 km wide would have boiled the ocean away in its entirety. Smaller objects obviously generate less steam on impact, but would still generate spikes in the Earth's atmospheric temperature as the strength of the moist greenhouse temporarily stepped up.

Frightening though such impacts would have been, their effects would have been transient. Even a brutal 300 km wide impactor (something that would leave a scar up to 3000 km wide on the ocean floor) would only par-boil the surface for 10,000 years or so: certainly long by our standards, but nothing compared to the length of the geological period. Water would rain out of the initially hot (80 °C) atmosphere at a terrific rate—perhaps equivalent to a global depth of one meter per year.

However brutal all this sounds, there is sturdy geological evidence that the Earth was relatively balmy by as little as 100 million years after the Moon formed. Tiny crystals of zirconium silicate (zircons) were produced in sufficient abundance that they eventually became preserved in 3.3 billion year old rocks in Australia. The Jack Hills zircons tell of a time when the Earth was able to form granite. Granite only forms when there is sufficient liquid water in the mantle to allow the separation of the rocks minerals (including the zircons) from the hot brew. As the mantle is relatively dry, the major source of water comes from the subduction of oceanic crust. Moreover, the trace elements in these zircons indicate that the granite formed in part from melted clay-like minerals and these only form in the presence of liquid water. Although the manner in which this crust was delivered to the mantle is a matter for considerable and quite contentious debate, it must have happened in some form so that enough water and other minerals were present to broil the granite out and form early continental crust. The implication is, that even as little as 100 million years after the planet formed, the Earth hosted at least some aerial land masses and experienced erosion, which formed sediments like those of today.

By contrast the crust of Venus is rather telling. Most of the crust is basaltic and we know this from two sources: the multiple Soviet Venera landers of the 1970s and early 1980s; and more recently from Venus Express. Radar and other measurements of the surface show a primarily dark, flat rocky bowl and the likeliest volcanic rock that would produce this is basalt. Basalt is abundant on all of the rocky planets, the asteroids, and probably Io, too. Indeed, basalt may be the most common extrusive volcanic rock in the universe. Granite needs water to form and the only likely areas for granite on Venus are two prominent highlands, Ishtar Terra and Aphrodite Terra. If we assume that these are continents then we can make some useful inferences about the climate under which they formed and how long it lasted. On Venus although most of the surface lies within 1 km of the lowest points around 14 % can be classified as highland. Again, if we assume that these regions are all equivalent to terrestrial continents (or the inert remains of associated regions called island arcs) then we can infer how long this took to form. If plate tectonics was about as active on Venus as it was on the Earth—and their similar mass, gravity and composition would make this a reasonable assumption—then it is possible to infer that plate tectonics likely persisted for around 500–1000 million years. On Earth the preserved Archean continental crust is limited in extent and of comparable global coverage to the highland regions of Venus. There is certainly a lot of hand waving going on here, but it's not unreasonable. If, as on Earth oceans supplied the water needed to make the granite we can assume that there were oceans for at least this time. On Earth, within ten million years of subduction ceasing volcanism and the production of granite also stops: by this point the mantle above the subducting plate has dried out and no more granite can be produced. The highland crust of Venus, therefore, forms a kind of clock that can be used to infer the presence and persistence of its oceans.

When the late heavy bombardment ceased 3.9 billion years ago, the Earth was able to settle down into a more sedate period. During the next 300 million years it was able to convert organic slime into recognizable bacterial fossils and the story of life took off. Yet, this transition from hell to heaven was not quite that straightforward. For one the proliferating life conspired with the steady growth of continents to remove carbon dioxide from the atmosphere. With the Sun still faint, planet Earth was vulnerable to global freezes. The earliest detectable freeze—the Huronian Ice Age—occurred just over 2.2 billion years ago. Details are scant as little continental rock remains to produce a legible tale.

Shortly before the Huronian freeze, the Earth's atmosphere began its dramatic final transition to its modern composition. Sometime around 3.2 billion years ago cyanobacteria learnt to process water into free oxygen gas and usable hydrogen. The latter was already used to convert carbon dioxide into glucose, but until this time the bacteria were unable to harvest sufficient energy to split water. Instead they used the abundant hydrogen sulfide. When water became usable, suddenly one of nature's most reactive gases, oxygen, was free to trample all over the planet's geology and pre-existing life. For around 800 million years oxygen was soaked up by hydrogen sulfide and metals in the oceans and the atmosphere saw little of this noxious element. There came a time when those bacteria that were able to produce oxygen broke free of whatever geological chains were present and took over the oceans. Iron sulfide was converted to iron oxide rust and rained out onto the ocean floors. Billions of years later we would dig this up to ferment our industrial revolution. Most bacterial life in the oceans would have been quickly exterminated, with what remained, forced to adapt to this new poison or retreat to the muddy depths of the ocean floors. Oxygen's rise undoubtedly eliminated huge swathes of life on Earth but in doing so, cleared the way for the rise of more complex multicellular life.

At some point during the terrestrial Archean or early Proterozoic, Venus and the Earth diverged. While the Earth was vulnerable to extreme cooling, Venus was moving in the opposite direction. The ever-brightening Sun would have taken Venus past a tipping point where its warming oceans would have first given up much of their store of carbon dioxide then their bulk water through evaporation. In concert, these greenhouse gases would have begun to rapidly warm the planet. Somewhere between 3.5 and 1 billion years ago the last drop of Venus's oceans drifted into the atmosphere. Most likely this would have been sooner rather than later. With its moderating oceans gone and an atmosphere stuffed to the hilt with greenhouse gas, Venus began its thermic ascent to its current position as the hottest body in the Solar System, other than the Sun. The once twins were no more identical.

The history of the Earth's atmosphere has shaped the evolution of life as well as the deeper machinations of the planet. Our Goldilocks' world owes its benign quality to the interaction of the deep interior with the wet and now oxidized surface. Liquid water remains because the pace of geological change broadly matches the opposing change in the luminosity of the Sun. Moreover, changes to the surface, brought about by plate tectonics, have facilitated the preservation of liquid water. An early Earth was deficient in terra firma was darker and absorbed more sunlight keeping it warm. As the land rose and carbon dioxide was drained from the air, the Sun grew brighter at just the right pace. Plate tectonics probably also maintains the planet's geodynamo by helping stir the pot of molten iron in its core. This is a two-way street as plate tectonics—at least in its present guise—is likely only possible because we have a wet, cold surface. Remove the water and not only would the plates gum up, because the mantle would be more viscous, but the steady addition of continental crust would end as well. The planetary interior and exterior are fundamentally entwined and it is this that maintains the habitability of the Earth.

This state of affairs will not last forever. Although the Sun has been a benign accomplice in the evolution of the Earth and its atmosphere, this relationship will break down in the not too distant future. It is to this eventual divorce, and its lethal consequences, that this chapter now fatefully turns; but not before we encounter one last—and possibly rather long—freeze.

The Last Icehouse

The current planetary climate is controlled in large part by the greenhouse gas carbon dioxide. Although exact measurements are uncertain, the amount of this gas may have been 100 times greater when the Earth formed than it is now. Over geological time, despite blips like the Permian and Eocene, the overall concentration of this gas in our atmosphere has been in decline. Although photosynthesis has played a significant role in enhancing the rate of decline, the primary driver has been a series of reactions called the carbonate-silicate cycle.

In essence the carbonate–silicate cycle is a process driven by two competing forces: erosion and burial versus volcanism and uplift. When silicate rich rocks are exposed to carbon dioxide, particularly in the presence of water, chemical reactions convert the silicate rocks to carbonates and free silicon dioxide. This lowers the amount of carbon dioxide in the air. The greater the rate erosion occurs, the greater the rate of decline in the amount of carbon dioxide in the atmosphere. This, as you may recall, was the critical link between the rise in the Himalayas and the decline in atmospheric carbon dioxide levels that would ultimately take the planet into the most recent Ice Age (Chap. 2). However, the Earth is a dynamic planet, where the crust is buckled, twisted and in places drawn down into the mantle.

Where the temperatures exceed several hundred degrees, the carbonate-rich rocks decompose: the process is grandly known as thermal decomposition. Primarily, on Earth, the process happens where oceanic crust is subducted. But wherever temperatures are suitably high, carbonate rocks will cook, decompose and release their store of carbon dioxide.

Over time, geological activity has declined on Earth, as the planet's interior has cooled down. Moreover, more and more of the planet has become covered in continental rocks, which are too light to subduct. Consequently, when carbonates become washed onto their margins, it is more difficult for these rocks to fall into the ocean depths where they can then be subducted. As a result more and more carbon dioxide is being lost from the system, forever trapped as limestone and other carbonates on the continental margins.

Thus, human activities aside, carbon dioxide is becoming (or would have become) a very insignificant player in the global greenhouse over the next several tens of millions of years.

The Rise of Amasia

In 250 million years time, the Earth's continents will have reassembled into a twisted mirror of the Permian Pangaea. Although this may not be the planet's final resting place for the continents' vast bulk, their assemblage will coincide with a turning point in carbon dioxide levels. The Sun will be around 2–3 % more luminous than it is now—noticeable to our eyes—and, consequently, global temperatures will have risen from an average of around 15.5 °C now to around 20 °C. Carbon dioxide levels will have dropped from 400 ppm now to around 50 ppm. At this critical point, broadleaved plants will no longer be able to collect sufficient carbon dioxide to successfully photosynthesize. Those that can't adapt will become extinct. The grasses, on the other hand will continue unaffected as they can collect sufficient carbon dioxide down to levels of approximately 10 ppm. Perhaps the giant forests of today will be replaced by thick stands of bamboo. Regardless, existing floras will begin a final and irreversible decline as carbon dioxide levels become critical.

What will conditions be like in this future Pangaea, or *Amasia* as some have christened it? In general, the popular literature paints a grim picture resembling the end of the Permian with hot deserts dominating the interior and a much more stormy sea battering its shores. In part this may be true but it is, like it was for the Permian Pangaea, unlikely to be the whole story. For one, with an extra 5 °C or thereabouts in global temperatures, much of the planets ice is likely to have melted, but probably not all. Remember an extra 5 °C is something we might experience in 100–200 years with continued consumption of fossil fuels.

Amasia is most likely to reside over the planet's North Pole (Fig. 3.10). The most recent models show the future Pangaea will form through the closure of the Arctic Ocean and part of the North Atlantic basins. This will create a large landmass—possibly a rather mountainous one over the North Polar Region. The North Pole, itself, is likely to lie roughly along the eastern edge of Greenland. Land, being more reflective than water and with a much lower heat capacity (Chap. 2) will chill rather nicely. Thus, even with globally higher temperatures, there is a good prospect that some form of polar cap will be present in the north, particularly as much of this region will be fairly mountainous following the continental collision that closed the basin.

Likely extending towards the south will be the remnant Atlantic Ocean. The most recent models have this looking very much like the Tethys, but turned 90° on its side (Fig. 3.10). This pattern, if it is produced should facilitate the movement of cold waters from the North Polar Region southwards. Thus, unlike the Permian, the oceans could still be thoroughly mixed, if sufficient ice is retained in the north. Indeed, a Tethyan-like set-up would have subduction zones extending along one or both of the Atlantic margins, generating yet more mountain building. All of these high terrains would readily allow glaciation along the edges of part of the supercontinent. Certainly, this is speculative but definitely not unreasonable. We can only really know by surviving and observing this future world directly, which is rather out-with our current technological level…

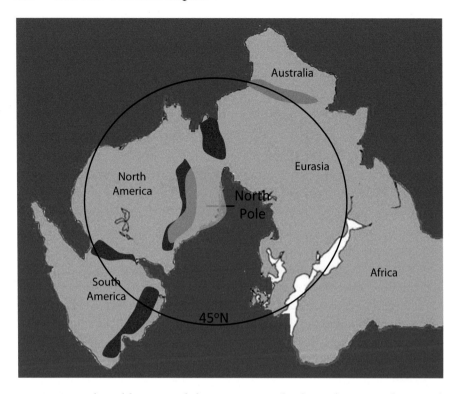

Fɪɢ. 3.10 A plausible view of the continents looking down on the North Pole, 250 million years, hence. *Darker browns* indicate mountains produced by the continental collisions that formed Amasia; *lighter brown*, the somewhat older belt produced by the collision of Australia with Indochina. A plausible ice cap is shown in *translucent lighter blue* overlying the northern mountains and extending somewhat over Greenland to the North Atlantic. Further mountains along the western Atlantic seaboard represent new belts formed by likely subduction of the Atlantic ocean crust under the Americas

Meanwhile, the situation of the South Polar Region is a bit more complicated. The most recent models do not clearly reveal where Antarctica will lie. The island continent could equally move north and close the Indian Ocean, leaving a body of open water over the South Pole. Alternatively, other models also suggest that Antarctica stays roughly where it is now, stranded. However, given that by 250 million years from now, Antarctica would be surrounded by a vast expanse of old, cold ocean crust, one would have thought subduction would become extensive around at least some of its shores, causing it to move northwards to warmer climes. If it does, and the Indian Ocean becomes its target, then the Indian Ocean may be reduced to a relatively small inland ocean, while Antarctica finally thaws out after 280 million years in the freezer.

Given its isolation from the rest of the planet's oceans the interior ocean, if it forms, could well become anoxic because its warm waters will be unable to mix with the greater body of circulating water on the planet. Regardless, the Indian Ocean or its descendent will be a source of moisture to the supercontinent's interior, along with the remnant Atlantic Ocean. Moreover, in a warmer world, with less water likely stored as ice, sea levels will likely be higher than present, which will leave open the prospect of much of the low-lying continental interior being flooded. Exactly where this will be depends on many factors which are, once again, hard to predict.

Given this set-up, what will the prospects be for future life? On the face of it probably rather good. Species will have to adapt to a radical change in plant life, but the supercontinent is unlikely to be a bland desert, given the predicted arrangement of the continental masses. Thus there should be plenty of different ecological niches for life to occupy.

One of the more dramatic assumptions about the climate of Amasia is its propensity for ravaging by tropical storms. Many published climate models for Amasia assume the greater heat from the Sun will drive the formation of proportionately more brutal tropical storms than are found on Earth today. In reality this depends on a lot of unknowns and assumptions. Firstly, although the planet will be warmer overall, it is the distribution of heat—and in particular the warmth in the oceans—that is critical. Future Earth might well generate more tropical storms rather than more brutal ones. Or the layout of the continents or of intervening cold ocean currents might mean that few access the shores of the future landmass. We can still get a few clues about the planet's likely tropical targets. Looking at the likely geography of Amasia one can speculate as to which areas would be prone to the effects of tropical storms. The least-favored (most heavily afflicted by tropical storms) would be the western shores of Amasia—areas that are now Eastern Africa, Eastern South America and, further east, southwestern Australia: these will all likely face the future Panthalassa. A good stretch of warn water is just what a tropical storm needs to develop and the set up of the continents will facilitate a thorough battering of these parts of the globe. The Northern oceans are largely subdued in extent, although the Atlantic basin might spawn a smaller share of tropical storms, with these affecting what is now Northern Canada and northern-most South America (the area that is currently Venezuela).

Will Amasia be the last supercontinent? Again, that's hard to predict but given the secular cooling of our planet, the continents should eventually cease wandering sometime between 250 million and 800 million years into the future. Once the mantle becomes too cold and viscous to allow subduction the internal works of the planet will begin to gum up. The timing of this phase is important as when subduction ceases—or becomes intermittent, the return of carbon dioxide to the atmosphere will drop further. Moreover, less water will cycle into and out of the mantle and that may have profound consequences for the way the atmosphere is maintained and the supply of nutrients to organisms. Again, this is speculation.

Looking beyond the bounds of Amasia and into the more distant future we will begin to see the futures of the Earth and Venus converge once more. Evan as Venus and the Earth began life in a very similar state, but then diverged, the ever brightening Sun will bring the climate of these two worlds back together once more. Before we examine the fate of the Earth and Venus in Chap. 5, we will branch off into more speculative territories. Chapter 4 will look at the weird and wonderful atmospheric phenomena on Earth, but we close this chapter with a view to the distant future of the Earth. In particular we ask: can humanity or its descendents hold back the tide and keep the Earth habitable until the Sun consumes us?

Geo-Engineering the Future

Life on Earth, in terms of its diversity, has likely peaked—indeed, it probably peaked during the Carboniferous, around 300 million years ago. Although life will continue to evolve to fit its new surroundings, over the ensuing billion years, or so, the trend for life on Earth is irrevocably downwards. That is if we let nature take its course and let the planet heat up.

Is there another way for the Earth—beyond the next few million years to at least until the Sun becomes a red giant? Could this idea be extended to look for intelligent life elsewhere in the universe? Think about it: conditions for life on Earth may become intolerable once the Sun is a mere 10 % brighter than it is at present. If you could block out that extra 10 % and keep the shield up for as long as you required it, then the planet could remain positively balmy, even as the other planets began to overheat. This could be maintained right through the entire main sequence lifetime of the Sun. There are two obvious possibilities for shielding, neither cheap and neither cost-free in terms of the environment, but compared with the alternative, a costly move elsewhere in the solar system, then a spot of geo-engineering might just come in handy.

The first option is to put sulfates into the atmosphere (or something similarly reflective) to block out the radiation from reaching the planet's surface. That might work, but you would need to constantly replenish the sulfate or other atmospheric shield as they would tend to wash out or precipitate out of the atmosphere. That might get tricky. Sulfates also alter the distribution of rainfall, so you would need to plan ahead where your crops grow and people live.

Likely more expensive, but a better long-term option would be to physically build a reflective mesh around the planet. Undoubtedly something of an engineering nightmare—and hardly cheap—such a feat would be possible with a bit of asteroid mining and space-based construction and it might be cheaper than building a fleet of exodus craft to take you to the nearest habitable world. Over time, such a shield would need repair and enhancement, but it could provide a long-term strategy to keep the planet habitable. With less sunlight now reaching the surface, solar furnaces could be used to cook carbon dioxide out of rocks to replenish the atmosphere and maintain plant life.

If we wanted to look for a truly advanced civilization elsewhere in the cosmos we might want to look for the signature of such a planetary shield. Depending on the planet's composition we could search for enhanced reflectivity (an unusually high albedo) for a planet that lay just inside the orbit at which one might expect a runaway greenhouse to commence. Now, Venus already has this and no one would suggest Venus harbors highly intelligent life.

However, something a bit more subtle—and with a chemistry not dependent on sulfuric acid—might just imply, if not definitively prove, intelligence lay behind the high albedo. Markers would be an oxygen-rich atmosphere implying photosynthetic life and a spectrum compatible with metallic or other reflective shielding. Free metals are not commonly found high in planetary atmospheres.

Conclusions: The Big Picture

In this chapter we have built up a broader picture of the Earth's climate system, importantly illustrating how the planet's climate has changed over evolutionary time. Although there have been some monumental swings the planet's in-built regulatory systems have brought the climate as a whole back into line with habitability repeatedly. The underlying players are the Sun and carbon dioxide.

While carbon dioxide is by no means intelligent, its ability to respond to changes in the amount of radiation the planet receives from the Sun is crucial to the well-being of the planet. For most of the planet's history carbon dioxide has been the regulator of planetary temperature. It was only in the last 34 million years that the level of carbon dioxide fell so low that the underlying astronomical cycles began to show their effect. Previously, these cycles gave small boosts to the global temperature but never took it low enough for global glaciations to be an issue. From here until around 1800 A.D. the level of carbon dioxide was sufficiently low that it then responded to the change in temperature. This meant that as temperatures fell, the hydrological cycle slowed and with it the rate of erosion. This in turn allowed more carbon dioxide to build up and persist in the atmosphere, which then warmed the planet more. When the Sun was brighter, as a result of both changes to the Earth's orbit and angle of tilt (the Milanković Cycles) or to its secular brightening with age, the rate of erosion stepped up and more carbon dioxide was lost from the atmosphere. This in-built thermostat has operated successfully for 4.5 billion years until human activity uncoupled the gas pedal from the clutch and the brake.

Indeed, the accelerating rate of melting of the northern and southern caps is unprecedented in geological history. Even in the closest match, the Eocene, the rate of temperature increase was roughly ten times slower than the currently observed rate. No natural cycle in recorded geological history (barring asteroid impacts and a Yellowstone-style super-volcano eruption) has changed the temperature and the mass of carbon dioxide so quickly. Even the relatively speedy Milanković cycles operate on longer timescales measured in tens of thousands of years: not decades.

Putting this in context, if we assume that we are not alone in the Universe, it is likely that other intelligent species face the same issues elsewhere in the cosmos, assuming that all life is governed by the same underlying laws. Species have to fit within the broader canvas created by the geochemistry and geophysics of their home world. There are going to be many themes that transfer from the Earth to other celestial bodies as clearly the underlying astrophysics is universal. Thus, if we find a planet with roughly the same mass and composition as our world orbiting a G-class main sequence star at a distance of 150 million kilometers, expect it to look rather like the Earth, if not as it is now than as it once was or will be. However, don't forget that this exoplanet might not look like the Earth does now, simply because it shares our physical location. Our planet has changed significantly and with it the advent of life has radically altered the planetary atmosphere. The temporal dimension is as important as the spatial dimensions when we think about habitability or simply the gross features of the planet's atmosphere. Were the Earth lifeless, there would be little free oxygen and the atmosphere likely more noxious. Without human activity, carbon dioxide levels would be on a downward trend. It is from the perspectives laid out in chapters one and two that we can now move on to examine the workings of the atmospheres of the other planets that we know about. This is a growing family and thus far few are clones of the Earth. With improving technology and our ability to resolve greater and greater details, expect a clone or three of the Earth to pop up soon, providing another planet on which to observe climate.

References

1. Zahnle, K., Arndt, N., Cockell, C., Halliday, A., Nisbet, E., Selsis, F., et al. (2007). Emergence of a habitable planet. *Space Science Reviews, 129*, 35–78.
2. Fernando S. Paolo., Helen A. Fricker., Laurie Padman (2015). *Volume loss from Antarctic ice shelves is accelerating.* http://www.sciencemag.org/content/early/2015/03/31/science.aaa0940.
3. Svensen, H., Planke, S., Malthe-Sørenssen, A., Jamtveit, B., Myklebust, R., Eidem, T. R., et al. (2004). Release of methane from a volcanic basin as a mechanism for initial Eocene global warming. *Nature, 429*, 542.
4. Leconte, J., Forget, F., Charnay, B., Wordsworth, R., & Pottier, A. (2013). Increased insolation threshold for runaway greenhouse processes on Earth-like planets. *Nature, 504*, 268–271, doi:10.1038/nature12827.
5. Som, S. M., Catling, D. C., Harnmeijer, J. P., Polivka, P. M., & Buick, R. (2012). Air density 2.7 billion years ago limited to less than twice modern levels by fossil raindrop imprints. *Nature, 484*, 359–362.
6. Rosing, M. T., Bird, D. K., Sleep, N. H., & Bjerrum, C. J. (2010). No climate paradox under the faint early Sun. *Nature, 464*, 744–749.
7. Mitchell, R. N., Kilian, T. M., & Evans, D. A. D. (2012). Supercontinent cycles and the calculation of absolute palaeolongitude in deep time. *Nature, 482*, 208–211.
8. Hansen, J., Ruedy, R., Sato, M., Imhoff, M., Lawrence, W., Easterling, D., et al. (2001). A closer look at United States and global surface temperature change. *Journal of Geophysical Research, 106*, (D20) 23947–23963.
9. Travis, D. J., Carleton, A. M., & Lauritsen, R. G. (2002). Climatology: Contrails reduce daily temperature range. *Nature, 418*, 601.
10. Pollard, D., & DeConto, R. M. (2009). Modelling West Antarctic ice sheet growth and collapse through the past five million years. *Nature, 458*, 329–333.
11. Naish, T., Powell, R., Levy, R., Wilson, G., Scherer, R., Talarico, F., et al. (2009). Obliquity-paced Pliocene West Antarctic ice sheet oscillations. *Nature, 458*, 322–328.
12. Kossin, J. P., Emanuel, K. A., & Vecchi, G. A. (2014). The poleward migration of the location of tropical cyclone maximum intensity. *Nature, 509*, 349.
13. Michael E Mann. (2002). The Earth system: Physical and chemical dimensions of global environmental change Volume 1, (pp. 504–509). In Michael C MacCracken & Dr John S Perry (Eds.), *Encyclopedia of global environmental change*. New York: John Wiley & Sons. ISBN: 0-471-97796-9 Editor-in-Chief Ted Munn.
14. Hoffman, P. F., Kaufman, A. J., Halverson, G. P., Schrag, D. P., Hoffman, P. F., Halverson, G. P., et al. (1998). A Neoproterozoic snowball Earth. *Science, 281*, 1342–1344.
15. Peterson, T. C., Gallo, K. P., Lawrimore, J., Owen, T. W., Huang, A., & McKittrick, D. A. (1999). Global rural temperature trends. *Geophysical Research Letters, 29*(3), 329–332.
16. DeConto, R. M., Galeotti, S., Pagani, M., Tracy, D., Schaefer, K., Zhang, T., et al. (2012). Past extreme warming events linked to massive carbon release from thawing permafrost. *Nature, 484*, 87–90.
17. Wignall, P. B., & Twitchett, R. J. (1996). Ocean anoxia and the end Permian Mass extinction. *Science, 272*, 1155–1158.
18. Daniel H. Rothmana., Gregory P. Fournierc., Katherine L. Frenchb., Eric J. Almc., Edward A. Boyleb., Changqun Caod., & Roger E. Summons. (2014). Methanogenic burst in the end-Permian carbon cycle. *Proceedings of the National Academy of Sciences*, 111 no. 15 > Daniel H. Rothman, 5462–5467, doi: 0.1073/pnas.1318106111.

4. Weird Weather

Introduction

In principle this chapter concerns luminous phenomena, some of which are natural, while others might well be tricks of the light. Around 2000 thunderstorms are active within the Earth's atmosphere at any one time. Most of these occur in the afternoon when the land or sea is at its hottest. Although almost anyone reading this book will be familiar with thunder and lightning, thunderstorms generate a diverse family of other phenomena, which will be explored in the following pages.

Note that whereas the other chapters in this book are built on reliable and verifiable evidence, this one is altogether different. While most of what is said is verifiable, there will be some stories which are, perhaps, more anecdote than hard science. Yet embedded within this raft of diverse and peculiar tales are some scientific gems that lie at the frontiers of our understanding. A few will be misidentifications and will be debunked, but others will develop into truly odd manifestations of our planet's atmospheric and internal engine. There is a splattering of new science and a lot of wonder in these phenomena, some of which might just have counterparts on other worlds that we are discovering today. Read on and imagine.

Strange Lights from Thunderstorms

Ball lightning is a good, as in verifiable, starting point. Ball lightning was known and recorded either in drawings, painting or in stories for hundreds of years. Figure 4.1 reportedly shows a painting made by an unknown artist around 1888 of three luminous balls traveling along a small ravine during a thunderstorm. In those pre-UFO ages we can be pretty sure the Russia artist was documenting a natural phenomenon and was less likely to be letting his imagination wander.

© Springer International Publishing Switzerland 2016   159
D.S. Stevenson, *The Exo-Weather Report*, Astronomers' Universe,
DOI 10.1007/978-3-319-25679-5_4

Fig. 4.1 The first known image of ball lightning. This painting was published in La Physique Populaire in July 1888. The artist remains unknown but this work is believed to have been painted it following a thunder-storm near St Petersburg, Russia. As with any artistic rendering how much "artistic license" was employed is unclear, but differences in the color of different balls are obvious

Descriptions of ball lightning vary considerably. Some describe it as moving upwards, others downwards, sideways or unpredictably in different directions. These can even be against the direction of the prevailing wind. Some ball lightning appears to move towards metal objects, or along the surfaces of walls, while others indicate free movement in the air. Some ball lightning appears capable of penetrating solid objects, including glass, wood or masonry and in some instances appears to pursue people. In general it is luminous white to yellow or even red in color and adopts round or oval shapes, with some more cigar-shaped or flanked by sparks. In most cases the balls are opaque but a few accounts exist that suggest balls are more translucent in nature. Some balls are associated with the smell of sulfur, ozone or nitrogen dioxide, but this is uncommon. As you can see there are almost as many descriptions of their appearance as there are accounts of their presence. This diversity of these accounts has led to a lot of confusion and suspicion that ball lightning only existed in the imaginations of the observers. Indeed, one working hypothesis for the phenomenon was that electromagnetic fields in thunderstorms made people hallucinate. Although this sounds outlandish, the effects of electromagnetic fields on our perceptions are well known and are suspected to lie behind many alien abduction stories. Therefore, maybe, just maybe, thunderstorms do cause hallucinations.

Despite the many differences in the reported incidents, there are a few clues that might help scientists uncover the secrets of ball lightning. Aside from the clear association with lightning—in particular cloud-to-ground lightning—and thunderstorms in general, there are a few other reports of ball lightning or something like it being reported in quite different contexts. In World War II submariners often reported consistent accounts of small, luminous balls of light emerging when battery banks were switched on or off. This suggests one of two things: either the balls were produced directly by electrical discharges that ionized the air, or the electrical discharge caused an electromagnetic pulse—a wave of microwaves that energizes the air indirectly. Both are plausible in thunderstorms.

Another interesting observation was made by pilot Lieutenant Don Smith during a routine flight to Hawaii in the 1960s. His cargo plane was flown with its radar turned up to its maximum intensity while the plane penetrated dense fog. Two horns of St Elmo's fire appeared on the random (the radar cover). St Elmo's fire is a luminous electrical discharge seen around pointed surfaces in the presence of strong electrical fields. St Elmo's fire is now very well understood, however in this instance, what was more

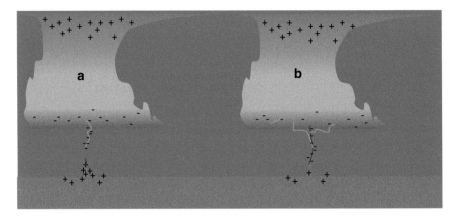

FIG. 4.2 A simplified anatomy of lightning. Convection within a cumulonimbus cloud allows charge to separate. Although not clear, it appears formation of ice is key. During the formation of ice, small particles break off and carry positive charges upwards, while negative charges remain with the larger ice particles and water droplets lower down. During a cloud-to-ground strike (shown) the cloud's strongly negatively charged bases induces a positive charge in the ground underneath by repelling negatively-charged electrons. These charges are most strongly concentrated near the tips of objects (**a**). A faint, low current stepped leader stroke advances relatively slowly towards the ground while corresponding positively charged "feelers" (*red*) move upwards from any suitable point. When one makes contact with the stepped leader the circuit is completed and the cloud discharges a powerful electrical current to the ground (**b**). Where there is a large spread of negative charge in the base, further areas of the cloud discharge in succession, leading to a prolonged spider-like display. Intra-cloud discharges drain positive charge from the cloud top, along with rain or hail

interesting was the subsequent appearance of ball lightning in the cockpit between the two horns of St Elmo's fire. Clearly there was some form of "communication" between the two phenomena suggesting the strong electrical field that had generated the St Elmo's fire also initiated the ball lightning event.

Slightly creepier was the appearance of ball lightning in another plane flying over Alaska near a thunderstorm in 1962. The crew of a P3 Orion were joined, mid-flight, by a red translucent ball of light. The crewmen reported that the large ball appeared near the plane's rear bulkhead before slowly passed down the galley of the plane towards its front. After a few moments the basket-ball sized orb reached the front section of the plane only to

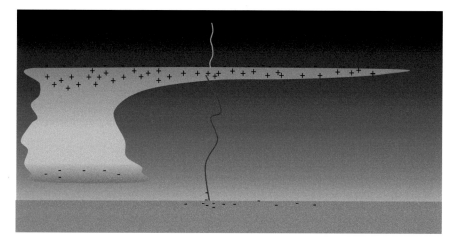

Fig. 4.3 Positive lightning. In some instances, during particularly powerful thunderstorms, cloud-to-ground lightning can originate directly
from the positively charged anvil of the thundercloud. This requires a
much higher voltage than conventional lightning as the distance the
bolt needs to travel is greater. The very strongly positively charged anvil
again induces the separation of charge on objects on the ground. However, in this case the ground becomes negatively charged. The same process happens as before with a positive leader moving downwards then
making contact with a negative feeler. Electrons then flow up the path,
rather than down it, to cancel out the charge in the cloud. Interestingly,
in these storms the connection with the ground appears to result in further electrical discharge between the cloud and the upper atmosphere
(*light blue squiggle*). This process appears to require the presence of a
negatively-charged cap above the top surface of the positively charged
anvil—although the details are far from clear

disappear through the fuselage near the plane's electrical distribution center.

Both of these observations suggest a link between ball lightning and strong electromagnetic fields generated naturally or
through inadvertent man-made activities. But there are other
observations that suggest a connection between this enigmatic
phenomenon and the direct contact of lightning with surfaces.

Two dramatic accounts both involving churches in renaissance England appear to be the first documented instances of ball
lightning in the world. The first occurred in Wells, Somerset,
England, in 1596. Here, during a violent thunderstorm, a luminous ball of light entered the church through a window during a
sermon. The ball—about the size of a soccer ball—passed along

the wall of the church near the Pulpit, before violently exploding and scattering parishioners across the ground. The explosion was hot enough to melt the metal parts of a nearby clock and leave odd star-shaped burns on some parishioners' clothing. Aside from that, and structural damage to the interior of the church, no one was harmed.

The second event, which is now the more famous—and is consequently often erroneously referred to as the first recorded ball lightning observation—was made on October 21st 1638. This vivid account, similar to that in Wells, tells of the Church of St Pancras in Cornwall being struck by an almighty bolt of lightning during a powerful thunderstorm. Incandescence then entered the church, through the window, where 300 parishioners were attending a service. The ball of light was hot enough to burn several of those attending the service, in some cases underneath apparently un-marked clothing. Soon thereafter the ball was said to have violently exploded throwing many of the churchgoers to the ground, one so violently as to smash his skull apart on impact with the ground. In all four were killed and 60 injured; with severe damage caused to the structure of the building. Accompanying the storm, a small tornado appears to have subsequently done away with a dog that ran from the explosion in the church. The evidence of this unusual event is still visible on boards within the church.

Another tale comes from 1944 Uppsala, Sweden, where a thunderstorm is reported to have created a ball of light that penetrated a glass window, leaving a 5 cm-wide hole. But it was reports from the 1960s that ultimately led to ball lightning being taken seriously by the scientific community.

Perhaps most peculiar in ball lightning folklore is the manner by which the phenomenon went from fantasy and delusion to accepted scientific fact—or at least onto the record of Nature's letters page in a 1969 issue (Nature 224, 895). Professor R. C. Jenninson of the Electronics Laboratories at the University of Kent was flying in an Eastern Airlines Jet from New York to Washington when ball lightning apparently "emerged" in the gangway of the plane, lazily bouncing down the aisle past the scientists and other passengers, before winking out. The main problem with ball lightning despite these first-person accounts, as it is with many of the other luminous phenomena described here, is that they have proved to be impossible to recreate convincingly in the laboratory. This tends to inflame skepticism in the scientific community. Various explanations that are touted tend to err on the side of science fiction and fantasy, rather than reality. Explanations of ball lightning vary from the unlikely (antimatter meteors, mini-black holes or spinning electric field vectors) to the surreal (aerogels, heavy neutrinos or self-confined light)—it seems that there are more explanations than there are validated observations of the phenomenon. That said a few models do appear to hold water. We're going to look at a couple of these, one of which has some observational evidence to support it.

The two front-runners are the silicon vapor hypothesis and the microwave cavity hypothesis, with a third known as the lightning ball hypothesis a close third.

The silicon vapor model posits that when lightning hits the ground in some instances the composition of the soil allows the formation of a ball of combusting gases. The soil has to have the following constituents: silica glass (sand) and a source of organic carbon, such as dead leaves, animals, bacteria and fungi. When the lightning strikes all of this mulch of material is vaporized and the organic material reacts with the silicon dioxide to form carbon dioxide and silicon vapor. All of this happens in an instant and the hot silicon vapor that is produced then begins to burn in the air, reforming the silicon dioxide. In this hypothesis, it is this chemical reaction that makes the ball lightning glow. Though this couldn't readily explain ball lightning seen onboard aircraft, the theory did gain some backing from observations made of the phenomenon during a thunderstorm in China.

In 2012 a group of Chinese scientists from Northwest Normal University in Lanzhou, China were observing a thunderstorm on the Tibetan Plateau. During the storm a cloud to ground strike generated a short-lived luminous ball. Using a spectroscope attached to their camera they managed to obtain the chemical composition of the lightning and the accompanying ball. As expected the lightning showed the presence of oxygen and nitrogen from the air, while the spectrum of the ball showed silicon, calcium and iron, as well as carbon, oxygen and nitrogen (Fig. 4.4). The 5 m wide ball itself had a lifetime of less than 2 s but it did manage to move around 15 m before winking out. Interestingly, the ball's behavior

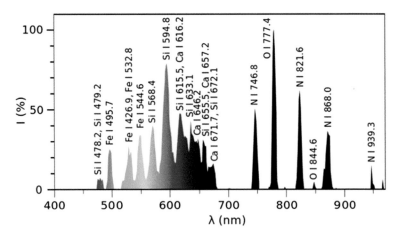

Fig. 4.4 A representation of the first spectrum taken of ball lightning. The spectrum was captured during a Chinese thunderstorm in 2014 using a modified camera. The spectrum is of the emission type—indicating that it is a low density, ionized gas. Chemically, the gas broadly matches the silicon-vapor model, which suggests ball lightning is produced when lightning hits and vaporizes silica-rich soil. Elements: *Si* silicon, *Fe* iron, *Ca* calcium, *N* nitrogen, *O* oxygen; with wavelengths in nanometers indicated. Interestingly, carbon, which is a prediction of the vaporized silica model, is missing. Image courtesy of Wikipedia commons universal license author: Olli Niemitalo

showed that it was influenced by nearby power lines. There was a clear 50 Hz flicker that was associated with the oscillating electric field in the cables. This suggests that the ball contained ionized gas that the electric field from the power cables could influence. Thus clearly, the ionized silicon vapor theory does work, at least for some ball lightning. Both in the laboratory and in this instance, in the field, these balls are very short lived. So, are there any other means by which ball lightning could come about?

The next most favored theory involves radio or microwaves. In the 1920s Nikola Tesla is said to have produced ball lightning routinely while investigating sending electrical energy through the air using radio waves. Similarly, scientists have created ball lightning–like phenomena in the lab using beams of microwaves. Y. H. Ohtsuki and H. Ofuruton succeeded in making ball lightning by interfering opposing beams of microwaves. It is not unreasonable to think that lightning could generate pulses of microwaves that might do the trick to create ball lightning. A derivative of this is the Handel Maser-Soliton theory in which the microwave source is a maser with a diameter up to 5 km. Masers are microwave-wavelength equivalents of optical lasers where molecules in a gas, typically water or carbon dioxide can amplify a microwave signal producing a coherent beam of radiation. If that sounds outlandish bear in mind astronomers have been aware of masers hundreds, thousands or even millions of kilometers across that lurk in places like the atmospheres of aging red giant stars, collapsing nebulae or the dusty torus of gas that surrounds many super massive black holes. Thus, unlikely as it sounds, there are clear precedents for such phenomena in nature: our atmosphere might just behave like an enormous microwave laser on occasion, and those 5 km wide sources might just be thunderstorms.

One of the best observational sources comes from a 1954 ball lightning sighting, or as Domokos Tar (Swiss Federal Institute of Technology) likes to call his observation, a lightning ball. During a thunderstorm Tar was stuck in Budapest. Lightning struck the ground close to where Tar was standing. Domokos Tar later wrote, *"Immediately a very strong wind began to blow. I saw the bushes bend to the right under the wind. After the lightning, it got relatively dark, because of the dark clouds. In the background there was a dark building. The bushes were still moving in the wind and on the left wet leaves and grasses were whirling through the air. After about 2 sec of darkness, suddenly a very beautiful sphere with a diameter of about 30-40-cm appeared about 1.2 m above the ground, at the same distance (about 2.5 m) from the lightning impact point as the bush, but in the opposite direction, to the left. The Lightning Ball (LB) was very brilliant, like a little sun. It was spinning counter clockwise. The axis of the rotation was horizontal to the ground and perpendicular to a straight line drawn between the bush (trees), impact point and the ball. The LB had one or two plumes or tails. The plume was not as brilliant as the sphere, and reddish. Very strange was that the plume was not on a perpendicular plane to the ground, but to the north of this plane. In other words, the plume had a component in the YZ-plane. After a very short time about 0.3 s the plume merged in the ball. The LB was moving at slow, constant speed to the left on straight line mentioned. It had absolutely sharp contours. Its brightness was constant across the entire surface. At that moment, I no longer saw any rotation."*

Tar's observations were recounted at an annual symposium on ball lightning in 2006 that provided a fascinating overview of the phenomenon. Tar's lightning ball model only holds about a hundred or so Joules of energy so approximates a domestic light bulb. This is shown in Fig. 4.5. The model itself seems perfectly reasonable and Tar even suggests that such a collapsing system might be applicable to the generation of energy by nuclear fusion. There are a few variants of this model around involving rings of plasma generated through lightning strikes, but Tar seems keen to emphasize that his lightning ball model is distinct from ball lightning because it operates at lower energies. Whether this distinction is real or not is as open to question as the model or models for the observed balls that are identified. It seems obvious that the vaporized silicon model could work in concert with the ring models, or each could operate separately. Likewise the maser hypotheses could work with either, as well. Either way ball lightning, no matter how peculiar it seems, is an established occurrence.

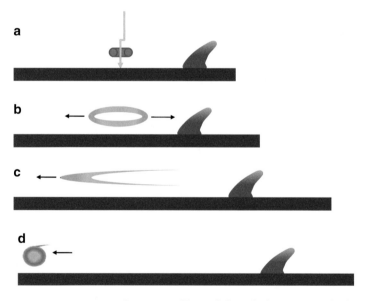

Fig. 4.5 Dominic Tar's Lightning Ball Model. A lightning strike (a) generates a rotating charged ring of air. This ring expands outwards until on one side it encounters an object (b). The ring breaks at the contact point with the object then collapses along its length (c). Eventually, the ring collapses to form a rotating ball (d). This glowing mass of air is the lightning ball. A hybrid ball involving vaporized soil particles could also give rise to the silicon vapor identified in the Chinese observations in Tibet (see text)

Why do some thunderstorms appear to make ball lightning and not others? This might come down to the nature of the storm. Most thunderstorms produce negative lightning bolts. A precious few, five in every ten million, lightning bolts are much grander and likely bear a positive, rather than a negative charge (Figs. 4.2 and 4.3). Such super-bolts last around ten times longer than conventional negative lightning and carry proportionately greater charge. It's certainly possible that ball lightning only forms in those rare situations where positive cloud-to-ground strokes occur.

In support of this is the curious tale of the 1978 "Bell Island Boom". Bearing some connection with the Seneca Guns described later in this chapter, this loud explosion was heard over 100 km from Bell Island, a small island near Nova Scotia. Residents of the island reported a loud explosion (or possibly three closely spaced explosions) and one resident reported seeing a bright flash followed by the appearance of a luminous ball. This faded after a few seconds. There was considerable damage to nearby electrical equipment, suggesting a large electrical discharge had occurred. Many interesting conspiracy stories were kicked up by this, including Soviet woodpecker radio signals or a Tesla device that beamed radio waves into objects to induce massive electrical discharges. Using data from the Vela gamma ray satellites, which were used to monitor nuclear tests, John Warren and Robert Freyman from the Los Alamos National Laboratory (then called the Los Alamos Scientific Laboratory) detected a superbolt over the island at about the time the reports of the explosion or explosions came in.

Earthlights or Headlights?

Across the globe there is a bewildering array of other types of lights besides ball lightning that appear to come and go at will in particular locations. These "other" Earthlights can be nagar lights, igniting marsh gas or earthquake lights, described later, but there are also less well explained ones that appear when the weather is largely fair and in non-seismically active areas. Burning marsh gas, a mixture of phosphine and methane, rises upwards as one would expect since it is hot and moving by convection. Swampy areas with related decomposition of plant material give rise to these flammable gases. The Naga fireballs on Thailand's Mekong river are light, small rapidly rising bursts which appear at the end of the annual Buddhist lent, khansa, in October. Their recorded history is spotty but they are observed by many each year at the festival season, whether originating as tracers set by humans, as phosphine gas, or as plasma orbs.

The Paulding Lights are another intriguing case. Believed to be supernatural in origin by some in the region where they occurred, a group of students at the Michigan Technological University Society of Photo Optical Instrumentation Engineers (SPIE) used a spectroscope to find the scientific explanation. How these lights came to be understood says rather a lot about our perceptions and what we are prepared to believe.

In order to understand the Paulding Lights you need to picture the scene. They appear in the distance along the line of an old dirt track flanked on its right by power cable-bearing pylons. The lights are mostly white or off white; but red lights are common, as well. To the unaided eye the lights magically appear flicker then disappear, only to reappear once more. There are several very good videos on *YouTube* replete with people whooping and crying at the seemingly magical appearance of the lights. With a small pair of binoculars an observer at one end of this track will see lights appearing before moving up or down, in part hidden by the trees. These observations are always indistinct and taken from a bridge overlooking the path or on that section of the path under the bridge. On a good day many observers will gather, sometimes even film crews, all trying to catch a glimpse. So far so good: the Pauling Lights are very well documented and are clearly a genuine phenomenon. The problem is what is their cause?

UFOs and psychic energy are front runners in some quarters, as well as lights associated with tectonic strain (earthquake lights) and ball lightning. However, clearly none of the latter natural phenomena fit the bill. Enter the SPIE team from Michigan Tech University armed with a powerful spectroscopic telescope. Training the scope towards the direction of the Paulding Lights they could clearly see the junction of the main interstate (Highway 45) and a side-road that runs continuously with the end of the dirt track that the lights are seen on.

As dusk descends the team continues to spy on the junction until the famous lights appear. Again, by naked eye the same configuration of white and occasionally red lights are seen. Some flicker; some appear to descend a short distance then wink out. However, under the gaze of the telescope all becomes clear: the lights are definitely head and tail lights of cars. The single, flickering lights are resolved into pairs of headlights or less frequently red tail lights on cars moving onto or off the interstate in an exact spectroscopic match.

In 2005 students from the University of Texas carried out a similar investigation with the even more famous Marfa Lights. Once again, under the gaze of a spectroscopic telescope the Marfa Lights are resolved into car headlights or camp fires, distributed by refraction through the atmosphere. In both cases the lights are only clearly visible to distant observers under specialized conditions. In these instances a layer of warm air sits above the ground with a cooler layer below. Such a layer readily forms in warm weather when the sun goes down and the ground cools quickly through radiation. Light from distant cars (up to 8 km away) is bent by the change in the density of the air as it moves towards the observer. In some instances this causes it to bend enough so that it is visible over longer than normal distances. What is obvious about many of these light phenomena is that they appear to exist only in the distance, near the horizon. They never seem to appear near the observer. In a few documented instances observers report that the lights appear to play tag with them, keeping pace with their movements. This can be explained if cars are driven by law-abiding citizens that are keeping to the speed limit and hence remain equidistant. The same phenomena has been observed, and explained, at the so-called Spooklight near Hornet Missouri; again, spectroscopic investigation showed that these were refracted car headlights.

Refraction can occur in two ways. Where the air is warmest above the ground, but colder upon it, light rays from an object bend downwards towards the observer. This is known as a superior image. The converse, with warmer air near the surface is called an inferior image and is responsible for conventional mirages that are seen on hot days. In the case of the Paulding and Marfa lights, the colder air at the surface has this effect leaving light rays apparently coming from higher up in the air than they truly are. At sunset the Sun's rays are refracted through the air in a similar way, with colder air at the surface. This allows the Sun to appear as though it is still above the horizon when in fact it has set.

In the opposite situation, the warmest air lies at the surface. This most commonly happens on hot summer days. Light rays are then bent upwards through the hot overlying air. This most commonly forms mirages, when there appears to be water lying on a dry surface. This water is simply a reflection (or rather a refraction) of light from the sky above.

More complex superior mirages, called Fata Morgana, can also occur when the air has a more complex structure, but is still predominantly coldest near to the surface. Named after the Arthurian sorceress Morgan le Fay, these were first noted in the Straights of Messina. According to legend this wicked sorceress hung out in the straits, luring sailors to their deaths by leading them to believe land was near. In a typical Fata Morgana image, the original object, which has its image refracted, is rather distorted and unrecognizable.

In Australia, Professor Jack Pettigrew examined the famous Min Min lights of Queensland. Pettigrew confirmed that the lights there, similar in appearance and behavior to those in Marfa and Paulding, had an origin in refraction, which in turn was caused by changes in the temperature of the air with height. Pettigrew observed that a great number of natural and man-made lights can contribute to the observed Fata Morgana and that these are not always evident in the light of day. However, these odd light phenomenon are not confined to distant lights.

Fata Morgana or Something Else?

The Hessdalen lights are perhaps the most famous Earthlights on the planet. This is certainly down to their frequency, at one time in the early 1980s being visible 10–20 times per week. They have been the subject of prolonged scientific investigation involving an international array of investigators.

Hessdalen is located in a broad valley around 120 km south of Trondheim, Norway. For decades unusual lights have been reported regularly at various locations within the valley, sometimes near to or apparently on the ground, with others higher up in the air. Most are white or off white in color and may appear randomly or appear to move at high speed.

The lights associated with this small, isolated village have become something of a national phenomenon for Norway and are certainly the most researched Earthlights on the planet. Their nature as of yet defies simple explanation; many appear in mid-air and can be tracked on radar before becoming visible. Most impressively many of these lights appear to be moving at very high speeds, typically 10,000–20,000 m per second, or over 2–5 million miles per hour. Whatever the radar is detecting must be very low in mass otherwise it would suffer spectacular drag and frictional heating in the Earth's atmosphere. A simple bundle of electric or magnetic field lines could do this and reflection of radio waves by aurora has been known for some time. In the late 1960s George Millman (General Electric Corporation) wrote about the reflection of radar by magnetic fields at low latitudes in The Journal of Geophysical Research, Space Physics.[1] While these are largely visible phenomena, radar reflections from some form of magnetic field is certainly possible as an explanation for those phenomena that appear to have unreasonably high velocities and are at least initially invisible.

Other suggestions for the phenomena involve a diverse array of phenomena. These include: natural batteries in the valley that involve iron-rich rocks on one valley side and copper-rich rocks on the other; piezoelectric discharges from strained rocks; even odd combustion reactions involving the element scandium that is somehow kicked up in clouds of dust from the valley floor. Piezoelectricity could explain the phenomenon, but the rocks of Hessdalen are not under geological strain, so how that process would work in this location is unclear.

The problem with the Hessdalen Lights is the likely diverse nature of their origin. Although many Earthlights will turn out to be either man-made lights or perhaps mirages of distant lights, such as those at Paulding or Marfa, some, such as those in Norway, are likely to be genuinely novel phenomena and worthy of thorough investigation. Often from such interesting science comes something unexpected and useful.

---

[1] Journal of Geophysical Research, Space Physics, Volume 74, Number 3 (1969) Wiley Online

Earthquake Clouds and Lights

Moving on from the genuinely odd, or somewhat suspect, phenomena of Earth Lights we enter still mysterious but extremely well documented earthquake lights (EQLs). Like Earthlights these lay on the fringes of mainstream science for decades but their existence is now becoming more widely accepted. There now exists a very well populated library of accounts of atmospheric lights accompanying, preceding or following a minority (perhaps 0.5 %) of major earthquakes. Indeed, some very thorough scientific reports accompany their sightings before, during and after earthquakes such as the 2009 L'Aquila earthquake in Italy or the 2008 Sichuan quake in China; however, many more are known.

Moreover, many of these odd and rather beautiful lights are accompanied by the formation of clouds which may, on some instances, be visible from space (Fig. 4.7). Just like the Hessdalen Lights, the manner in which these lights and clouds are generated appears to be utterly mysterious. Figure 4.8 illustrates some earthquake lights that have been recorded on film. From the ground some of these clouds appear to be luminous and somewhat similar to aurora (upper right Fig. 4.6). While others appear to be just light rather than illuminated clouds (lower left Fig. 4.6).

Although it is certainly believable that earthquakes generate luminous phenomena, you have to be rather careful about which are genuinely connected with quakes and which are simply coincidental observations. The main problem with the phenomena is the broad dispersion in their timings and their relative rarity. There is no doubting the photographs and occasional video evidence are real. The question is are they associated with the earthquake or are they misidentifications of man-made or natural phenomena, or just a very interesting and unusual coincidences? Given the broad nature of the observations, from balls, to lightning-like discharges, to diffuse glows, and that some of these lights appear well in advance of the earthquake or even several weeks afterwards, proving any connection is fraught. Some luminous phenomena are clearly iridescent clouds, a well-known and very beautiful rainbow-like pattern caused by sunlight refracting and scattering within ice clouds. These iridescent clouds could have connections

Fig. 4.6 Earthquake lights and clouds from different perspectives. *Upper left* (a) shows a photograph taken of a luminous ball near L'Aquila on June 20, 2008—10 months before the quake. Photograph by Bruno Chiarelli. Photograph (b) is one of many ground photographs taken by an observer around 30 min before the deadly 2008 earthquake in Sishuan Provence in China. Photograph (c), *lower left* was taken near the Sakurajima volcano in Japan. As well as the prominent light near the foot of the volcano there are numerous glowing balls (one is *arrowed*). When the brighter light "turns off" in the YouTube video this is taken from, the balls disappear. The wide spread in the timing of lights leaves open the door that they are unconnected to earthquakes

with the quakes or they could just be coincidental in their appearance. Iridescent clouds are certainly not that rare. Other flashes during earthquakes could be downed and blowing transformers.

Observations of earthquake lights made around the time of the 2009 L'Aquila earthquake are certainly unusual and suggest some interesting phenomena is going on. Whether these are truly linked to the earthquake is the point in question. At L'Aquila, before the quake, countless observers reported flame-like eruptions from the ground and flashes as well as increasing reports of ball-lightning-like phenomena (Fig. 4.6). These tended to give way to broader atmospheric glows both during and after the main shocks had passed. However, some observers reported lightning-like discharges coming from the ground after the after the main shocks had passed. Other observers noted odd clouds; many like those seen in the moments before the Sichuan quake, looking like aurora, while afterwards odd, stratified clouds with a violet complexion covered the ground near the region's mountains.

How might we explain these phenomena, first of all assuming that all have connections with earthquakes? As with earth-lights we enter territory with fairly wild speculation. Explanations included the piezoelectric effect, suggested also for Hessdalen, which involves shocked or strained quartz. Alternatively, some suggest earthquakes cause disruption of the Earth's magnetic field leading to flashes or other luminous phenomena.

Robert Thériault and colleagues have suggested a new phenomenon to explain EQLs (Fig. 4.7). This is similar in some regards to the piezoelectric effect, but somewhat more radical. In their

FIG. 4.7 Cross-cutting basaltic dyke at Innellan, western Scotland, slicing through older, metamorphosed limestone. The left edge of which is indicated in *blue*. Is this 60 million year old feature the sort of rock that generates earthquake lights? If Robert Thériault and colleagues are correct, the inhabitants of this British village could be in store for some "UFO activity" should the earth violently shudder

new model it isn't quart-filled rocks like granites that are responsible but instead their more iron-rich cousins, basalts, dolerites and gabbros. In particular, where these rocks are arranged in vertical or near vertical structures called dykes, the build up or release of strain associated with earthquakes generates ionized gases within the rock. These are then captured by the Earth's magnetic field and funneled along the length of the dyke at high speed, eventually erupting out of the rocks and into the atmosphere above. The source of all this activity is defects in the formation of iron-rich crystals within the rocks. Positive charges uncouple from their surroundings when the crystals are stained then escape along the axis of the channel. This eventually brings them to the surface. Although it seems quite convincing that some EQL phenomena, such a erupting blue jets or lightning-like discharges, would seem to be explicable in the context of this model, others, such as iridescent clouds, diffuse glows or ball lightning-like phenomena would be harder to ascribe. That *some* luminous phenomena are associated with violent underground events seems hard to refute. And the figure of 0.5 %, mentioned earlier? This low figure, you may recall, is the proportion of earthquakes to which earthquake lights are associated through observation. The low number, Freund and colleagues suggest, is down to the geology of the local rocks. Only in those places where there are steeply-dipping layers of basaltic rocks will earthquake lights be found. This should be a testable prediction as most of the planet is sufficiently well-mapped to identify those rocks the authors suggest are associated with earthquake lights (EQLs).

What of earthquake clouds? Figure 4.8 shows a train of clouds originating above the epicenter of the 2003 Bam earthquake that trail in the prevailing winds east southeastwards towards western India. Other such clouds have been documented for nearly 400 years. Like the EQLs definitively linking clouds to earthquakes will be fraught with risk. In particular the definition of what constitutes an earthquake cloud seems to vary and the mechanism of their formation is speculative, to say the least. It is suggested that such clouds may form when ground water percolates into rocks that are cracking ahead of the main earthquake. As the rocks crush and grind together the water is heated and escapes as invisible clouds of steam. These then condense downwind when they have

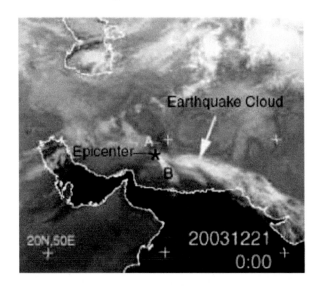

Fig. 4.8 An earthquake cloud? This satellite photograph taken shortly before the 2003 Iranian earthquake centered on Bam shows a large cloud originating above the earthquake epicenter and drifting downwind across southern Iran and Pakistan. The cloud was visible for the best part of a day. Photo taken by the IndoEx satellite/University of Dundee

risen to suitable height. In some instances steam is seen to escape from some areas, however, this is not common. Alternatively, clouds could form in more complex ways when crushed rocks release clouds of dilute ions into the air. These ions then attract water molecules which condense to form clouds. None of this is scientifically unreasonable: the question is, is any of it true?

What this area of research needs is a mechanism that can be directly tested on site. Given the sporadic nature of earthquakes this is a bit like looking for a needle in a haystack. That said, 20 years ago supernova hunting was in a similar state. The introduction of automated searches soon revolutionized the science; opening the door to much more systematic, observational astronomy while providing a far clearer understanding of the diverse phenomena called supernovae.

For now, it is fairly positive some of the EQL phenomena are real, like the lightning-like discharges and blue jets erupting from the ground as well as some of the broader luminous displays. Are the others really associated with the earthquake or are they just happening coincidentally? For those which we have a reasonable degree of certainty are we seeing a new connection between the planet's interior and atmosphere, one far more subtle but no less impressive than volcanic eruptions or other geothermal activity? If these observations are confirmed, we shouldn't be surprised to find similar lights in the atmospheres of Mars or extra-solar worlds. Will these worlds be oddly illuminated by fleeting phosphorescences?

Sprites, Jets and Other Luminous Atmospheric Phenomena

Another observation dismissed for decades was the appearance of brief blue or red discharges above thunderstorms. A lot of pilots' sightings of these enigmatic lights were grouped with UFOs and consigned to the dustbin. As there are some glorious sightings, photographs and videos of these illuminations, the combined body of evidence became overwhelming in the early 1990s when they were finally added to the list of phenomena associated with thunderstorms. Figure 4.9 shows a particularly wonderful photograph taken of sprites above a tropical thunderstorm taken from the International Space Station.

The origin of sprites is partly understood and appears to be linked to particularly large thunderstorm cells with an extensive anvil structure. Sprites appear simultaneously with cloud-to-ground strikes and in particular the discharge of the positively-charged core of the anvil or upper cumulonimbus cloud. The upper portion of most thunderstorms carries a strong positive charge, while the corresponding negative charge is found lower down, where the air is warmer.

Around the outer surface of the positively charged anvil is a halo of negative charge that is produced by induction. The strong positively charged ice particles cannot conduct to the air as it is a strong insulator so instead they attract negative charges towards them in the surrounding air. Thus the entire upper portion of the cloud is shielded from the stratosphere by a shell of negative charge. What happens next is unclear, but the trigger for these novel luminous phenomena appears to be the discharge of the positively charged portion of the cloud to the ground below.

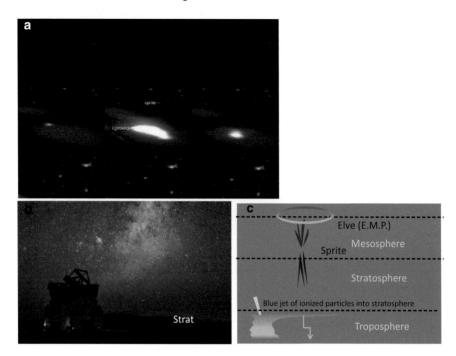

FIG. 4.9 A *red sprite* (central image, above) fluorescing 60 km above a thunderstorm over Myanmar near the Indian Ocean (**a**). The sprite appears over the center of the thundercloud's anvil just as lightning discharges below. Image courtesy of NASA/ISS. Lower image is a sprite caught by the Southern Observatory in Chile (**b**). The sprites are located above very distant thunderstorms over Brazil. Below (**c**) sprites (*red tendrils*) tend to form above the positively charged anvil when it discharges directly to the ground (*yellow forked arrow*). Elves (*the blue ring*) are rapidly expanding blue rings of light caused by electromagnetic pulses associated with the sprites reaching the base of the thermosphere (Chap. 1)

This, in turn, triggers two other processes. In the first, a breakdown occurs between the upper, positively-charged region of the storm and the screening negative charge that surrounds it. This process appears to trigger the formation of smaller blue jets that are seen above the anvils of some storms and extend into the stratosphere. In some storms, there is instead an upward directed collapse of the negatively-charged region found in the middle of the storm. In these events, the bubble of negative charge explodes upwards through the top of the cloud and into the stratosphere, eventually reaching the base of the mesosphere, a good 30–80 km higher up. This generates much larger blue jets, unimaginatively called gigantic jets.

Sprites are better understood. These are directly associated with the discharge of the positively charged anvil to the ground and are presumably a response of the upper atmosphere to the rapid change in the voltage (the potential difference) between the top of the anvil and the ground underneath; or to a change in voltage between the top of the anvil and the screening charges that surround it. They are exclusively mesospheric phenomena (40–90 km up) while the jets are confined to the stratosphere and appear to behave in a similar way to cloud-to-ground lightning with leader and return stroke components.

On closer analysis the gigantic jets appear to be a hybrid between the smaller blue jets and the red sprites. Smaller blue jets give rise on occasion to gigantic jets when the smaller blue jets have initiated the breakdown of the electrical resistance through the full depth of the stratosphere. Where they carry sufficient oomph, the discharge of the blue jet triggers a sprite, which expands upwards towards the mesosphere. This, in turn, creates an ionized pathway for further electrical discharge, which extends from the cumulonimbus cloud, directly to the base of the ionosphere. With the insulating layer disrupted, a negatively charged bubble of gas erupts out of the cloud and punches through the stratosphere. A gigantic jet is born.

Finally, there is one further mesospheric phenomenon known as elves that is fairly well understood. These are rare, faint concentric shells of expanding luminosity that can be seen at the base of the thermosphere (Chap. 1). Elves form when an electromagnetic pulse, associated with the lightning discharge, reaches the base of the thermosphere and expands outwards through the highly ionized gas. The electromagnetic pulses associated with lightning also heat the upper layers of the mesosphere and layers of ionized gas in the ionosphere. These pulses can be up to 20 billion Watts—a rather formidable output. As well as generating gamma ray flashes (below) there is the suggestion that such electromagnetic pulses have sufficient energy to accelerate electrons from the atmosphere out into the surrounding space. Here, they are captured by the Earth's magnetic field and travel through the Van Allen radiation belts. Electrons from here can then be scattered back into the upper atmosphere, completing a vast circuit that links our atmosphere and interplanetary space.

Gamma Ray Bursts on Earth

The repertoire of phenomena associated with terrestrial thunder-storms took a major leap up when, in 1991, the Compton Gamma Ray telescope detected millisecond-long pulses of gamma rays originating within the Earth's atmosphere. Although gamma ray bursts were well known from astronomical sources (and from terrestrial nuclear tests) it came as something of a surprise when the orbiting telescope detected these highly energetic (20 MeV) bursts. The bursts, known as Terrestrial Gamma Ray Flashes (of TGFs), were as brief as they were violent with each lasting between 0.2 and 3.5 milliseconds. The origin was clearly connected to thunderstorms with the few detected bursts at the time coming shortly before lightning was detected from the ground.

Although the precise mechanism remains elusive, the likely scenario involves the involvement of cosmic rays. These highly energetic particles bombard the Earth's atmosphere minute by minute and will generate sprays of secondary particles in the stratosphere. Some of these liberate electrons and it is these electrons that are the eventual trigger for the gamma rays. What happens next is a powerful lightning strike close to where the spray of electrons has formed. Some researchers have the electrons sourced within the top of the thundercloud while others require only that it appears close to a storm. When lightning discharges it generates a powerful electrical field that accelerates the cloud of electrons, which then whiz upwards towards space. As these interact with the nuclei of neighboring atoms, the electrons rapidly decelerate and release what is known as Bremsstrahlung, or braking radiation. This can include gamma rays if the electrons have enough oomph to begin with.

More amazing still was the discovery in 2011 that thunderstorms can also liberate jets of antimatter. The scenario is rather well understood, but their detection relied on a bit of luck and help from the Earth's magnetic field. The antimatter electrons, known as positrons begin their life as a gamma ray. These gamma rays originate in thunderstorms through the mechanism that was discussed above. Once generated, these gamma rays can interact with further atomic nuclei in the atmosphere. In the right circumstances these produce electron–positron pairs. Ordinarily these pairs of particles might wander around before the positron interacted with, and then annihilated, a nearby electron. However, in the rarefied confines of our upper atmosphere, both the electron and positron can escape. This involves these charged particles being rescued by our planet's magnetic field. Born within our planet's magnetized umbrella each particle is whisked away from its partner and shot out into space. Normally at that point they would either escape the Earth in its entirety or become trapped in our magnetic field, but thanks to our technology some have a grizzlier fate. High above the Earth, some of the positrons happen to whack into the orbiting Fermi Telescope. When they do so, they immediately annihilate. Their demise results in the formation of a characteristic pair of gamma rays, each with the energy of 511,000 eV. The plucky antimatter particle evades many hazards throughout its flight and escape is just on the horizon when they are annihilated in an instant. This convoluted journey involved a thunderstorm, a cosmic ray, and the Earth's magnetic field. As with the discovery of gamma rays by Compton Observatory, the detection of antimatter by the Fermi Gamma Ray Observatory was entirely fortuitous. Given the convoluted path to make antimatter, does ball lightning seem so odd by comparison?

Terrestrial gamma ray bursts and the formation of antimatter are also interesting because only a minority of thunderstorms seem able to produce these. Over 1000 storms may rage across the planet mostly in the tropics each day, but less than 100 TGFs have been detected over this period. This might be because only a minority of thunderstorms are energetic enough or perhaps tall enough to launch gamma rays through the overlying atmosphere into space, or it might be because the gamma rays are focused into beams rather than broad sprays of radiation. Most of these beams simply evade our detection. More work is clearly required.

Aside from the dramatic displays of lightning associated with thunderclouds that can deliver 10,000 Amps of current in every strike, the atmosphere also delivers a steady stream of charge, amounting to 1000 Amps across the entire surface of the planet.

This is delivered through a voltage (potential difference) of 120 V per meter. One thousand Amps would be rather unpleasant if focused on one spot, but across the entire face of the planet this 1000 Amp flow amounts to a drizzle of 10 trillionths of an amp per meter square. Clearly, this is not enough to light even the dimmest of bulbs. Measly though this is, this arm of the circuit is merely the closest part of an even larger circuit that extends through the ionosphere to the magnetosphere and ultimately outwards to the stars through the solar wind. Within this vast electrical track thunderstorms play their part by using the Sun's energy to charge up the ionosphere from below. When lightning discharges a torrent of electrons to the ground, the cloud can then discharge positive charges upwards towards the base of the ionosphere. This process helps maintain the voltage between the Earth's surface and the ionosphere above.

While thunderstorms are a dramatic player in the global electrical circuit they are also a profuse source of radio waves. Each lightning bolt generates an electromagnetic pulse, as we have seen, and this is detectable as annoying interference on our televisions and radios when storms are nearby. Storms also generate much lower frequency radiation as part of this repertoire. The electromagnetic pulses that race away from lightning discharges have a range of wavelengths. Those in the 50 Hz range are able to propagate right around the Earth numerous times before dying away. As they do these waves can generate so-called Schumann Cavity Resonance. This phenomenon is caused by the electrically insulating properties of the lower atmosphere, which separates the charged ionosphere from the chargeable surface of the planet. Schumann Cavity Resonances are effectively standing waves that oscillate, or vibrate, through the depth of the stratosphere and troposphere. These vibrations release radio waves at various low frequencies that can be detected by radio antennae. Because lightning can excite the lower atmosphere in the right range of wavelengths, it can generate these Schumann Cavity Resonances. This allows this low energy radiation to be used to monitor the frequency of lightning strikes across the globe as a hum of radio waves. Observations suggest that the overall frequency of lightning is increasing and at least in theory this could be direct consequence of rising global temperatures.

Sky Quakes

One further phenomenon is truly global in nature and just as mysterious. So-called sky quakes have been identified in various locations, dotted around the globe, each with its own regional name. Along the banks of the Ganges in Bangladesh the booms are called *Barisal Guns*; along the eastern shores of the North Sea in Belgium and the Netherlands they are called *mistpoeffers*; while in Seneca and other parts of the south eastern US they are referred to as *Seneca Guns*. In Connecticut Valley they are called *Moodus Noises*, while in the Philippines the name *retumbos* is used. In each location these wonderfully evocative names describe the rumbling booms that are sometimes seconds long that occur without warning and at any time in the day. They are said to resemble distant thunder or artillery fire and may be accompanied by low rumbling which shakes buildings. Although this is clearly reminiscent of earthquakes scientists trying to assign the reported noises to seismic activity often (but not always) come up short.

Although some of these booms have been definitely linked to the sonic booms of military aircraft, others are clearly not manmade with accounts going back over 100 years in locations as far apart as Victoria in Australia and Seneca in the US. So, what causes the *Seneca Gun* to fire?

Meteors entering the atmosphere might work as a cause, but the frequency and their seeming preference for some geographical locations would make that hypothesis unlikely. Gas erupting from underwater is possible and would tie with the frequency of these lights near the coast or lakes. Similarly, collapsing underground limestone caverns might work as well. In some instances, in much like the manner Fata Morgana is a mirage of light caused by atmospheric refraction, the *Seneca Guns* and related sky quakes might be distant thunder refracted over the observer's horizon. In at least some instances this idea has been investigated using regional meteorological reports and there is no evidence that any thunderstorm lies close enough to the purported sounds to have created it. That said thunder refracted from further afield could explain some of these sky quakes.

In a few instances where small earthquakes appear to be associated with similar noises the arrival of the earthquake P-wave may, on occasion be registered as a rumble by observers. However, none of the current explanations quite fits the bill. The Seneca Gun or sky quake might be a combination of different noises, including gas burps, collapsing caverns and super-sonic military jets. At Seneca some booms are definitely accompanied by visible rumbling which would indicate either atmospheric sonic booms or earthquakes. Hopefully, time will tell what causes these kinds of phenomena and whether any are related to geological turmoil or atmospheric disturbances.

The Red Rains of Kerala

Finally, turning from the surreal to the sublime, we look at the red rains of Kerala. In the southern Indian province of Kerala blood-red rain fell sporadically across much of the region during the late summer of 2001. Some reports tied the appearance of this lurid precipitation to the sound of thunder, which was taken by some to imply an atmospheric boom, perhaps caused by an incoming meteor. Analysis of the rains revealed they were colored by cellular structures. The assertion that these had an association with the boom and supposed meteor led to the idea that this was some sort of panspermian event with alien cells having arrived from outer space. Early analysis suggested that the spores had an unusual composition. However, later investigation confirmed that the red cellular material was some form of algal spore.

In the same region black, green and yellow rain has also been reported. The same phenomenon was recorded in Kerala in 1896 and again in 2012, suggesting an interesting local source of these spores. Similar rains have been reported to the south in Sri Lanka, implying a localized but generally widespread geographical origin. At present the source is fairly mysterious, but there is no doubt that the coloration is down to terrestrial rather than extraterrestrial sources.

As areas only meters apart can have red rain or clear and quite normal rainfall, the source of the spores must be very localized and somehow remain isolated within the atmosphere so that it doesn't all mix into a bland baby-pink. Quite how this is achieved could have implications for issues as diverse as the circulation within the atmosphere and, potentially, the spread of pathogens by air. The red algal spores could thus be used by scientists as a natural tracer for air movement presenting a novel and quite colorful means to probe the dynamics of the Earth's atmosphere.

Flying Spiders

Even animals can take advantage of atmospheric irregularities. Charles Darwin observed the phenomenon of "flying" spiders onboard HMS Beagle in the early 1830s, and found hard to explain how spiders could launch themselves from objects to fly horizontally when the air was apparently still. Convection would tend to lift the spiders vertically, not sideways. Confounded by his observations Darwin went as far as to suggest that the observed fanning out of fibers from the spinnerets might best be explained by electrostatic repulsion. Indeed, it is now thought that the pervasive 120 V per meter field, permeating the Earth's atmosphere might be the force at work.

On a calm, clear day, as we've seen there is a persistent field generated by various phenomena. This field and the accompanying charge separation forms the basis for the formation of thunderstorms under differing circumstances. Spiders might well utilize this field to launch themselves, for only 30 nanoCoulombs (30 billionths of a Coulomb of charge[2]) is sufficient to provide enough repulsion to drive uplift of the spiders silk fibers for a spider that is 1–2 mm across.

As Darwin noted it is certainly suggestive that when spiders produce the threads for "ballooning", as this process is known, the fibers immediately fan out and rise upwards, which would imply they are repelled by themselves and the underlying surface. This implies that they have acquired the same electrical charge as the underlying surface. Thus, it seems reasonable that spiders are initially launched from the surface by electrostatic repulsion, before any light air currents can catch and transport them further.

---

[2] 1 Coulomb is one unit of electrical charge that is equivalent to one amp of current flowing in one second: $Q = It$, for the math fan; where "Q" is the charge in Coulombs; "I" is current in Amps and "t" is time in seconds. It is often more customary to define the Coulomb in terms of electrons, where 1 C is equivalent to the charge on approximately 6.241×10^{18} electrons.

Conclusions

While the general outlay of meteorological and related phenomena is clearly fairly well understood, there still exists a battery of observations on Earth that retain a considerable air of mystery. Some, like the lights at Marfa and elsewhere, clearly fall into the realm of misidentified human activity, but there are clearly others that have truly natural origins that we still do not understand. Similarly, we are bound to encounter phenomena on other planets that simply don't fit current theories. Getting a better grip on the weird end of terrestrial weather will certainly aid our understanding of phenomena on truly alien worlds. Look at Chap. 6 (Mars) or Chap. 7 (Jupiter and Saturn) for a clear indication that this supposition is true. The universe throws all sorts of apparently inexplicable phenomena at us. The beauty of these observations does not lie in presupposing that these are mysterious and inexplicable, but instead reveling in the joy of unpicking them and finding out how they work. Meteorology and climate science are universal phenomena and we can only understand the climate of our world and of alien worlds if we focus on both. It will be interesting to complete the full inventory of weird and wonderful atmospheric phenomena: undoubtedly it will be far greater than we imagine.

References

1. Dimitar Ouzounov, Sergey Pulinets, Alexey Romanov, Alexander Romanov, Konstantin Tsybulya, Dimitri Davidenko, Menas Kafatos, Patrick Taylor. (2011). *Atmosphere-Ionosphere response to the M9 Tohoku Earthquake revealed by joined satellite and ground observations. Preliminary results.* Retrieved from http://arXiv.org/pdf/1105.2841.pdf.
2. Dijkhuis, G. C., Callebaut, D. K., Lu, M. (Eds.), (2006). Observation of lightning ball (ball lightning): *A new phenomenological description of the phenomenon.* In Domokos Tar. Proceedings of the 9th International Symposium on Ball lightning (ISBL-06), 16–19 August 2006, Eindhoven, The Netherlands, pp. 222–232.
3. The Society of Physics Students at the University of Texas at Dallas. (2005). *An experimental analysis of the marfa lights progress report* (Submitted). Available as a PDF online.
4. Pettigrew, J. D. (2003). The Min Min light and the Fata Morgana. An optical account of a mysterious Australian phenomenon. *Clinical Experimental Optometry, 86*(2), 109–120.
5. Castelvecchi, D. (2015). Rogue antimatter found in thunderclouds. *Nature, 521*, 135.
6. Fidani, C. (2010). The earthquake lights (EQL) of the 6 April 2009 Aquila earthquake, in Central Italy. *Natural Hazards and Earth System Sciences, 10*, 967–978.
7. Park, S. K., Johnston, M. J. S., Madden, T. R., Dale Morgan, F., & Frank Morrison, H. (1993). ELF precursors to earthquakes in the ULF band: A review of observations and mechanisms. *Reviews of Geophysics, 31*, 117–132.

8. Zhonghao Shou, Darrell, Harrington. (2000). *Bam earthquake prediction & space technology.* www.earthquakesignals.com/zhonghao296/copies/harrington.pdf.

9. Stolyarov, A., Klenzing, J., Roddy, P., Heelis, R. A. (2005). *An experimental analysis of the Marfa Lights.* Progress Report submitted by: The Society of Physics Students at the University of Texas at Dallas. Retrieved from http://www.spsnational.org/wormhole/utd_sps_report.pdf.

10. Domokos Tar. (2008). New revelation of lightning ball observation and proposal for a nuclear fusion reactor experiment. In: Vladimir L. Bychkov & Anatoly I. Nikitin (Eds.), *Proceedings of the 10th International Symposium on Ball Lightning (ISBL-8),* Kaliningrad, Russia, pp. 135–141.

11. Hayakawa, M., Nakamura, T., Iudin, D., et al. (2005). On the fine structure of thunderstorms leading to the generation of sprites and elves: Fractal analysis. *Journal of Geophysical Research, 110,* 27–35. D06104.

12. Siingh, D., Singh, A. K., Patel, R. P., Singh, R., Singh, R. P., Veenadhari, B., Mukherjee, M (2009). *Thunderstorms, lightning, sprites and magnetospheric whistler-mode radio waves.* Retrieved from http://arxiv.org/abs/0906.0429.

13. Thériaulta, R., St-Laurentb, F., Freundc, F. T., & Derr, J. S. (2014). Prevalence of earthquake lights associated with rift environments. *Seismological Research Letters, 85,* 159–178.

14. Tavani, M., et al. (2011). Terrestrial gamma-ray flashes as powerful particle accelerators. *Physical Review Letters, 106,* 1–7.

15. Gorham, P. W. (2013). *Ballooning spiders: The case for electrostatic flight.* http://arxiv.org/pdf/1309.4731v2.pdf.

16. Schneider, J. M., Roos, J., Lubin, Y., & Henschel, J. R. (2001). Dispersal of *Stegodyphus dumicola* (araneae, eresidae): they do balloon after all! *The Journal of Arachnology, 29,* 114–116.

5. Venus

Introduction: The Twin That Isn't

Take two planets of roughly the same mass and composition and place them at similar distances from their parent star. You'd clearly expect them to end up as near twins. However, what the Solar System shows is that it doesn't take much of a butterfly to drastically alter the fate of worlds.

Venus undoubtedly was bequeathed with the same amount of volatile materials as the Earth and it is supposed that early on the two planets might have looked rather similar: immense water worlds under a thick, hot atmosphere of water vapor and carbon dioxide (Chap. 3). However, Venus received just that little bit more radiation from the Sun than the Earth did and never quite cooled off the way the Earth did. What happened next would set the fate of Venus and turn it into the noxious hell it is today. This chapter describes how a world much like our own can end up so very different to ours despite a similar start; and how the fate of its atmosphere became tied to the fate of the planet as a whole.

The Venusian Day

Venus is a thoroughly odd planet. A Venusian sidereal day—the time it takes a fixed, distant object like a star to culminate each rotation period is 243 Earth days: moreover, Venus does this backwards (its rotation is retrograde). Consequently, the Sun rises in the west on Venus and sets in the east. The Earth, by contrast, has a sidereal day of 23 h 56 min and rotates on its axis in a prograde direction, matching the direction of its orbit around the Sun. Moreover, while the Earth rotates on its axis at 1670 km per hour at the equator, Venus crawls around at 6.5 km per hour—a modest walking speed.

© Springer International Publishing Switzerland 2016
D.S. Stevenson, *The Exo-Weather Report*, Astronomers' Universe,
DOI 10.1007/978-3-319-25679-5_5

Meanwhile, the Venusian year is 224.7 Earth-days long. Therefore, the retrograde rotation on its axis means that the Sun rises and sets on a period considerably less than the sidereal day. Instead of 243 days, the Venusian solar day is only 116.75 Earth-days long, as on every rotation on its axis, the retrograde spin brings the Sun to the same point of culmination far earlier than it takes the planet to spin on its axis. These odd characteristics are as stable as they are normal. In the 16 years separating the arrival of the Magellan and Venus Express Probes, the Venusian sidereal day lengthened by 6.5 min for every sidereal day that had elapsed. This means that the length of a Venusian sidereal day was 156 min or so longer in 2006 than it was in 1990. How did this strange state of affairs arise and why does it continue to change? Just imagine what it would be like if we gained the best part of 3 h a day every decade.

The likely principle factor directing this odd state of affairs is likely the shear bulk of Venus's atmosphere and its relative proximity to the Sun—40 million kilometers closer than the Earth. The Sun exerts strong tidal forces on Venus's thick carapace of gas and on its (probably) plastic mantle. Many calculations show that this should have been sufficient to tidally-lock Venus to the Sun within a few hundred million years of its formation. If this had happened, Venus would now present one face permanently to the Sun, while the other resided in darkness: this is something that we shall look at in Chap. 10 when we examine the planets that orbit red dwarfs. Venus, however, is not (quite) tidally locked to our star. So, how do we explain the odd rotation pattern of Venus?

Venus's massive atmosphere has clearly had an enormous influence on the evolution of the planet as a whole. When one looks at both worlds you can see that the atmosphere on Venus and the Earth carry vastly different amounts of momentum and this, in turn, grossly affects the interplay of energy between the spinning planet and its atmosphere. Although the solid part of each planet is around one million times more massive than the Earth's atmosphere, this figure is around 100 times smaller for Venus. Although the masses of atmosphere and underlying planet are still very different, the bulk of Venus's atmosphere is still sufficient to have significant effects on the rotation of the planet. Clearly, over time this massive atmosphere must drag on the planet, both through the bulk gravitational pull and through the effect of friction. Moreover, with the atmosphere acting as a giant ocean of liquid-like material, the solar tides can also exert a significant pull, which in turn influences the planet's spin.

For many decades it was thought that the retrograde spin of Venus's surface, relative to its orbit around the Sun, implied that Venus was likely hit by a giant impactor, much like the one thought to have formed the Earth and Pluto's large moons. However, although such an impact is possible, and perhaps even probable, the chance that it would have hit Venus in just the right way to flip it over is unlikely. Instead more recent calculations place the origin of Venus's retrograde spin squarely with its massive atmosphere. In 1978 Andrew Ingersoll and Anthony Dobrovolskis (Caltech) showed that solar tides acting within the massive atmosphere were more than sufficient to slow the planet's spin to the point that the planet now completes an orbit of the Sun faster than it rotates on its axis. This process was dependent on a number of other factors, including the pull of the Earth. However, in principle tidal forces could slow the rotation of the planet to its current, decelerating, level. Venus may be in a brief window where it the interplay of forces have led the planet to abandon tidal locking. Thus, if you were to look at Venus one billion years or so ago it might have been locked, facing the same way to the Sun for hundreds of millions of years. However, repeated gravitational nudges from the Earth, or other planets, accelerated Venus out of its locked position. Now, it may be that Venus is slowly returning to a life presenting only one face to the Sun. This, too may last for another few hundred million or few billion years before repeated nudges from the Earth allow Venus to slip free once more.

A Noxious Vision of Hell

When Mikhail Lomonosov first spotted the planet's atmosphere during a solar transit in 1761 it might have been hoped that Venus, with its permanently obscured face was some sort of tropical paradise—if a rather cloudy one. Indeed, this idea persisted until relatively recently when the atmosphere began to be probed by Earthly machinery.

However, set against the idea of a tropical paradise were calculations made by Rupert Wildt in 1940. Wildt showed that the amount of carbon dioxide in Venus's atmosphere would lead to temperatures far in excess of the boiling point of water. Recall the climate skeptic argument (Chap. 3) that there is some sort of upper limit to which carbon dioxide could heat the Earth when its absorption bands become saturated. However, as the concentration of greenhouse gas rises, the range of wavelengths, over which the carbon dioxide (or any greenhouse gas) will absorb radiation, will increase in kind. Such collisional broadening (Fig. 3.7) ensures that Venus, and presumably the early Earth (Chap. 3) are (or were), very hot indeed. All of this thanks to the absorption of infrared radiation by carbon dioxide gas. Wildt's calculations were confirmed by the fly-by Mariner probe in 1962. When the Soviet Venera 4 arrived in 1967 it confirmed that carbon dioxide and its associated massive greenhouse effect dominated the climate of Venus. The idea that Venus might be some form of lush tropical jungle was thoroughly incinerated.

The vision of hell was confirmed when the Soviet Union landed probes in the late 1960s and early 1970s. During their short, unpleasant lives the Venera landers indicated a barren volcanic hell, where a dim light from the Sun refracted into an odd apparition of Hades. Venus is a planet boiled and crushed under an atmosphere 92 times more massive than that surrounding our world. So dense was the atmosphere that the pressure at the surface was roughly equivalent to being under a depth of 1 km of water. Were you able to survive the temperature (and suit yourself up appropriately) you would be able to swim in the dense carbon dioxide-rich atmosphere, which behave much more like a dense liquid than a gas under Venusian conditions. Even the modest 10 m per second breeze would be sufficient to knock you off your feet were you attempting to walk.

More interesting are the effects of refraction. Refraction is the bending of rays of light (and all electromagnetic waves) as the radiation travels from one material to another. In Chap. 4 we saw how refraction can generate some odd and quite confusing effects, such as the Marfa Lights. If Venus's atmosphere was transparent, from your position on its surface, the horizon would appear to bend upwards around you. Moreover, were you at a height of 35 km, light would be totally internally reflected around the planet, so that Venus would appear to extend infinitely in all directions. However, sadly in practice atmospheric haze prevents such an optical wonder. So, although the bowl-like effect would be apparent, you wouldn't see as far before the light was scattered and absorbed.

Disappointing though this is you might still see an interesting effect after Sun set. On Earth the atmosphere still remains light for perhaps half an hour after the Sun had gone down. This is because the light from the Sun reflects off particles in the atmosphere and because light refracts around the edge of the Earth. On Venus the latter effect is far stronger as a result of the thick sea of carbon dioxide on the surface. As a result—haze dependent—the Sun might appear as a thin sliver of light all through the night, until it rose again the following morning. In the case of Venus, remember that this night is more than 121 Earth days long. Finally, this thick layer of carbon dioxide is thought to behave as a super-critical fluid, meaning that although strictly speaking a gas, it has many of the properties of a denser liquid. Of these possible properties the key one is its ability of the dense carbon dioxide atmosphere to conduct heat very rapidly. This property ensures that there is little difference in temperature between the day and night sides of the planet at the surface, even at the surface where winds are light.

The temperatures at the surface of Venus are sufficient to melt lead, tin and zinc; and probably also induce fairly vigorous chemical reactions between the rocks. Clearly, the extreme heat will also drive out any moisture that was present in these rocks when they reached the surface. After all, on Venus it is only when you reach around 50–60 km up that temperatures fall consistently below the boiling point of water and whatever moisture is present would be able to condense.

The Structure of the Venusian Atmosphere

As on the Earth the atmosphere can be divided into distinct layers. The lowest, extending up to around 60 km, is equivalent to the terrestrial troposphere. This has a lid at 60 km defined by a small temperature inversion (a region where temperatures rise

Table 5.1 Temperature and pressure within Venus's atmosphere at various heights above its surface

Height above surface (km)	Temperature (°C)	Pressure (as a fraction of that at the Earth's Surface)
0	462	92.1
10	385	47.4
20	306	22.5
25	264	14.9
50	75	1.0
55	27	0.60
70	−43	0.037
100	−112	0.000027

Of note at about 55 km the pressure and temperature are suitable for human survival

with height) that separates the troposphere from the mesosphere above. As with the Earth air circulates from the equator polewards in two large Hadley Cells (Fig. 5.1). The slow rotation of the planet ensures that these reach far closer to each pole than they do on the Earth. Fully 120° of latitude is covered by these vast, slowly over-turning flows of air. This is double the latitudinal range that is seen on the Earth, leaving only a relatively thin sliver of surface covered by a different mode of air flow (Table 5.1).

The fast and counter-clockwise flow also means that most of the atmosphere moves in a retrograde direction, like the surface underneath, but that it moves faster than the planet rotates. Such motion is known as super-rotation. This phenomenon appears to have slightly different origins, depending on the planet that is examined. In the case of Venus the atmosphere completes a circuit of the planet in far less time than the planet takes to rotate on its axis: 4 Earth days compared with 243 Earth days, respectively, or one sixtieth the length of the sidereal day. Compare this to the mean wind speed on Earth which is roughly one tenth the rotation speed of our planet.

While super-rotation in itself is remarkable, most of this super-rotation is accomplished above 50 km where winds blow up to 400 km per hour in a band south of the equator. How is this feat accomplished on a planet that rotates so slowly? Led by Héctor Javier Durand-Manterola, scientists from the Universidad Autonoma de Mexico have analyzed winds flowing across the terminator of Venus at altitudes of 150–800 km. These trans-terminator flows

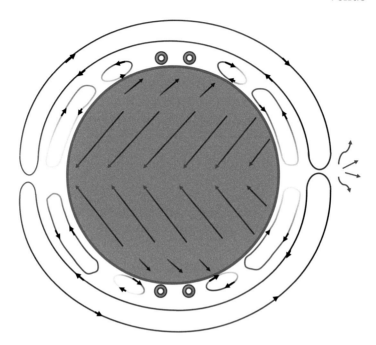

FIG. 5.1 Atmospheric circulation on Venus. Two broad patterns are discernible. Nearer the surface the atmosphere is broken into a number of large cells which shrink in size towards each pole. The largest of these is equivalent to our Hadley Cells but extends across twice the latitude the equivalent cells do on Earth (0–60° on Venus). A smaller "polar collar" may be thought equivalent to our Ferrell Cells (Chap. 1). Meanwhile, nearest to the poles are two anticyclonic vortices shown as blue donuts. Overlapping these in the mesosphere and lower thermosphere (90–150 km up) is a simple cell that transfers air at height from the sub-solar point to the anti-solar point on the night side. This thermal tide is driven directly by solar heating and ultimately leads to oxygen loss to space on the night side (*brown arrows*). Some of this ends up captured by the Earth

contain very little mass but move at several kilometers per second—far higher than the air movement below. Thanks to differences in the speed of these winds in the dawn and dusk sides of Venus, complex pressure waves are formed as the faster moving dusk terminator flow pushes across the night-side of the planet and into the outgoing dawn terminator flow.[1] As a result of the collision atmospheric waves are generated that transfer energy and momentum into the lower atmosphere. Similar waves are exam-

---

[1] Measurements made by Venus Express confirm that atmospheric gravity waves (chapter 10) transfer momentum from the ionosphere to the cloud decks and drive the super-rotating flow.

ined in Chap. 10. However, in this case these *atmospheric gravity waves* transfer momentum from the Earth's surface to its upper atmosphere.

The scientists speculate that it is this transport of energy and momentum has accelerated the atmosphere of Venus to dizzying speeds. Compare this mechanism to that proposed for the generation of super-rotation within the atmospheres of hot Jupiters in Chap. 10. Although both planets might appear superficially similar, with atmospheres dominated by super-hot gases, the origin of this heat is quite distinct. In the case of Venus heating of the upper atmosphere is effectively delivered from below, as the thick carapace of greenhouse gases captures the Sun's radiation. In the atmosphere of a hot Jupiter, the heating is strongest at the top of the atmosphere, where direct heating from the parent star is most intense. Thus, it shouldn't be surprising that both planets have super-rotation but that it is driven by such different mechanisms.

Returning to Venus, ultimately the origin of the pressure waves that drive super-rotation can be traced back to the solar wind. Thus, if correct, it is the Sun's hot breath that forces the atmosphere of Venus to move so swiftly. A similar mechanism may operate out at Titan (Chap. 9) causing its atmosphere to super-rotate as well.

Air speeds are low at the surface but increase to around 100 m per second (360 km per hour) at the tropopause. Thick clouds of sulfuric acid, with a smattering of sulfur, sulfur dioxide and water vapor fill this layer and obscure the surface almost in its entirety. Above this region the cloud layer rapidly clears, leaving a haze of sulfuric acid in a layer perhaps 20 km thick that can be thought of as the mesosphere. Unlike the Earth there is no intervening stratosphere where temperatures briefly rise again with height. This is because Venus lacks a distinct ozone layer that would absorb solar energy and cause warming.

In the mesosphere temperatures fall much more slowly with height and there are slight blips in the overall trend before we enter the thermosphere above. Indeed the lowest part of the mesosphere, which extends from 65–120 km above the ground. Here, the temperature remains roughly constant in the lowest 8 km of this layer, at around –43 °C. However, as we continue upwards and clear the upper cloud deck we enter the upper mesosphere where temperatures continue downwards to around –110 °C. On the dayside of Venus, this is the coldest region in the atmosphere. Above 95 km we cross the mesopause: the broad transition region between the mesosphere and thermosphere. Temperatures rise slowly in this region to temperatures between 27 °C and 120 °C at 120 km up. However, the Sun plays a very big part in controlling the temperature of the thermosphere. In the shade of the planet, the thermosphere in the night-side of Venus is even colder than the mesosphere with temperatures lowering towards –120 °C.

The layers above the tropopause also circulate from the sun-lit side to the dark side of the planet. There is very little north–south movement of air at all and the entire process is driven by solar heating which drives gases away from the area that is most strongly heated. In this layer carbon dioxide is broken down to carbon monoxide and oxygen, some of which later escapes to space at the anti-solar point on the planet's night side.

In the troposphere the circulation immediately above the surface, poleward of 60°, the Hadley cells give way to two smaller cells. The first cell, known as the polar collar is broadly equivalent to terrestrial Ferrel Cells and is a zone of upwelling gas on Venus. Astronomers can track the air flow here by the distribution of all that noxious carbon monoxide which is concentrated within these polar bands. Clouds generally rise highest in these zones, driven by convection from below. Presumably the cooler air sinking at either end of the Hadley cells drives the upward movement of air in the polar collar a bit like a cold front kicking up warm air to its front. Near the northern and southern flanks of the Hadley cells the Coriolis Effect is sufficient to drive airflow into two circumpolar jets that circulate at roughly 140 km per hour at 50–60° north and south of the equator.

In the late 1970s the Pioneer probe detected an anticyclonic storm located over the North Pole. Consequently, it was not entirely surprising when Venus Express detected the corresponding circulation over the South Pole in 2006. However, while the North Pole revealed a single storm that over the South was altogether more striking and unusual: an s-shaped pairing of two vast anticyclones, swirling air back downwards to the planet's surface. Each swirling eye counter-rotates around the other, forming a pair of hurricane eyes. However, like Jupiter's red spot, each Venusian anticyclone is an area of relative calm in an otherwise turbulent sea of poisonous gas. Each vortex rotates around their common center over a period of about 3 days in the same general direction as the rest of the atmosphere: i.e., from west to east at this latitude. Within and around the storm edges winds drop from around 50 m per second around the eye to zero within the eye, where winds are descending. In many regards these storms are analogous to the Polar anticyclones that form over Antarctica (and to a lesser extent) over the North Pole on Earth during the winter in each respective hemisphere.

Within the polar collar clouds can form at the top of the troposphere or within the mesosphere on a very rapid and short term basis. Venus Express saw this kind of event in 2007 and was thought to have happened when a blast of sulfur dioxide was injected into the mesosphere by a storm lower down. Such nacreous clouds, composed principally of sulfur dioxide and sulfates, are fairly rare phenomenon on Earth but are probably relatively common in the middle atmosphere of Venus, where such gases are abundant.

As you ascend higher through the atmosphere you reach the ionosphere. Here, as on the Earth and other planets ultraviolet, x-ray and gamma radiation from the Sun, together with a barrage of cosmic ray particles, blast electrons from the nuclei of the abundant oxygen and carbon atoms. This breaks carbon dioxide down to carbon monoxide and free oxygen. Gases continue to flow away from the sun-lit side towards the opposing darkness. Meanwhile at the point where the solar wind reaches the ionosphere a turbulent boundary forms called the ionopause. Here the solar wind is largely deflected by a combination of increasing gas pressure and the effect of magnetic fields created as the wind meets the ionized top of the atmosphere. This, and a bow-shock formed somewhat further upwind, towards the Sun, gives Venus some limited protection from the erosive power of the solar wind. However, as the European Space Administration's Venus Express showed, a considerable mass of atmosphere is still removed on a continuous basis by the constant stream of gases from the Sun. In all Venus loses roughly 1000 kg (1 metric ton) of oxygen per hour: however, the surprising thing is that this is roughly the same as the amount of mass lost by the Earth—and by Mars. An absence of a magnetic field is apparently no great problem for Venus.

No Layer but Yes, Ozone Around Venus

Soon after its arrival in 20006, Venus Express began sampling the atmosphere of Venus at different altitudes. It used measurements in the infrared and microwave portions of the electromagnetic spectrum. These were complemented by an array of measurements at radio wavelengths, which among other things monitored the planet's magnetic field and lightning.

Above the Earth ozone is formed in significant quantities from molecular oxygen. Ultraviolet light interacts with the normal two atom diatomic form of the gas that is produced primarily from bacterial photosynthesis in the oceans. The diatomic form of the gas is converted into the unstable three atom combination we call ozone. Ozone is proficient at absorbing ultraviolet light in the window at which this would otherwise penetrate the air and reach our planet's surface.

On Mars and Venus there is no photosynthesis (as far as we are aware) so the oxygen needed for the formation of ozone comes from the photochemical break down (photolysis, or light-splitting) of carbon dioxide gas, which both planets have in abundance. On Venus carbon dioxide is split by light to form carbon monoxide and a free oxygen atom. These free atoms are then swept around to the night side of the planet by Venus's thermal tide (Fig. 5.2), where they can then combine with one another to form first diatomic oxygen (the stuff we breathe) or triatomic ozone. High above the obscuring haze and clouds, in the mesosphere of Venus carbon dioxide gives birth to ozone with some help from the Sun.

At no place are concentrations sufficient to shield the atmosphere below, but if there were life to be had on Venus, its thick sulfuric acid clouds and haze would do the trick and prevent the ultraviolet from harming it. Ozone or not, the place to look for any kind of life will therefore be within the clouds, noxious to our senses, though they are.

The Edge of Space

Continuing 100 km above the layer containing the maximum concentration of ozone we reach the edge of the atmosphere. Immediately under the Sun's glare at the subsolar point the solar

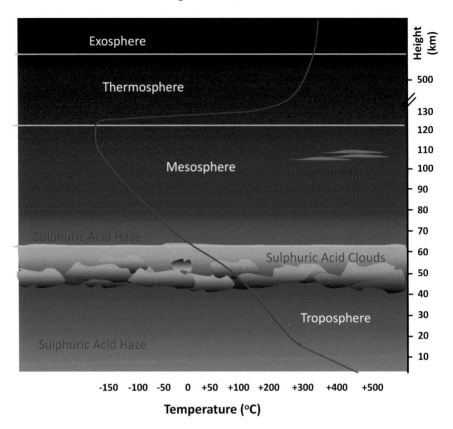

FIG. 5.2 The structure of Venus's atmosphere. Overall it has a much simpler structure than the Earth's with a steady drop in temperature and pressure to a height of over 120 km. There is no comparable stratosphere with a rise in temperature as is found above the Earth. Above 120 km temperatures rise again as we leave the thermosphere and enter interplanetary space. The height of the boundary between the thermosphere and exosphere depends on which side of Venus faces the Sun, being lowest on the Sun-lit side at around 220 km above the surface but rising to 350 km on the night side. Moreover, while on Earth most of the thermosphere is ionized this is only true of the sunlit side on Venus

wind blows the bow shock inwards to 1900 km above the surface of Venus, which is roughly one third the radius of the planet (Fig. 5.3). However, when Venus Express made this measurement it was at the solar minimum in 2007 when the solar wind was at its weakest. Between the bow shock and the ionosphere lie two further boundaries: the ionopause, which marks the top of the ionosphere; and the magnetopause, which is the upper boundary of the induced Venusian magnetic field. The former lies at roughly

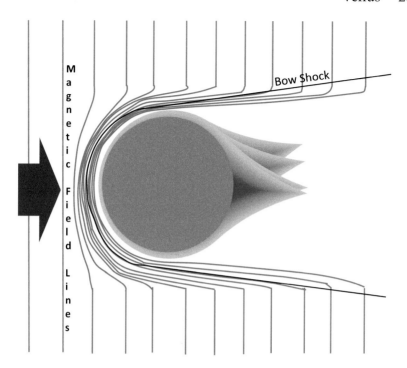

Fig. 5.3 The gross magnetic field and ionopause around Venus. Interactions of the solar wind (*red arrow*) with the planet's ionosphere result in the formation of a bow shock (*black*) and a "magnetic barrier" between the ionosphere yellow-orange". When the solar wind is stronger the bow shock approaches within 1900 km of the surface. However, when the wind is weak the ionosphere balloons outwards and forms a long comet-like tail in the wake of the planet. Oxygen atoms, with others, escape in the tail

250 km but is absent on the night side of the planet. The magneto-pause lies immediately above this at 300 km. Again, like the bow shock, these distances vary with solar activity, somewhat counter intuitively being highest when the Sun is most active. Between these two boundaries there lies a magnetic barrier which prevents any of the solar wind penetrating more deeply into the ionosphere.

Beyond the planetary terminator, on the night side of Venus, the magnetic field of Venus is drawn outwards downwind to radii exceeding ten times that of the planet (Fig. 5.3). It is here that energized oxygen ions, neutral oxygen gas and hydrogen ions, as well as a dose of helium venting from the planet's interior, ultimately escapes into interplanetary space. As Fig. 5.3 shows gases from the ionosphere can leave Venus on the night side of the planet. When the Sun has very low activity and the speed of the solar wind is low, gases within the ionosphere balloon outwards and more escape into interplanetary space. At higher solar wind speeds the ionopause wraps more fully around the planet and less gas is removed. This is mostly oxygen that has been released from the break-down of carbon dioxide, but hydrogen (from water and sulfuric acid) and helium also escape here.

One of the surprises from the Venus Express mission was the discovery of "reconnection events". Here, within the magneto-tail—the region downwind of Venus where the field is dragged out—the magnetic field lines can flip backwards and launch particles back towards the night-side of Venus. Here, they can induce air glow or even aurorae as happens at the Earth.

One problem with these observations is the obvious difference in the amount of water on each planet. While the Earth loses approximately the same amount of oxygen as Venus per second, this rate may not have been the same in the past. On Earth most of the oxygen comes from water vapor that is wafted up into the stratosphere and mesosphere. Here ultraviolet light splits it and releases first hydrogen, then by a complex route, oxygen. On Venus the present loss of oxygen comes from the splitting of carbon dioxide. Almost all of the water is gone. Carbon dioxide is much denser than water vapor and clings more closely to the planet's surface. This property makes it harder to shift than lighter water vapor. At best this implies that although oxygen losses are comparable, in the past the lighter water would have been much easier to dispose of than the carbon dioxide is today.

Overall the dominant process by which Venus loses atmosphere is through a mechanism with the grand title electric force field acceleration. High in the atmosphere, electrons are energized by ultraviolet and x-ray photons from the Sun. As electrons are less massive than other particles, they are less tightly held by Venus's gravity and thus more likely to escape from the top of Venus's ionosphere. As these move outwards and captured by the Sun's solar wind and magnetic field, they leave behind a net positive charge. This, in turn causes the remaining ions to repel one another, causing the upper atmosphere to balloon outwards and ultimately driving the ions outwards, where they too can be captured by the Sun's rasping wind. This, plus the effect of photochemical reactions releasing oxygen and the drag of the solar wind, led to the steady loss of Venusian air. However, as we shall see later, the rate of loss is not as dramatic as it first seems.

Changes to Wind Speed at Venus

One of the surprising discoveries made by Venus Express where long and short-term changes in wind speed. These indicate that for utterly mysterious reasons the average wind speed at 50 km has been increasing over the 7 years Venus Express was observing the planet. When the probe arrived in 2006 the average wind speed was 300 km per hour. By the demise of the craft in 2014 wind speeds had increased to nearly 400 km per hour. Igor Khatuntsev (Space Research Institute in Moscow) and lead author of the Russian-led paper to be published in the journal *Icarus* carried out one study using the data from Venus Express, while an independent Japanese group carried out a separate set of observations made from the Earth. Khatuntsev's team calculated the changes in wind speed by recording how cloud features in images moved between frames. In all over 45,000 features were methodically tracked by hand and more than 350,000 further features automatically compared using a computer program. The Japanese team used a similar process, using an automated process to monitor the position of different cloud features over successive image frames and using some simple math to derive the velocities.

Over shorter periods of time both teams also recorded rapid variation in cloud velocities with some features circumnavigating the globe in under 4 days, while a few orbits later the speed had decreases so that over 5 days were needed for clouds to cover the same distance. Such short term changes must reflect alterations in the amount of momentum shared by Venus's core, mantle, crust and atmosphere. Changes in Venusian wind speed might be caused by an exchange of momentum between the accelerating atmosphere and the decelerating planetary spin. Perhaps more surprising is the possibility that the atmosphere may be becoming spun up by an exchange of momentum between the Earth and Venus as they pass one another in their orbits around the Sun. As such, the Earth also exchanges momentum with its atmosphere.

Therefore, changes to the wind speed on Venus might even be caused by changes in the circulation of our atmosphere. What a weird thought.

Finally, as Venus was being observed the Sun was moving from solar minimum to maximum: perhaps the increase in the strength of the solar wind was pumping more energy into the upper atmosphere. From here, the mechanism that is believed to pump up super-rotation (and is described above) might simply have been transporting the greater momentum that was present in the solar wind to the lower atmosphere. If that is true, expect the speed of the atmospheric winds to decrease with the decline in the Sun's output. However, if the pattern is linked to an exchange of momentum with the planet's surface, expect a much more gradual change in the spin of the planet that runs independently of the solar cycle. However, maybe we should think again: could the spin of Venus (as a whole) change with the solar cycle? Does the Sun's magnetic cycle drive changes to the movement of its atmosphere, which ultimately exchanges momentum with its surface? Wouldn't that be an idea: Venus—a planet driven by its weather.

Snow on Venus?

In 2012 Venus Express reported that along the day-night terminator of Venus temperatures could fall as low as –175 °C at an altitude of 125 km (Fig. 5.4). Scientists, led by Arnaud Mahieux (Belgian Institute of Space Aeronomy) determined the atmospheric pressure and carbon dioxide concentration before calculating the temperature. This very cold layer is sandwiched between two warmer layers which should lead to the trapping of any carbon dioxide ice particles and perhaps lead to the formation of clouds analogous to noctilucent clouds, here on Earth. On occasion Venus Express noticed bright patches of cloud along the limb of Venus and it's possible that these are the proposed carbon dioxide ice clouds. Future explorers might just see snow on Venus—or more accurately high above Venus. Clearly such a frosty phenomenon wouldn't stand a snowballs' chance in hell anywhere else on this planet… (Fig. 5.4).

Yet, despite Venus's renowned extreme heat, there is another possible location for snow on Venus, and this is *on Venus*. When

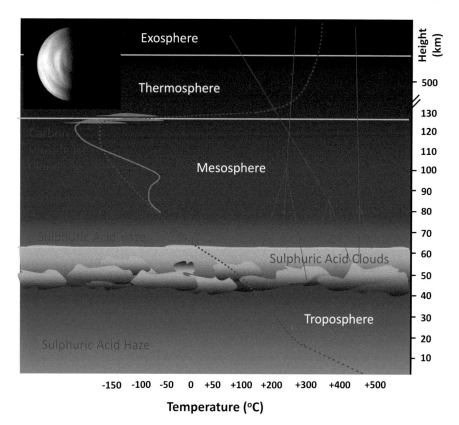

FIG. 5.4 Unexpected phenomena in the Venusian atmosphere. Near the day–night terminator (*photo inset*) temperatures fall to low enough temperatures for carbon dioxide to freeze out at around 125 km up. Here, carbon dioxide ice clouds might be dense enough to allow the formation of dry ice snowfall. *Purple straight lines* indicate cosmic rays streaking through the atmosphere. Although not yet clear these may create ionization in the sulfuric acid clouds that ultimately charges them enough to create lightning. Venus image from Venus Express (ESA)

Magellan mapped the surface during the early 1990s it detected some highly reflective material on Venus's highest peaks. Although no one would suggest conventional snow and ice—even at the relatively *cold* surface temperatures of 380 °C found at the summits of these mountains—a peculiarly Venusian snow might just be possible. Although you'd be hard-pressed to go skiing on this pressure cooked (45 bar) toasty material, what might be either elemental Tellurium or perhaps lead sulfide, would make the mountains seem rather terrestrial in appearance. Of course, all of this has to be envisaged through the peculiar magnifying lens of the lower Venusian atmosphere, but nonetheless if you could see Maxwell Montes rising upwards in the distance, its top gleaming in the faint, reddened sunlight, you might at least be temporarily reminded of the Earth (before Venus crushed, fried and dissolved you).

Lightning on Venus

On the Earth there are over 40 flashes of lightning per second, amounting to over 1.4 billion flashes per year. On a particularly stormy, early autumn day across Europe, in 2015, there were over 230,000 strikes. This rather large figure puts the Earth in one of the most electrically active planets known. However, within the Solar System at least three other planets are known to host thunderstorms. Jupiter has been known to host thunderstorms since Pioneer 1 soared past in the 1970s. Saturn also hosts a large population of thunderstorms, primarily associated with specific large storm systems that periodically dot the planet's atmosphere (Chap. 7). It is also likely that Uranus and Neptune will join the fraternity but Venus remained a curious outsider. Well, it turns out that it isn't (Fig. 5.5).

Fig. 5.5 *Left above*, a typical thundercloud (cumulonimbus) moving briskly eastwards over Crieff in Scotland. The fibrous top part of the cloud consists of ice particles, which develop a positive charge (Chap. 4). The base of the cloud, which is largely hidden by the trees, retains a negative charge, although this is grossly simplified. Above snow is falling from the anvil of another thundercloud, on a chilly late May day in Scotland. *Photographs by author*

How do we know that Venus has lighting? Well, aside from the possibility that it might be visible through a telescope or through the eye of an orbiting satellite, most lighting would be invisible as it occurs within Venus's dense, acidic and highly reflective clouds. However, what clouds hide in visible light they cannot hide at radio frequencies. Lightning produces very characteristic short bursts of radio waves, lasting roughly 1 s. The effect is quite unmistakable (Chap. 4). Using such radio frequencies, the Soviet Venera probes obtained data that suggested Venus does indeed have lighting. This was subsequently confirmed by observations made at similar frequencies by Magellan and later Venus Express. Venus Express also detected lightning through its effects on planetary magnetism. Each discharge generated bursts of magnetism associated with the movement of electrons along the lightning bolt, and through the effects of magnetic fields associated with the electromagnetic pulses (EMPs) lightning generated (Chap. 4). Interestingly, around the Earth such EMPs also affect the magnetosphere surrounding our planet (Chap. 4). Lightning discharges shove the terrestrial Van Allen radiation belts upwards creating safe zones at altitudes of a several hundred kilometers. Although Venus does not have the equivalent radiation belts, as it lacks the equivalent magnetic field, it does still generate its own micro magnetic shield as we have already seen. To recap this is formed by the interaction of the solar wind with the thermosphere, which generates a bow shock around the planet. This region is not only influenced by the strength of the solar wind, but thus will also be affected by EMPs generated deeper down within the clouds of the planet. As Chap. 4 suggested, terrestrial and other lightning links the atmospheres of planets to the broad galactic and intergalactic magnetic fields. The effect may be slight but it is still a fundamental link between the cosmologically tiny (a lightning bolt) and the vastness of the cosmos as a whole. Just imagine the beating of the wings of a butterfly on Earth affecting the formation of a distant star, by the subtle but measurable effect it has on terrestrial lightning and henceforth the planetary magnetic field.

That Venus had lightning was something, but the question remained how a planet with an apparently dry atmosphere produces the separation of charge needed to synthesize lightning? On the Earth, Jupiter and Saturn (Chap. 7) lightning forms in a similar way with charges separating within large cumulonimbus clouds. However, this mechanism cannot happen in quite the same way on Venus as dry Venus does not host an icy cloud structure in its main cloud deck: temperatures are just too high. For example at the cloud tops in Venus temperatures are just about freezing and the clouds are almost exclusively made of sulfuric acid. However, the underlying force that produce lightning is still convection as it is on the Earth, but with a generous hand from cosmic rays (Fig. 5.4). Christopher Russell (University of California, Los Angeles) and colleagues have analyzed Venus's electrical displays from data acquired by Venus Express. This reveals that the Earth and Venus are rather similar in terms of the strength of the lightning. As on Earth the majority of lightning on Venus occurs on the day lit side of the planet and convection within the reflective cloud layer is probably required to drive charge separation. However, although cosmic rays certainly help generate electrical charges within terrestrial clouds, they probably play a subordinate role to charge separation on ice particles and water droplets. On Venus, without ice, cosmic rays probably play the dominant role in generating charge differences in Venus's clouds—perhaps with an additional input from ultraviolet light (Fig. 5.4). Convection may simply drive clouds to different altitudes, with those reaching the greatest heights becoming the most strongly ionized. Recall that the energy of cosmic rays attenuates (weakens) with decreasing altitude. Therefore, cosmic rays have the greatest energies at the greatest altitudes in the atmosphere.

Most lightning on Venus also occurs in its tropics (within the 120°-wide strip flanking the equator). The principle difference is that convection probably does not originate at the ground, as it does on Earth, which is almost uniformly hot and poorly illuminated. Instead the relevant convection is likely confined within the cloud layer itself taking energy conducted and convected from the dense, fluid-like layer of carbon dioxide below. The sulfuric acid that forms the clouds is of low density in the deepest layers of the atmosphere as here temperatures are so high that the compound evaporates before it reaches the ground (Fig. 5.6).

On the Earth around 90 % of the lightning within the tropics is between or within clouds, but this value drops as you approach the poles. Over Norway roughly half the detected lightning is of this variety, with the other 50 % striking the ground. The proportion is

Fig. 5.6 Fall-streak virga—a.k.a. "jellyfish clouds". On a warm summer's day ice falls from mid-level altocumulus castellanus and trails behind the clouds, which are moving on a modest south-east wind. The ice sublimates long before it hits the ground. Such clouds might be common on Venus where sulphuric acid rain fall is both less dense than the carbon dioxide sea into which it is falling and boils at 337 °C, a temperature reached above the summits of Venus's highest mountains. *Virga photographed by author*

set mostly by the level at which water can freeze. This is clearly going to be higher over the Tropics than it is nearer the Poles. On Venus we have no real idea where the lightning is, but the majority is likely to be of the intra-cloud variety, in part because of the effect described above but also because the cloud bases are so much higher than they are on Earth. To strike the ground a much larger voltage would be needed to overcome the insulating effects of the greater depth of air. We know that on Earth, the much more powerful super-bolts, which strike from higher up, are also much, much rarer (Chap. 4).

Moreover carbon dioxide is a very poor conductor of electricity, meaning that to strike the ground a bolt of electricity that was generated within the clouds 40 km up would need to be extraordinarily long and thus of an extraordinary voltage. This assumes that Venusian air does not carry some form of highly conducting gas or other particle and this is not easy to determine. However, from terrestrial observations of our storms the effect of temperature should make most (if not all) Venusian lightning of the intra-cloud variety.

Why Does Venus Have So Much More Carbon Dioxide than the Earth?

Appearances can be deceptive. When you look at Venus you see an atmosphere dominated by carbon dioxide with very little nitrogen, compared to the Earth. However, this is by comparison of percentages alone. Remember that Venus has over 90 times the mass of carbon dioxide than the Earth has, so that value of 3.5 % nitrogen actually corresponds to nearly four times the mass of this gas in Venus's atmosphere compared to that found on Earth. As for the carbon dioxide, recall that on Earth most of this is taken up as carbonate rocks (Chap. 3) leaving only 1/250,000 the total mass of carbon dioxide as free gas. Solid carbonate rocks form when carbon dioxide and water interact with dissolved ions—mostly calcium and magnesium. Thus, on the Earth this process, known as sequestration, has accounted for the loss of most of our atmosphere to the surface. Of course, sequestration has also removed water vapor and carbon dioxide. On the Earth if you could resurrect this carbon dioxide as gas, the Earth would have roughly three times the mass of carbon dioxide in its atmosphere as is found on Venus: the Earth would have a thicker atmosphere than Venus has.

Constructing a Dry Planet

When we look at water and the Noble Gases something else becomes obvious. Clearly Venus is very dry indeed and the different noble gases reveal that much of Venus's original quotient of some of the noble gases such as Argon. Quite apart from assuming that the Earth and Venus might have started out roughly the same, with the same budget of water as our world, there is clear evidence that Venus has lost almost all of its water and would therefore have originally been a rather wet planet just like the Earth. The clue to this planetary dehydration is the ratio of two isotopes of hydrogen. If you recall from Chap. 2 chemical elements come in a variety of flavors called isotopes. Some are heavier than others and thus easier or harder to remove by evaporation or by other processes. For example as Chap. 2 showed, oxygen comes in three flavors of different masses: oxygen-16, oxygen-17 and oxygen-18. Oxygen-18 has two more neutrons than oxygen-16 making it 2 g heavier for every 600,000 million trillion atoms that are present. Likewise hydrogen comes in three flavors: hydrogen-1 (boring old hydrogen); hydrogen-2 (or deuterium); and hydrogen-3 (tritium). Tritium is unstable and radioactively decays in a short time, so we shall ignore it; but deuterium is completely stable and roughly twice the mass of bog-standard hydrogen-1. Because it is heavier it is harder to shift from the atmosphere than hydrogen-1.

Now, deuterium and hydrogen are normally found in a very similar ratio throughout the universe—or at least in water. Therefore, if that ratio is very skewed in favor of deuterium it tells you straight away that there was a lot more of the original hydrogen-1 that has now been lost. Quick calculations indicate that Venus has lost more than 99.9 % of its original store of hydrogen—and most of this would have been found as water. The conclusion is clear: Venus was just about as wet as the Earth but then lost almost all of its water. At present, if we assumed that the amount of oxygen we observe leaving Venus is from water, as it is on Earth, and to a lesser extent on Mars, then Venus would only have lost about 8 cm depth of water if we assume an ocean had completely covered its surface. For comparison the Earth would have lost about 9 cm in depth, with Mars losing around 30 cm: the difference in depth of water lost is a reflection in the difference in surface area of each planet. The problem is this isn't enough. Venus has lost far more water than the Earth and clearly something much more dramatic than the current rate of loss must have occurred.

In Chap. 3 we compared the early histories of Venus and the Earth. Both planets likely began life in the same way and both likely had identical (or near identical) conditions for the first hundred million years: a hot pressurized greenhouse. From here on two models are possible. In the "hot early" scenario Venus stays hot and any early ocean is unstable and soon succumbs to evaporation. In the "cool early" scenario Venus, like the Earth becomes cool enough, for long enough to form stable oceans and run plate tectonics.

In the hot early model water vapor is held in the atmosphere as a hot gas for tens of millions of years, with much of it rising upwards from a high tropopause to the stratosphere. Here the more intense ultraviolet light of the Sun (both because Venus is closer to the Sun and because the early Sun released more energetic, ultraviolet-rich flares and a stronger stellar wind) steadily broke the water vapor down to liberate its hydrogen. This light gas steadily boiled off into space leaving a more oxygen-rich atmosphere. After maybe, 100–500 million years all of the water was lost—the hydrogen to space and the heavier oxygen largely retained. What happened to that gas remains a mystery, but presumably it reacted with materials on Venus's surface or sulfur-rich gases in its atmosphere. What was left was an increasingly dense and hot carbon dioxide atmosphere. The final straw for Venus might have ironically come from the failure of plate tectonics.

Measurements taken by the Visible and Infrared Thermal Imaging Spectrometer (VIRTIS) on Venus Express made in the infrared portion of the spectrum shows that the rocks on the Phoebe and Alpha Regio plateaus are lighter in color and look old compared to the majority of Venus's surface. On Earth, such light-colored rocks are usually granite and form continents; and because granite is a lot less dense than basalt it floats higher in the mantle than basalt. The presence of granite would naturally explain why these regions are the only noticeable highlands on Venus's surface. As was suggested in Chap. 3 that parts of Venus's crust appear to be granite implies that Venus held oceans for at least a few hundred million years. That such continental crust covers only a trivial proportion of Venus's surface also implies what oceans Venus once held must have been transient and probably lost within one billion years. This hotter version of Venus might well have hosted early life, giving rise to the possibility that early in the Solar System's history there were three Earth-like worlds: a steamier Venus, a Goldilocks' porridge-like Earth and a more frigid and possibly effervescent Mars.

However, by one billion years, Venus was likely overheating; it's oceans on the wane and the atmosphere steadily filling with water vapor and carbon dioxide gas. Whether this led to a thermal runaway—a thoroughly unpleasant phase where water vapor drive the temperature to over 800–1000 °C—is unclear. As temperatures rose to over 70 °C then 100 °C the remnant oceans would have boiled away while the air filled with a torrid sea of broiling thunderclouds. Extending up to 100 km in height, these monstrous storms pumped water vapor high into the stratosphere where ultraviolet light broke it down. While the Earth maintained a cold and stable lower stratosphere, Venus's was overwhelmed by vigorous convection extending from the hot surface below. Although ozone would have been forming from the oxygen that was released from the split water, this would not have been sufficient to shield the water below. Once so much water is split the hydrogen that is released limits the stability of ozone and the stage was set to dry out the planet. The escaping hydrogen can also drag the oxygen (and ozone) away with it in a process called hydrodynamic drag.

Thus by 3.5–4 billion years ago Venus was drying out and perhaps initially cooling off a little. As water vapor split the greenhouse effect may have waned slightly and the clouds cleared revealing a hot desert-like planet. However, such relief would have been temporary. While much of the planets carbon dioxide was likely locked up in carbonate rocks on the planet's surface, the steady output of volcanic activity would have once more caused the levels to increase in the air. With every extra gigaton the temperature would have increased, even as the abundance of water vapor declined.

The question mark lies with the state of Venus's volcanic activity. With a similar mass and energy budget to the Earth it should have retained active volcanism. However, as the oceans vanished and the crust and mantle steadily dried out the process of tectonic activity would have soon ground to a halt. What happens next is speculation. With so much internal heat, volcanism can't simply stop as not enough heat would have been lost through the crust by conduction. While hot spot volcanism would have continued, an unstable situation was developing that would finally condemn Venus to the state of a torrid hell. Without plate tectonics to cool the mantle it would soon have overheated. As well as causing extensive melting of the uppermost layers, the crust would have begun to soften and buckle. At least once in the intervening three billion years the thick Venusian crust appears to have foundered. On Earth plate tectonics depends on the formation of a rock called eclogite. This metamorphic rock forms from basalt that is heated and compressed around 100 km under the surface. This transformation only happens nowadays when the crust is cooled, thickened and buckles downwards. On Venus this process is unlikely as the crust is too hot and dry. However, if the crust thickens up the base of it might become transformed into eclogite at its base and the resulting denser slab of rock would then sink directly into the mantle. Such a process might have happened on the early Earth when the mantle was much hotter and volcanic activity led to the formation of a thicker crust.

Why is this important? Well, if Venus developed a thick crust that then foundered into the mantle the process would have rapidly consumed the entire basaltic crust of the planet—probably leaving the island continents that formed early on. Just imagine the scene unfolding. Initially volcanic activity steps up as the mantle begins to overheat. Copious volcanic activity begins to resurface the planet, adding a lurid red glow to the already superheated world. After a few million years the crust has thickened up and begins to founder under its own weight, plunging catastrophically into the mantle. Piece by piece the crust would crack and whole segments would begin a rapid subduction under a fresh, thinner, molten sea of lava. With the atmosphere already hot, the extra input of heat would raise temperatures further. Worse still, as volcanic activity increased the amount of water vapor and carbon dioxide, temperatures would have begun to sky-rocket. However, that wouldn't be the end.

On the surface of the foundering basaltic crust would have laid thick deposits of carbonate rocks, created while Venus still had oceans. As this was dragged down into the mantle with the foundering basaltic crust, it would have broken down releasing its store of carbon dioxide, as well. At this point the ocean's worth of carbon dioxide would have been vented to the atmosphere. While initially, the wet greenhouse might have exerted roughly 100 times the surface pressure, this would have decreased to much more manageable levels as sunlight broke the water vapor down to oxygen and hydrogen and the oxygen became chemically bound to rocks through chemical reactions. However, as volcanicity stepped up, and in particular the massive amount associated with resurfacing got under way, the atmospheric pressure would have increased once more as massive amounts of carbon dioxide accumulated. In turn this would have enhanced the greenhouse effect until temperatures may have once again reached 1000 °C. If they did get this high, any remaining carbonate rocks would have decomposed and released their store of carbon dioxide, adding a further wave of greenhouse gas to the atmosphere.

The only saving grace for Venus would have been the increasingly dense burden of clouds. If Venus ever had a largely transparent atmosphere (to visible light) it would have rapidly clouded over once all of this volcanic activity kicked in. Clouds would have reduced the amount of solar energy that made it through and allowed temperatures to fall to their present levels. At some point around 500 million years ago, most of the volcanic activity declined to its current, apparently, low levels. The atmosphere then stabilized at around 95 bars of pressure with a temperature around 450–500 °C, remaining there until the present day.

Today, the largely dry planet is still venting hydrogen. Although much of this may come from the photolysis of sulfuric acid some of it is coming directly from the break-up of the small amount of free water found in the atmosphere. The Venus Express magnetometer instrument (MAG) detected hydrogen gas being stripped from the day-side. Until Venus Express arrived at the planet, this process had been suspected but there was no real evidence that it occurred, with most hydrogen thought to exit the planet through its magnetotail. It is clear that Venus is still drying out today—most likely this is residual water from the mantle that was dumped there during the planet's formation or subsequently during the brief era of plate tectonics.

So, how much mass is Venus losing from its atmosphere now? And perhaps more pertinently how long will it take the solar wind to strip poor old Venus bare? You might imagine that the relatively exposed Venus might not be able to hold onto its atmosphere very well. However, as we've seen the combination of high atmospheric mass with a dominance of the relatively weighty carbon dioxide, and the protective effect of the bow shock and ionopause means that Venus currently loses only one metric ton per hour (roughly 8,700,000 kg per year). That might sound a lot but with an atmospheric mass of roughly 500,000 million trillion kilograms it means that Venus can retain a reasonable atmosphere for the best part of 55 trillion years. The Earth has a similar mass loss rate and if nothing else were to change the Earth would be stripped bare in a comparatively trivial 591 billion years. Of course, long before that happens both planets will be vaporized by the expanding Sun…

Life in Hell?

Here's an interesting idea: Venus is still habitable—albeit not on its hellish surface. Recall, that at about 55 km up the temperature is around 25–30 °C with a pressure comparable to sea level on Earth. Moreover, photolysis of carbon dioxide gas (the chemical break up of carbon dioxide caused by ultraviolet light from the Sun) also releases fairly reasonable amounts of oxygen. In principle life could survive in these conditions as long as they didn't mind the low pH and the relative dryness. On Earth bacteria are known to survive and reproduce within our clouds—and many species of bacteria survive hot, acidic conditions. Thus, maybe, just maybe, there is life in the middle atmosphere of Venus.

Future Venus: The Earth and Venus, Twins Once More

Although Venus and the Earth are very different now, they both began their lives with hot, carbon dioxide and water vapor-rich greenhouse atmospheres. Perhaps for the first 50–100 million years these encased a molten ball of rock and metal. However, soon after 100 million years had passed the surface would have solidified and lowering temperatures would have allowed water to begin falling from the skies (Chap. 3). Under a dense carbon dioxide-rich atmosphere, water began to accumulate in oceans that would then have begun to pull down the carbon dioxide.

As we have already suggested, perhaps for a few hundred million years, both the Earth and Venus presented a deep blue, clouded face to the universe. However, as soon as Venus began to lose its oceans the planet would once again have clouded over and the long, dark separation of these two worlds would have begun. Yet despite this schism, the parting is only going to be temporary. At some point in the next two to three billion years Earth will pursue its twin along a path to hell. Before the Earth succumbs to this fate it will pass through a process of slow decline where falling carbon dioxide, but rising temperatures force the demise of first land plants then land animals.

800 million to 1 billion years into the future carbon dioxide levels in the Earth's atmosphere will have fallen because the gas will have been absorbed into the oceans and into the planet's crust. Less and less carbon dioxide will be returned by volcanism, as this too declines with the secular cooling of the planet.

While photosynthesis in higher plants will cease when the concentrations of carbon dioxide fall lower than 10 parts per million (0.01 %), bacteria in the oceans will continue to operate photosynthesis and generate oxygen as a byproduct for a long time thereafter. As long as life can persist, feeding on these bacteria life on Earth will survive. I speculated in "Under a Crimson Sun" that plants could survive if they "mined" carbonate rocks for carbon dioxide. At present most plants (and their symbiotic fungi) actively secrete acids and other chemicals into their environment. In principle, any plant growing on carbonate-rich rocks, such as limestone, or lime-rich soils could capture carbon dioxide from the rock using acids secreted around its root system. There certainly would be evolutionary pressure for such a move, but whether it would be sufficient to maintain a terrestrial biosphere is anyone's guess. A more grisly alternative is that plants take up carnivorous lifestyles, obtaining their carbon from animals. In terms of the biosphere, they couldn't all adopt this change as this would be akin to a living perpetual motion machine with animals surviving by eating plants and plants surviving by eating animals. Thermodynamics will step in and prevent such practice becoming widespread amongst plants.

More importantly for our planet will be the ever-brightening Sun. At one billion years from now the Sun will be 10 % brighter than it is now. This will be sufficient to raise the temperature of the planet to around 30 °C, or a good 7 °C higher than it's been in any previous greenhouse phase, such as the Permian-Triassic boundary. The planet will be then be ice free for the first time in around 5.4 billion years and much of the continental surface is likely to be under water. This state of affairs will be exacerbated if plate tectonics has ceased and mountains have largely eroded down. Venus, meanwhile will hardly notice the difference with its thick, highly reflective atmosphere keeping the increasing glare of the Sun at bay.

While Venus will remain largely unscathed for the ensuing few billion years, the Earth's biosphere and atmosphere will be forced to change. It's at this one billion year mark that two different fates emerge. It was thought that the most likely fate would be a rapid rise in temperature over a few tens to hundreds of millions of years. This could result in temperatures escalating to the point at which the surface of the Earth was completely melted. This was because if water vapor was allowed to trap sufficient energy the surface temperature would need to rise to the point at which it was hot enough to release energy at visible wavelengths. Then, and only then, would the atmosphere be able to release energy trapped by water vapor. However, in new work, described recently by Jérémy Leconte and co-workers (Laboratoire de Me´te´orologie Dynamique,) is that the Sun will begin a protracted but accelerating boil-dry of the oceans. This will likely take over a billion years to complete. With the planet at an average temperature of 35-45 °C water vapor will act much more effectively as a greenhouse agent. Remember that 45 °C is an average for the planet, which means that some areas will see temperatures well above 70 °C. As more water evaporates the planetary greenhouse will strengthen and so temperatures will rise, further accelerating the effect. In this scenario the oceans, and with them life, will be eliminated by 2-3 billion years into the future.

The fate of the water that boils away will mirror events on Venus perhaps five or six billion years before. At present the Earth's water is largely contained within the troposphere by the temperature inversion at the tropopause: although the number seems large, only a small amounts of water escapes into the stratosphere. This is then split by ultraviolet light to hydrogen and oxygen.

From here, somewhere in the region of 95,000 metric tons of hydrogen is then lost to space, per annum. Most of the accompanying oxygen retained because it is heavier. A comparison of terrestrial isotopes of hydrogen in some archean rocks with those in present day equivalents implies that the Earth has lost nearly a quarter of its water since it formed. Fortunately, above the Earth the stratosphere is cold and hence dry. Moreover, as we saw in Chap. 1, above the tropopause air begins to warm with height because of the absorption of ultraviolet light by ozone. This gas is concentrated within the stratosphere. Such a warming trend blocks convection from carrying moisture any higher and thus preserves most of our planet's riches in the troposphere, where it is shielded from ultraviolet light by the abundant ozone above.

However, as happened on Venus when the terrestrial troposphere warms the lid of the tropopause is forced upwards until the moist air below becomes accessible to much more abundant and energetic ultraviolet light. As increasingly energetic and intense ultraviolet radiation splits the Earth's water, hydrogen will drift off into space and be blown away in a majestic comet-like apparition—though one visible only if you could see in the ultraviolet. As with Venus, the Earth could, in principle develop an oxygen-rich atmosphere. However, whether this happens will depend very much on how much material is available on the planet's surface to oxidize. This will include dead and dying organisms, as well as any sulfur-rich compounds thrown out by our planet's declining volcanic activity. As with Venus a considerable amount of oxygen will also be lost from the upper atmosphere, either dragged out by escaping hydrogen, or split into single oxygen atoms and ions that can be grabbed by the solar wind and escorted into the deep, near vacuum of space. Recent work by James Kastings (Pennsylvania State University) and colleagues suggests that the critical surface temperature is 350 K (77 °C). At this point the troposphere leaks so much water vapor that the stratosphere effectively becomes saturated and warms up, thus allowing all of the planet's water to escape.

By the time the planet has aged a further two billion years, in this slow-cook scenario, the Earth will have boiled dry. Without water to sequester carbon dioxide gas, this greenhouse gas will begin to accumulate once more within the atmosphere. Although our water will be split by ultraviolet light at a pace sufficient to prevent the complete meltdown of the planet's surface, temperatures will begin to rise inexorably to levels approaching the contemporary Venus. Although this will avert a catastrophic greenhouse effect, any life that remains in the crust will still be cooked out of existence.

One interesting consequence of these newer models is that Venus was more likely to have held onto oceans for a significant part of its early life. Instead of an aggressive, early thermal runaway (Fig. 5.7), Venus would have had a more benign climate and possibly life for up to a billion years or so from its formation.

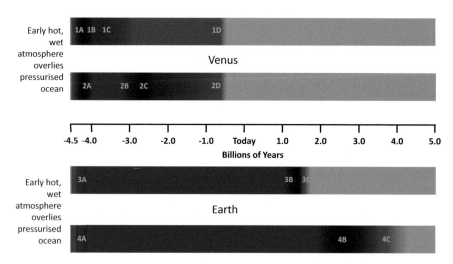

Fɪɢ. 5.7 Key events in the histories of Earth and sister Venus. *1A*—Transient deep oceans evaporate; *1B*—Runaway wet greenhouse; *1C*—Water lost to space and dry greenhouse begins; *1D*—global resurfacing event(s) fill atmosphere with carbon dioxide and water establishing modern Venus. In the alternative model early on Venus has cool oceans (*2A*). These survive for at least a few hundred million years before they begin to evaporate (*2B*). During this time plate tectonics makes Venusian highlands. In *2C* a hot, wet greenhouse is established before the water is lost to space; *2D* global resurfacing occurs as in *1D*. By contrast the cooler Earth maintains oceans after an early, hot, wet greenhouse (*3A/4A*). These persist until *3B* when a thermal runaway occurs or much later in *4B* when they have evaporated. In *3C* and *4C* a final hot, dry greenhouse is established, finally pairing the Earth and Venus once more

The Earth can maintain habitability for longer if the Earth's magnetic field falters in the next billion years—something that could happen when plate tectonics ends in the intervening time[2]—the Earth's atmosphere could be exposed to the erosive power of the solar wind. Now, we already know that this stellar breeze has minimal impact on the planet's atmosphere at present. If the rate of removal of gas (and the water vapor will do) can keep pace with the brightening Sun, the global greenhouse effect might be kept largely in check for a little longer, ensuring that the period from 2 to 2.8 billion years from now is more habitable than in the simplest scenario. With significant reserves of gases and liquid water, a steady loss of these volatile materials readily lowers the atmospheric pressure, which in turn weakens the greenhouse effect. You may recall from Chap. 3 that increasing the concentration of carbon dioxide enhances its greenhouse effect because at higher concentrations particles of gas collide and cause a broadening of the range of wavelengths over which the gas can absorb infrared (heat) radiation. Therefore, if we can remove gas from the atmosphere (mostly nitrogen) then the overall pressure will decrease and with it the strength of the planetary greenhouse.

During this extended phase the Earth would steadily dry out, but sufficient water could be maintained on the planet's surface to keep it habitable up to three billion years into the future. Beyond this point the atmospheric pressure would fall to less than a third its current level until liquid water became unstable at what would still be higher temperatures than are found on average today. What remained of the oceans would evaporate and the planet would finally run dry. Venus would have been unable to pull off this trick for two reasons. Obviously, being closer to the Sun it grew hotter faster than the Earth until its greenhouse became unstable. However, the geology of Venus also played its part. If Venus had been geologically inert with little volcanic activity, the brightening Sun might have split and driven off sufficient water to keep the planet cool and thus habitable for longer. However, while the young Earth stayed cool, with its greater distance to the Sun, Venus's atmosphere was soon overwhelmed by a combination of

---

[2] Plate tectonics may be essential for the maintenance of a global magnetic field. Dense cold subducting rock might stimulate convection in the planet's core, which in turn might drive the formation of the magnetic field.

evaporating water from its oceans and the constant addition of more carbon dioxide and water vapor from its abundant volcanoes. Venus never stood a chance.

Yet, if the Earth's volcanism is sufficiently subdued a billion or so years hence and the loss of water can keep pace with the brightening Sun, the Earth might cling onto a faltering habitability until 2–3 billion more years have elapsed. Certainly the Earth will be hotter and drier than we are used to, but life—quite abundant and complex life—could hang on.

For the sake of argument let's stick with the more benign fate for our planet as that gives us three billion years to play with. How will the Earth's climate systems alter—aside from getting warmer and drier; and at what point will the Earth and Venus come to resemble one another once more? Aside from the ever-growing influence of the Sun the next most important factor will be the spin of the Earth. This is coupled to the orbit of the Moon, with the exchange of momentum between the two driving the Moon outwards at roughly 4 cm per year. This process will continue until the length of a terrestrial (solar) day equals 1 month. At this point the moon will orbit the Earth around 540,000 km away. Meanwhile this "gain of momentum" by the Moon means that the Earth must be losing the equivalent amount, as momentum is always conserved. For our planet, the effect is causing the Earth to rotate more slowly: roughly 1.7 milliseconds are added to the length of the day each century. Were the Earth fully solid this value would be greater, but after the last Ice Age the mantle rebounded pulling mass inwards from the equator towards the poles, and this has caused a slight acceleration, partly off-setting the effect of the retreating Moon. Thus, the Earth will gain around 4 h to its day over the next billion years. Deceleration will continue after this point, until, as was said above, the length of the terrestrial day matches its month. However, the Earth will be truly dead by that point…

A decelerating Earth means a weaker Coriolis Effect and more sluggish west–east movement of the air in response to north–south differences in pressure (Chap. 1). When the Earth formed the planet likely rotated with a period of around 5–6 h. During the Proterozoic snowball (Chap. 3), a terrestrial day was approximately 21 h long. That is known with reasonable precision because tidal deposits are preserved in some sedimentary rocks that were laid down at the time. The to-ing-and-fro-ing of the water left very characteristic patterns in the deposits which allow the length of day to be established. Although the rate of slow-down decreases as the Moon pulls away from the Earth, the length of the terrestrial day will increase and with it the Coriolis Effect will decrease.

As the Coriolis Effect weakens the strength of terrestrial winds will decrease over time and will blow more north–south (on average) than they do now. This will facilitate a greater exchange of heat between tropics and poles, ensuring that the Polar Regions are more thoroughly heated than they are at present. This, coupled to a brighter Sun will make the planet more thoroughly cooked throughout. Embedded within the weaker west–east flow of air will be a grossly expanded pair of Hadley cells. A hotter planet with less spin will allow these to reach ever closer to the planet's poles, compressing the more westerly mid-latitude flow of air towards the poles. You can see this on Venus where the equivalent Hadley cells extend to 60° north and south of the equator—twice the width of the domain on the Earth. On Venus the belt of westerly winds is confined to a narrow collar around the Poles and this is the way the Earth's winds will come to resemble when the planet has grown hotter and spins on its axis more lethargically.

Inside the domain of the Hadley cells tropical storms will still generate assuming there is a sufficient width of ocean to supply them with moisture. Although tropical storms rely on the Coriolis Effect to gain spin, with more territory over which to form and more heat energy available than it is today, it might mean that much more of the planet will be afflicted by such tropical monstrosities. In Chap. 3 we saw how human activities already appear to be having this effect on the distribution of such storms, with a north and south migration of maximum storm intensity of 50–60 km per decade—this matching the steady expansion of our planet's Hadley Cells. Whether the tropical storms of tomorrow's Amasia will be the near-legendary mega-storms of the popular scientific media, remains to be seen, but the storm intensity will be a product of the temperature, distribution of land and the overall spin of the planet. More heat means more energy for the storm; while a slower spin means a weaker Coriolis Effect and less spin. Before the Earth boils dry tropical storms might just rule the meteorological roost over much of the planet.

From two to three billion years any residual volcanic activity would slowly refill the atmosphere with carbon dioxide and the greenhouse would accelerate. Although unlikely to get as hot as contemporary Venus, given that the Earth is further from the Sun, it would still be hotter than the boiling point of water. In this desiccated state the planet would remain until the Sun left the main sequence, expanded into a red giant then blew the atmosphere away. This process would take around 700 million years, until the swollen Sun's surface came so close to the Earth that tidal forces pulled the Earth in to be vaporized 200 years later. En route, the Earth may get a rather interesting atmosphere. When the Earth enters the last few million years of its existence, the solar surface will be less than one million kilometers from the Earth's surface. At this point, the Moon's orbit will take it half way to the surface of the expanding Sun. Perhaps it will come to resemble to a flaming molten orb, with a cometary tail of vaporized rock periodically showering the Earth as it swings to and fro around the molten orb of the Earth.

Although the dying Earth will rotate slowly it will still spin on its axis with a day length perhaps 100-200 h long. Finding a definitive figure for the length of the Earth's day is rather tricky—but were the Earth to survive the Sun's red giant phase the day length would eventually max out at 1000 h long in 50 billion years.

Immediately prior to the red giant phase the Sun will be around three to four times its current luminosity, depending on where you draw the line for the Sun's main sequence. This will be more than enough to make the Earth a close twin of the present Venus. Surface temperatures will easily exceed 500 °C. However, unlike the Venus of today the Earth's atmosphere could be relatively cloud free—if volcanism is very subdued on our rather decrepit planet and little in the way of aerosols are available. The slowly rotating Earth may have a dense but clear carbon dioxide-rich atmosphere that stirs slowly. This atmosphere, free of reflective clouds will absorb and retain much more heat than Venus does today and thus be proportionately hotter. Indeed, both Venus and the Earth should have broadly parallel geologies at this point meaning that Venus and the Earth could appear nigh on identical from afar. One significant difference is that the Earth actually has more carbon dioxide gas in the form of carbonate rocks than present day Venus. Most of these rocks are held at relatively shallow levels within the crust. As such, once global temperatures exceed 850–900 °C these carbonate rocks will break down (through thermal decomposition) releasing their store of greenhouse gas. As such there may come a time 3–4 billion years from now when the Earth is as hot, or hotter, than Venus at a comparable time. Rather than vying for habitability, Venus and Earth will vie for inhospitability. Aside from a few hardy fossils there will be no evidence that the Earth ever hosted life, never mind intelligent life.

However, if the Solar wind has had its way with our planet, the terrestrial atmosphere will remain relatively thin and the greenhouse effect proportionately weaker. Although still hot, it won't be hot enough to melt the surface. However, that (relatively) benign phase won't last much longer; for six billion years hence, the Sun will begin its final expansion and begin to melt the Earth and Venus.

Some seven billion years from now, the Earth and Venus will face-off with a Sun that is nearly 260 times its present diameter and around 2000 times as luminous. For the final five million years of its first red giant phase the Sun will expand most rapidly, vaporizing first Mercury then Venus. Both of these inner worlds will be directly swallowed, however, the Earth has a different fate. Although the Earth will likely be spared direct annihilation, the word "spared" will be a relative term. Basking, or perhaps basting, under a glare that raises the surface temperature to around 2000 °C might just permit the kind of hellish weather seen on *The Chronicles of Riddick*'s planet "Crematoria". If you haven't seen the film, Riddick has to escape his underground dungeon and run the surface to the nearest spacecraft, before sunrise. The emerging Sun not only blasts the surface with the hellish light, but also drives a superheated wind of vaporized rock and debris across the surface. Riddick, obviously escapes, but not before two of his colleagues have been turned into toast. Imagine sunrise: a vast red orb rises to fill nearly 70 % of the sky. The ground is a molten soup of rock, cast against a stark, black void above. As the Sun rises slowly to its zenith the molten stew at your feet starts to boil; not the whole surface but rather the lighter components. A thick dense wind of vaporized rock lifts upwards and flows in a massive thermal tide across the day–night terminator and into the dark side of the planet. Here, under a chill as deep as −180 °C the rock condenses and falls as a molten rain or a powdery, ashen snow.

Although unlikely, if the Moon has been ripped away at this time, then the Earth will tidally lock to the Sun. Then the constant erosion of the surface will strip away the crust on the exposed sunlit side. Progressively deeper layers of material will boil away, only to rain out on the dark, cold side of the battered Earth leaving the Earth resembling a lurid version of the contemporary Moon, with a thin molten crust on its Sunlit side and a thick ashen crust hidden in the perpetual night of its dark side. Unlike the present Moon, there really will be a dark side of the Earth. The Earth will be a bit of a mess, really.

Interesting though this fate might be much more likely is that the Earth will retain the Moon in a high, but, now, increasingly unstable orbit. As the Earth and Moon boil, the pair will have a brief (and fatal) final encounter. As the Moon orbits the Earth the denser solar wind blowing outwards will act like molasses on the orbital motion of our Moon. This will slow the Moon down, which, in turn, will cause it to spiral inwards. Interestingly as the two move ever closer, and the Moon speeds up, tidal forces between the molten Earth and Moon will spin the Earth ever faster. In the process, increasingly vast tides of molten rock will slosh across the Earth's surface. When the Moon reaches 11,400 km or so above the Earth's surface, the Earth's tides acting on it will tear it apart. For a brief moment a shower of molten rock will splash across the heavens. Under the glare of the incandescent Sun, this molten ring of material will gradually boil it away—a fate that mirrors that of the Earth, soon, thereafter. Once again, as with nearly every section of this chapter, there is an alternative. Should the Sun's bloated surface drag on the Moon more strongly than the Earth, the Moon might get pulled off its orbit around our world and into the depths of the Sun, before its orbit shrinks causing it to collide with the Earth. In which case, the fate of the Moon will simply presage the fate of the Earth a few million years later.

Earth's last "weather" a blisteringly hot rain of molten rock and ash will succumb to an increasingly dense, though no less torrid, fog. Timing is everything: Venus will last until around five million years before the peak of the first red giant phase; while the Earth will probably last for another 4.5 million years after Venus is vaporized. However, at some point, around 500,000 years before the end of the first red giant phase, the Earth will begin a 200 year-long death spiral into the Sun's outer layers. Temperatures will steadily rise until all of the rock begins to boil. Streaking, invisibly downwards towards the Sun's core, the cometary remains of the Earth will follow Venus and dissolve into nothingness, adding roughly 0.01 % extra heavy elements to its bulk. Only the sullen, molten face of Mars and the hot gaseous orbs of the outer giants will be left to witness the fate of the Sun's twins.

Conclusions

Venus and the Earth: the twins that aren't, but once were and will be once more. This terrestrial pair are so similar, in terms of mass and overall composition. However, thanks to a subtle difference in the location of their births, both worlds rapidly diverged early in their histories. Whether the Solar System hosted two Earth-like worlds (and possibly a third in Mars) remains unclear. There are tentative signs that Venus was cool and wet enough for at least a few hundred million years of its history to form small continental regions. However, it might be that the earliest ocean or oceans of Venus were simply pressure-cooked pans of boiling water that resided transiently before the greenhouse effect lifted them into outer space. What is clear is that by 2–3 billion years ago Venus was boiled dry and the modern greenhouse atmosphere began to take shape.

Contemporary Venus paints a portrait of the future Earth. At some point between one and three billion years into the future the ever brightening Sun will "catch up" with the Earth and drive its oceans into out space. Steadily rising solar luminosity and increasing carbon dioxide levels from residual volcanic activity will raise the global temperatures until they approach the melting point of the crust. Although this could be a fairly protracted affair for our world, if the Earth can hang onto enough water vapor its atmosphere will superheat much earlier (1–1.5 billion years from now). Either way, a global greenhouse effect will bring the atmospheres and histories of the Earth and Venus back into line with one another. For the remaining few billion years Venus and Earth have left, they will present a hot, but increasingly cloud-free face to the universe. These twins will continue to overheat until they both melt under an increasingly severe glare from our aging star.

Both worlds will then end their lives, with the Earth following Venus into the Sun's fiery maw roughly seven billion years from now. The twins will be no more.

References

1. Müller, N., Helbert, J., Hashimoto, G. L., Tsang, C. C. C., Erard, S., Piccioni, G., Drossart, P. (2008). Venus surface thermal emission at 1 micron in VIRTIS imaging observations: Evidence for variation of crust and mantle differentiation conditions. *The Journal of Geophysical Research, 113*, Issue E5, doi:10.1029/2008JE003118.
2. Ingersoll, A. P., & Dobrovolskis, A. R. (1978). Venus' rotation and atmospheric tides. *Nature, 275*, 37–38.
3. Leconte, J., Forget, F., Charnay, B., Wordsworth, R., & Pottier, A. (2013). Increased insolation threshold for runaway greenhouse processes on Earth-like planets. *Nature, 504*, 268–271.
4. Kasting, J. F., Chen, H., & Kopparapu, R. K. (2015). *Stratospheric temperatures and water loss from moist greenhouse atmospheres of earth-like planets*. http://arxiv.org/pdf/1510.03527v1.pdf.

6. The Wispy Weather of Mars

Introduction

Mars was born unlucky. Thanks to an unfortunate proximity to the proto-Jupiter, the debris from which Mars coalesced was severely depleted in mass. This left Mars something of an undersized runt, setting the tone for its future evolution and ultimate atmospheric demise. Unlike the Earth and Venus, Mars has a relatively slight gravitational pull, which is somewhat less than half that of its larger brethren.

The Mars of today is a cold, dry, barren world, but it wasn't always this desiccated. How it became so is an interesting tale that ties the deep interior to events happening millions of kilometers away. Much remains to be understood about the weather and climate on Mars, but there is already a wealth of interesting and mysterious phenomena that parallel many seen on Earth. Like the Earth, Mars has a diverse climate with an atmosphere structured much like ours. The large difference in the mass of our atmosphere and that of Venus and Mars means that the Martian climate is as different from Earth's as Earth's is from that of Venus.

The Structure and Formation of the Martian Atmosphere

The contemporary Martian atmosphere is a thin, dry affair with less than 1/100th the mass of Earth's. The atmospheric pressure averages around 6 millibar, although this figure varies considerably across the planet's surface. After all, there is a 22 km difference in height between the bottom of the Hellas impact basin and the summit of Olympus Mons. Pressure thus varies by more than one order of magnitude, being densest at lower elevations, where it is around 12 millibar. Meanwhile at the lofty summit

© Springer International Publishing Switzerland 2016
D.S. Stevenson, *The Exo-Weather Report*, Astronomers' Universe,
DOI 10.1007/978-3-319-25679-5_6

of Olympus Mons atmospheric pressure is only 0.02 millibars, or 150,000th of the pressure at sea level on Earth. As it turns out this is rather similar to the atmospheric pressure on Pluto (5 microbars or 100,000th of the pressure at the sea level on Earth).

Mars has such a thin atmosphere that vertical and horizontal heat transport by convection and advection is much more limited than it is in our atmosphere: on Mars direct transport by radiation dominates. That is not to say that convection and advection are not important, and convection is really rather rapid on Mars. With less mass to transport, and the atmosphere more transparent, overall, radiation is a far more effective means of getting heat from the surface to the top of the atmosphere than it is on Earth or Venus. This difference leads to some odd meteorological effects that we shall look at later. Most importantly for the idea of habitation, while the atmosphere is not as noxious as that of Venus, the pressure is so low that were a human to be exposed to it, the water in their blood would boil, leading to a rather unpleasant death. Occasional UFO conspiracy stories paint the surface as habitable if only we could stick a few plants there to make oxygen. Unfortunately the altitude would dessicate them unless they were in a pressure-controlled environment (as seen in the recent hit movie "The Martian").

Temperature-wise, Mars shows a lot of variability. The average temperature is around −20 °C, with surface temperatures that can be as high as 25 °C or more on the surface nearest to the equator. Meanwhile, in the long polar winters, temperatures fall well short of −100 °C, low enough for carbon dioxide to "precipitate" out as frost.

Figure 6.1 illustrates the general layout of the thin Martian atmosphere. Aside from the low overall pressure, perhaps the most striking thing is how dispersed it is. While on Earth the bulk of the terrestrial atmosphere's mass is found within 7 km of the surface that on Mars is far more diffuse. The troposphere extends up to 70 km, with a narrower stratosphere and mesosphere forming the upper layers of the Martian pancake. The troposphere is broader than that on the Earth because, on this occasion there is such a big temperature difference between the Martian surface and the tropopause. There is also no significant ozone layer. These factors allow the atmosphere to convect over a broader range of altitudes.

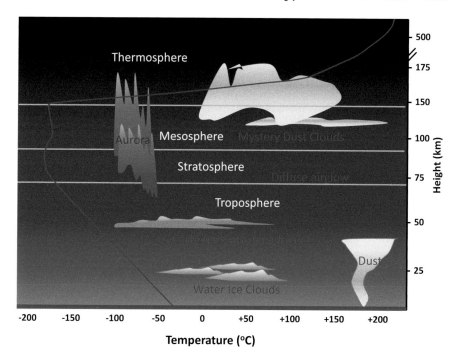

Fɪɢ. **6.1** The overall structure of the Martian atmosphere. Aside from being much colder than that of the Earth, the rarefied atmosphere is far more dispersed than above the Earth, with the troposphere, for example, spanning more than three times the height of the same layer above our planet. The newest feature is sporadic, seasonal clouds seen at 100–120 km above the surface. The nature of these are unclear. Diffuse airglow (low level aurora) occurs at 60–80 km

Overall, the temperature profile varies rather like that of Venus, but it is obviously a lot colder in the Martian troposphere. Above the troposphere, temperatures are rather similar to that above Venus (Fig. 5.1). This is largely set by two properties: the intensity of solar radiation and the composition of the air which is predominantly carbon dioxide. Carbon dioxide maintains the surface temperature on Mars as it does on Venus—but the far less concentrated greenhouse gas is less able to retain much heat. The uppermost layer, the thermosphere, has a temperature controlled by the intensity of ionizing radiation from the sun, but also by the still present, but even more dilute, carbon dioxide and carbon monoxide gas that still functions as a greenhouse gas.

Despite a Venusian temperature profile, surface temperatures excepted, and composition, Mars shares more climatic and meteorological phenomena with the Earth than it does with Venus. The more rapid Martian rotation and low temperature favors a terrestrial-like circulation driven by differential heating across the surface and by the Coriolis Effect (Chap. 1). But there are important differences. For one, the smaller radius of Mars (roughly half that of the Earth) means that the Coriolis Effect is weaker. Why? Well, imagine you are moving at 100 km per hour, northwards from the Martian equator. Mars rotates at about 300 km per hour around its polar axis. So, for every 100 km the air moves north, it also moves 300 km to the east. On Earth, the rotation speed is 1700 km per hour at the equator. Here, for every 100 km the air moves north, it moves 1700 km to the east. Thus air is going to be deflected by the rotation of our planet far more than it will be on Mars. Thus, although both the Earth and Mars rotate on their axis every 24 h or so, the Coriolis Effect is far stronger on the Earth than it is on Mars. Yes, air is still deflected in the same way as it is on Earth, but the effect is far weaker, so air moving northwards or southwards on Mars can travel far further (about five times further) towards the pole before being swung eastwards. Likewise, air moving towards the equator is deflected less strongly towards the west on Mars than it is on the Earth.

The low density and mass of the Martian atmosphere sets the overall theme for its weather and climate. Like a worn out sponge with many large holes, the Martian atmosphere allows most of the energy it receives from the Sun to pour back out into space. It has what atmospheric scientists call a low thermal inertia; and this creates some odd effects.

For one, although the Martian atmosphere can get rather warm at the surface these temperatures are restricted to the air closest to the surface. Ascend a few hundred meters and the temperatures will have fallen close to or below the freezing point of water. This strong gradient in temperature means that Martian air is very unstable and convects air rapidly. Thus strong heating near the surface can lead to some dramatic winds.

Moreover, when the Sun goes down Martian air loses heat very rapidly at night. This, in turn means that there are dramatic differences in temperature between the day and night hemispheres. With temperatures low enough for gases to condense out of the air, there can be rapid changes in mass of gas that is in the atmosphere. Rapid changes in temperature and in the mass of gas in the air result in large pressure differences between the day and night hemispheres—and between the summer and winter hemispheres (Figs. 6.2 and 6.3). These phenomena are largely unknown on the Earth: any effect thermal tides have is overwhelmed by the effect of our planet's rotation.

Mars thus, displays a very complex pattern of circulation driven by competing effects. The Coriolis Effect tries to organize the winds in a manner like the Earth (Figs. 6.4, 6.5 and 6.6), while the low density of gas results in flows from warm to cold (Figs. 6.2 and 6.3). The resulting pattern shows overlapping effects which work together in different ways in different seasons.

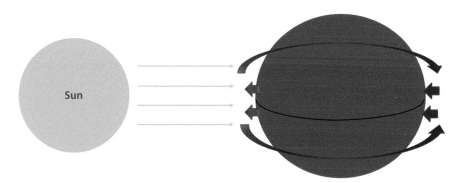

Fɪɢ. 6.2 Thermal tide flow in the Martian atmosphere. Thermal tides are airflows driven by strong differences in heating and cooling of air in the atmospheres of planets. Although Mars rotates at a similar rate to the Earth and has a modest Coriolis Effect, the atmosphere is so thin that radiation can easily escape. This leads to strong heating at the surface of the planet in the day and very strong cooling at night. This leads to a general flow of air from the day side towards the night-side of the planet: this is a thermal tide. The same process happens in the atmosphere of Venus but this is driven more by much stronger heating and a much slower rotation. As Mars rotates the area afflicted by cooling and heating migrates across the surface, interacting with surface features such as the Tharsis bulge, as well as with the Coriolis, and other, effects

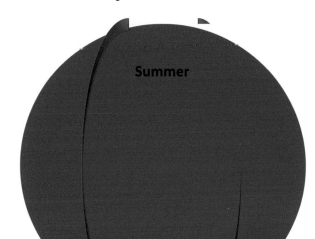

FIG 6.3 Bulk seasonal flow on Mars. Despite a modest Coriolis Effect and thermal tides driving air flow on Mars, there is an underlying seasonal flow of gas within the Martian atmosphere. As both carbon dioxide and water are at or below their condensation and re-sublimation points. This means that in the summer hemisphere both of these gases migrate from frost to the atmosphere, raising the overall atmospheric pressure. However, this freezes out once more in the winter hemisphere. This gives Mars a bulk, cyclical motion of gas from one hemisphere to another over the course of the Martian year. This is exacerbated by the Martian orbit which takes Mars more than 42 million kilometers further from the Sun during the Martian southern summer than it does during the Northern summer. Thus southern summers are warmer and northern winters colder than the corresponding season in the opposite hemisphere

For example, during the southern summer—which is also, currently when Mars is closest to the Sun—strong heating along the flanks of the southern highlands and Tharsis Bulge lead to strong updrafts. These updrafts draw in air from the northern hemisphere, which pours southwards, much like the strong Asian monsoon flow on Earth (Chap. 2). Bereft of oceans to add moisture these winds don't bring rain to Tharsis, but they can draw sand southwestwards towards the equator, then south eastwards towards the southern highlands. As Mars is a good 42 million kilometers from the Sun during the northern summer than it is in

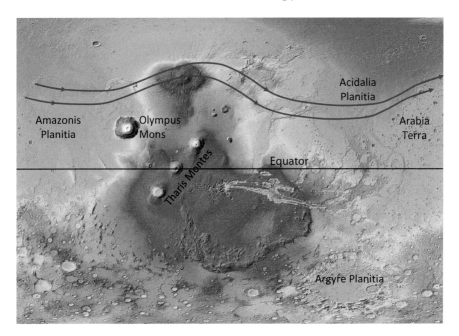

FIG. 6.4 Topography-driven Rossby waves within the northern winter hemisphere of Mars. Air is forced around and over the Tharsis Bulge and associated mountains. Just like the Rockies on Earth, winds dip south-wards over the Tharsis region and then northwards again to the east of the rise. Such topography driven waves are limited to 1–3 over the entire planet, while on Earth there may be up to six (Chap. 1). Such waves help to organize seasonal storms, including planet-wide dust storms that afflict northern Mars in this hemisphere's spring. Ground topography image: NASA

the southern summer, the northern Martian summer is therefore cooler and longer than that in the south. Consequently, the north-ern summer monsoon is much weaker than the southern summer one. This discrepancy leads to Martian sand dunes aligning in the direction of the prevailing winds of the southern summer.

The Martian atmosphere mimics that of the Earth in other respects. As well as organizing monsoon flows like those on Earth, the Martian atmosphere hosts large circulations like the Earth. Air rises strongly over the regions under the strongest illumination in the summer, moves Pole-wards at height, before descending over the Martian sub-tropics. These Hadley cells are directly analogous to those on Earth. North or south of these, at least in the winter and spring hemisphere, a second cell analogous to the terrestrial

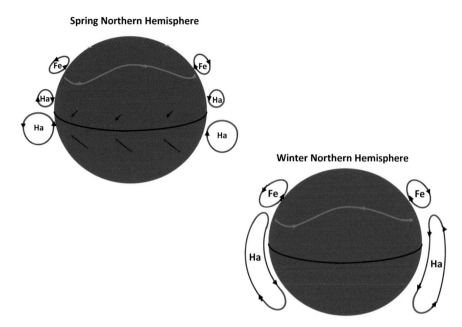

FIG. 6.5 The general, seasonal circulation on Mars. In the Spring, with the Sun over the Equator there are two Hadley Cells over the Martian Tropics. The Northern cell is weakest. Topography helps enhance the strength of the southern cell. Interactions between the westerly winds to the north; thermal tides (Fig. 6.2); topography (Fig. 6.6) and the reversing tropical Hadley cells can lead to the formation of planet wide dust storms that begin life as frontal-like disturbances within the Ferrel cells. In this season an easterly jet overlies the division between the Hadley cells as it does on Earth. The Martian atmosphere has a radically different structure during the northern and southern winter seasons. During the northern winter a single, large Hadley Cell (Ha) carries air from the warm southern hemisphere to the north while the northern Ferrel Cell (Fe) strengthens. Carbon dioxide is carried by this flow and precipitates out as frost on the northern ice cap along with much of the atmospheric water vapor. A westerly jet stream overlies the Ferrel Cell-Hadley Cell boundary much like the sub-tropical jet does on the Earth. This pattern reverses in the northern summer, with the Hadley cell carrying air southwards at height. As on Earth the Coriolis Effect bends the winds eastwards away from the equator and westwards towards the equator (Fig. 6.6)

Ferrel Cell organizes a belt of westerly winds. Above the axis of the twin Hadley cells lies an easterly jet stream; while above the junction of the Hadley and Ferrel cell lies a westerly jet stream, analogous to the sub-tropical jet stream on Earth (Chap. 1).

Although this is superficially like the terrestrial situation, Mars is constrained by its weaker Coriolis Effect and thin air,

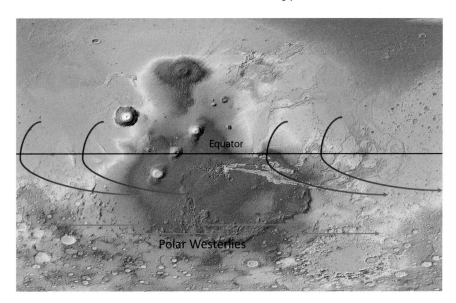

FIG. 6.6 Monsoon winds on Mars. As with the Earth, Mars has a seasonal monsoon (Chap. 2). However, unlike the Earth there are obviously no oceans, so the temperature gradient is entirely due to the position of the Sun. Like the Earth, Mars is closest to the Sun when the planet has its southern summer (*shown above*) and furthest from the Sun, currently during the northern summer. However, on Mars the effect is far greater as the Martian orbit is far more elliptical than ours. Consequently, the strongest monsoon winds blow to the south, an effect exacerbated by the large Tharsis bulge, which act much like the Himalayas and Tibet (Chap. 2). On Mars the Monsoon leaves its mark on the Martian surface in the form of sand dunes that align with the southern summer monsoon wind

which can't retain much heat. Thus, on Mars, air flows symmetrically around the equator in the spring when the Sun lies above the Martian equator. During the summer in the northern or southern hemisphere air flows through one cell to the winter hemisphere, leaving a single giant Hadley cell. The summer hemisphere also lacks a Ferrel cell, leaving Mars with a stripped down version of the terrestrial circulation system.

The transition from winter to spring has important consequences for the Martian circulation. Not only does a rising Sun begin to vaporize the ice cap and raise atmospheric pressure in the spring hemisphere, but it rapidly warms the ground increasing the instability of the air. The axis of the Hadley cell moves with the area of strongest heating and the Ferrel cell is overwhelmed.

In Chap. 10 we will see how this pattern of airflow might happen in an exaggerated form on an Earth like planet that was tipped over on its side. However, this never really happens on the Earth, primarily because the Coriolis Effect causes a far stronger deviation in air flow. However, we do see seasonal differences that are somewhat reminiscent of the situation on Mars. In the northern summer, in particular, as the land warms, the belt of westerlies becomes unstable and partly breaks up. The Asian summer monsoon only really kicks in when the westerlies running along the front of the Himalayas break down (Chap. 2). Over the northern oceans the jet stream buckles in response to the difference in temperature over the land and sea. During the northern summer the normally strong and broadly circular circulation around the Arctic also tends to break up into smaller cells. Although there is still a Ferrel cell per se, it is grossly weakened and is very erratic in its circulation pattern.

Dust Devils, Tornadoes and the Other Whirling Winds of Earth and Mars

On Mars the large differences in surface temperature from day to night and season to season generate large differences in atmospheric pressure. This leads to strong movement of air across the surface of Mars leading to some interesting weather.

At a local level, strong heating and limited retention of heat by the thin air leads to strong convection. On the Earth, such strong, localized convection combines with the effects of friction and airflow around obstacles to produce small, whirling columns of air. Although the Coriolis Effect has a role in these small disturbances, they are so limited in horizontal extent that this effect can be neglected. The terrestrial myth that the Coriolis Effect determines whether an emptying bathtub's swirl is clockwise or counter-clockwise springs to mind. Rather than the hemisphere being the determining factor in flow direction, minor differences in friction, water flow or the shape of the bathtub cause the water to preferentially spiral in one direction or another. With the right shaped bath and the plug withdrawn upwards, smoothly, the water will spin in either direction: friction does the work on its own.

The Coriolis Effect has no measurable effect unless you have a large tub and are very, very careful how you set up the experiment.

On Mars localized convection leads to spinning columns of air called dust devils, frequently seen on the deserts of the Earth. These spinning columns form from the ground up and grow as entrained debris are carried upwards by the spinning, ascending column of air. Rather more insidious and unpleasant versions of these dust devils form in terrestrial fires, where combustion rapidly heats the ground at the surface. Violent heating then leads to the formation of narrow columns of fire which twist in response to the movement of incoming air. Again, friction rather than the Coriolis Effect is at work. Such mini-twisters frequently assist in the spread of the fire across non-combustible surfaces, such as roads. Figure 6.7 shows such a mini-twister during an Australian bushfire.

Rather more violent than these miniature fire twisters are violent tornadoes that are spawned from thunderclouds, themselves generated by large bush fires with flame fronts several kilometers long. With a very large fuel supply, extensive fires can release so much heat and moisture that they form fully-sized cumulonimbus clouds. Such "pyro-cumulonimbus" clouds are fortunately rare, but these can generate hellish weather phenomena of their own. As well as lightning—which may start further fires, pyro-cumulonimbus have been known to produce black hail—soot encrusted and encased grape sized lumps of ice—and, as was mentioned, genuine tornadoes.

On the Earth, tornadoes and water spouts at sea are primarily associated with spinning cumulonimbus thunderclouds called supercells. Unlike dust devils, which plague Mars and the deserts of the Earth, convection plays an indirect role. Rather than growing tornadoes from the ground upwards, convection generates large cumulonimbus cells, and from these the funnels grow downwards. Growth of the funnel is a consequence of lowering atmospheric pressure that is associated with the development of the storm. As the condensation point of water decreases with pressure the decreasing pressure will cause moisture in the air to condense into a funnel-shaped cloud, the cloud densest where the pressure is lowest. As the storm intensifies and pressure continues to fall, the condensation level descends towards the ground.

FIG. 6.7 Meteorological effects caused by fires. In (**a**) a miniature fire devil joins the litany of catastrophe in a large bush fire in Australia. (**b**) shows a man-made fire devil at the Magna Centre in Sheffield. (**c**) (*above*) shows a pyro-cumulonimbus: a thundercloud formed above a large bush fire from the excessive heat and moisture released by combustion. In January 2003 this particular cloud hit Canberra, Australia. A short while after this photograph was taken the incendiary thundercloud generated a genuine tornado. As with conventional tornadoes this spinning vortex formed from the cloud down. Journalists, accompanying the Fire Service, filmed it flattening trees and buildings within the heart of the fire

Supercell storms don't spin because of the Coriolis Effect—and indeed, some storms have two cores spinning in the opposite directions. Spin is obtained when a horizontal swirl of air becomes entrained into the developing storm. In these systems winds moving at different speeds and at different altitudes, generates a spinning vortex of air. If this becomes drawn into the rising stream of air, within the storm, the spin becomes aligned vertically and the storm as a whole begins to spin. Where the spin becomes concentrated into a narrower spiral a tornado can result. Here, the collapsing column of spinning air conserves its momentum by spinning faster, just like an ice-skater. Supercells are common in many parts of the world, including Australia, central Europe, Russia, Argentina and southern Brazil. However, they occur most abundantly in the US, in the region known as tornado alley. In these regions there is the appropriate combination of meteorological conditions: a supply of warm moist air to fuel storm formation; a nearby source of warm dry air that caps convection for prolonged periods; and most importantly active cold fronts that can penetrate both air masses and displace their bulk rapidly upwards generating powerful storms. By virtue of these requirements, these phenomena only come together at certain times in certain locations.

Supercell thunderstorms have relatively long lifetimes as they can maintain a strong central updraft. Smaller thunderstorms tend to choke off their supply of warm moist air as soon as the rain begins to fall and cold air flushes downwards. This kills the storm within an hour of its formation. By contrast supercells can last for hours. In the case of the 2003 Canberra storm, the large bush fire generated the cumulonimbus cloud that morphed into a supercell storm. In turn, this supercell generated the tornado that repeatedly touched down in the suburbs of Canberra and causing further devastation. Just like a regular tornado, but unlike a dust devil or smaller fire devil, this vortex, colloquially called a "firenado", was formed from the cloud down. This firenado included fire in its arsenal of destruction, demolishing houses and incinerating much of their remains.

Returning to Mars, dust devils are most common in the Martian spring and summer when heating is strongest on the Martian surface. As with tornadoes on Earth, dust devils on Mars help redistribute the heat within the thin Martian atmosphere. They can combine in waves and deliver sufficient dust to the atmosphere that it blocks solar radiation from reaching the surface. This cools the surface and warms the atmosphere and, overall, makes the atmosphere more stable. In the process "gangs" of dust devils can generate larger Martian dust storms, although in general the largest of these—planet-wide affairs form when frontal disturbances form nearer the Martian poles in the spring.

Why is the spring the key season for dust storms? This is an interesting question that has only recently been solved. During the spring certain key atmospheric phenomena combine to generate ideal conditions for dust storm formation. Storms become global phenomena when frontal storms combine with the effects of Martian thermal tides to deliver extensive dust into the tropical Hadley Cells. Huiqun Wang (Harvard University) and colleagues combined observations of the 2001 planet-wide storm and computer simulations. Wang identified how different scale atmospheric processes collaborated in generating a storm that was to eventually blanket the entire planet (Fig. 6.8). Topography was key to their development. Dust storms develop and progress initially across the low elevation portions of the planet: Acidalia-Chryse, and some within Arcadia-Amazonis and Utopia. The arrangement of wide basins, flanked by regions of (broadly) north–south aligned topography helps generate Rossby waves (Chap. 1) within the northern atmosphere within the Ferrel cell. This is analogous to the effect the Rockies and Andes have on the westerly circulation on Earth. These topographic rises are critical to delivering energy and dust southwards.

In the southern spring and summer a number of different factors are coming together. The northern cap is cooling down and increasing in mass. Strong winds are moving mass (mostly dense carbon dioxide gas) into the cap and the dominant monsoon flow is flowing to the south of the equator. During the northern autumn, the Hadley cell is shrinking north of the equator and is eventually sub-served to a single cell extending from the south (Fig. 6.5). Thermal winds are still aligned southwards in the direction of the southern hemisphere since as this hemisphere warms, winds blow from the cold hemisphere towards the warm hemisphere at the surface, but away from it at height (Fig. 6.5).

FIG. 6.8 (continued) (b) On Earth most dust storms in the temperate regions and the sub-tropics begin their life like those on Mars. Cold fronts penetrate southwards during the autumn, winter and early spring. Here, over Africa and the Middle East (and similarly over the US in the spring or Australia during the winter) they encounter the dry continental interior. Abrupt changes in wind speed and direction can kick up enormous clouds of dust (brown arrows) ahead of the advancing cold front. Unlike the desiccated Martian storms, these cold fronts may have frontal thunderstorms associated with them and, in many cases, it is downdrafts from these thunderstorms that initiate the dust storms

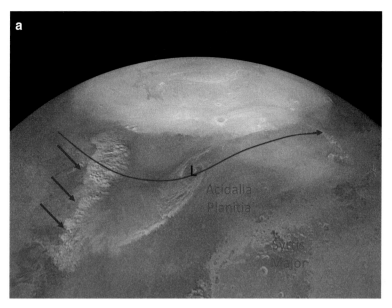

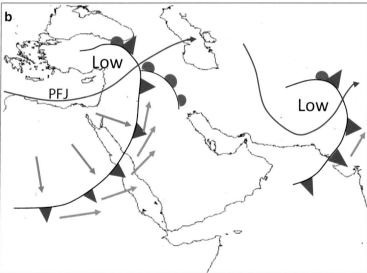

Fɪɢ. 6.8 (a) Developing dust storms on Mars in the Martian spring of 2002. A frontal storm (*blue arrow*) developing and moving south-eastwards along a dip in the Rossby wave (*purple*) towards the Martian equator. Although broadly analogous to a terrestrial cold front, the cloud along this boundary is dust, not water vapor. To the fronts rear a second frontal dust storm is developing (*dark blue arrows*). Where these frontal storms work together with the daily thermal tide and reach the tropical Hadley cell dust can be carried around the entire globe forming planet-wide storms. Underlying Mars photograph NASA/JPL (Mars Global Surveyor).

Periodically strengthening this bond are the Rossby waves that develop along the westerly jet stream that forms the divide between the Hadley cells and the westerly winds further north. As the north cools down, these waves become stronger as the temperature contrast grows between the northern polar cap and the rest of the planet. Like the Rocky Mountains on Earth, the Tharsis bulge promotes the development of waves that dive south over Acidalia Planitia (Fig. 6.4). Here, they drive the formation of low pressure areas and cold fronts, which migrate southeastwards across the flat plain towards Syrtis Major (Fig. 6.8). These fronts move rapidly: around 70–100 km per hour and are accompanied by smaller, regional storms. Such dust storms are frequent during the Martian spring and autumn (around 12 per season) but on the whole are limited in extent. Generation of a global storm requires these wave-generated dust storms to align with the daily tidal flow as the front is crossing one of the low elevation plains. When it does, dust is advected into the equatorial Hadley cell. "Successful" storms—those that generated planet-wide dust storms—were those that crossed 0° E between 9 am and 7 pm. This was when tidal winds across Acidalia-Chryse were also blowing to the south. At other times, the tidal flow was reversed and storm propagation was blocked.

In the northern winter, atmospheric waves are stronger and this, too, facilitated the progression of dust-bearing fronts to the equatorial regions. Interestingly, although the distance the cold fronts penetrate is a factor in getting dust near to the equator, this appears to have no direct link to the formation of global dust storms. Wang's group found the only factor that mattered was the speed of the front, which was solely linked to the tidal flow and the season. A deeply penetrating front that couldn't add its flow to the tidal flow wouldn't successfully deliver its load of dust to the tropical cells. Instead, most of the dust associated with these storms stays put, north of the equator. In the Mars Global Surveyor data the largest, global, storms were associated with a fortuitous movement of multiple frontal storms across Acidalia-Chryse. Such multiple introgressions give the atmosphere more chance align with the tidal flow and to deliver dust to the tropical circulation.

What happens next was demonstrated in research carried out in 1982 by Robert Haberle (currently at NASA Ames) and a decade later R. John Wilson (then at Princeton). Research indicated that there would be a dramatic change in the structure of the atmosphere. Dust absorbs heat energy far more efficiently than the thin carbon dioxide gas on its own. Able to hold and transport more heat the Hadley cell suddenly has a massive injection of energy. Just as extra carbon dioxide is slowly enhancing the strength and reach of the tropical Hadley cells on Earth, dust has the same but far stronger effect within the wispy air of Mars. Instead of extending from the southern mid-latitudes to just north of the equator, the southern Hadley cell will now extend all the way to the North Martian Pole, bulldozing the Ferrel cell out of the way. As soon as this happens dust fills most of the atmosphere. Surface temperatures drop, while the dust layer warms by 5–10°. Indeed, during the 2001 global storm the atmosphere warmed by a rather stunning 45 °C, causing it to puff upward, enough to threaten orbiting spacecraft.

Remember that on Mars, the atmosphere is so thin that it has what you might call a low buffering capacity: it cannot adjust smoothly to changes that occur within it. Were the Earth's atmosphere to be warmed regionally by dust, the circulation within its denser bulk would rapidly even out the differences in temperature and pressure. Were this the even denser, soup-like atmosphere of Venus, the atmosphere would hardly "notice" the difference in conditions.

There is an interesting link with terrestrial climate-skepticism, here. As Mars warned in the 2000s a small but significant number of skeptics suggested that this was evidence for the Sun driving our warming. After all, how could *terrestrial* carbon dioxide warm Mars simultaneously? Obviously, it couldn't: therefore, man-made global warming was a myth and the Sun must be warming both planets more. The problem, or rather problems, with that tale was that the Sun wasn't getting more luminous. Moreover, Venus, Jupiter and the other planets weren't getting warmer, either. So much for the Sun... Instead, Mars was simply getting warmer because of dust and changes to its albedo (reflectivity) at the same time as the Earth was getting warmer as a consequence of a greater burden of greenhouse gas. Unlike the Earth, Mars eventually cooled back down once the dust settled. Indeed, changes in Martian temperature, aside from seasonal differences, are largely caused by changes in surface albedo. The brighter Mars is, the colder it is and vice versa (Fig. 6.9).

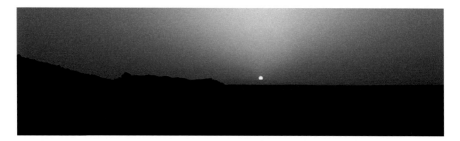

FIG. 6.9 Sunset on Mars, taken by the Late Spirit Rover at Gusev crater in 2005. It shows a couple of unique phenomena caused by atmospheric dust. The Sun is *pink* while the region immediately above the setting Sun glows *blue*. Both phenomena are caused by the scattering of blue light. Ubiquitous Martian dust scatters the blue light forward, with the greatest scattering at lower altitudes where the concentration of dust is highest. This creates a focus of blue light above the setting Sun. Scattering leaves red light projected directly from the Sun towards the Spirit rover's cameras. This combined with the color of dust in the air leaves a pinkish glow

Indeed, Lori Fenton (Carl Sagan Center), Paul Geissler (USGS) and Robert Haberle (NASA Ames) have examined more cyclical changes in Martian surface temperatures. Like the more drastic changes induced by global dust storms, seasonal changes in wind direction can enhance surface winds. Where the ground is brightest (has the highest albedo) the winds are lightest and where the ground is dark the winds strengthen. Wind scouring can remove lighter upper dust, exposing deeper, darker soils, which, in turn causes the winds to blow harder. These changes are also associated with temperature rises of up to 0.65 °C over the darker areas. Rises of this scale increase the formation of dust devils simply because the amount of convection increases. More dust devils mean that there is more transport of dust and heat into the atmosphere, which, in turn, increases the temperature of the lower atmosphere.

Aurora

Auroras are an inevitable consequence of having an atmosphere on a planet around a star. Therefore, all of the planets, bar Mercury and probably Pluto (if we can include this) host auroras. These luminous displays occur when charged particles from the Sun crash into the atmospheres of the planets. In the case of the Earth,

Jupiter, Saturn, Uranus and Neptune, such particles are funneled onto the planet's Polar Regions by the action of strong, internal magnetic fields. This usually isn't a direct dive from solar wind to atmosphere. Rather, magnetic fields first escort the particles into a large magnetized domain that surrounds the planet, called the magnetosphere. From here changes in the direction and strength of these field lines cause particles to be dumped periodically into the atmosphere of the planet, nearest the magnetic poles. Such "reconnection" events are fast and follow magnetic field lines from the Sun linking and re-working the planet's field.

Consequently, with the exception of Venus and Mars auroras are confined to the polar regions and it is only when the Sun ejects a particularly energetic burst that aurora can spread further afield cross the disc of the planet (Fig. 6.10).

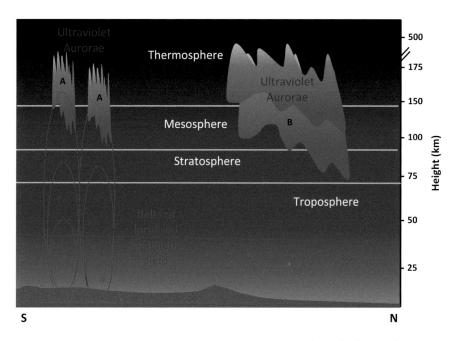

FIG. 6.10 Auroras on Mars. In 2004 Mars Express found ultraviolet auroras associated with localized regions of more intense magnetic fields in the planet's southern hemisphere (**A**). Such auroras are unique in the solar system. In December 2014 MAVEN found more widespread "Christmas Light" auroras over much of the northern hemisphere (**B**). Although spotted in the UV part of the spectrum these auroras may be visible as a diffuse, green glow in the night sky, as well. Moreover, Mars hosts a diffuse airglow, much like Venus, which is also found in the ultraviolet part of the spectrum on the night-time hemisphere of the planet

Venus and Mars are different—lacking an appreciable magnetic field of their own. The average present magnetic field at the surface of Mars is less than a few tens of nanoteslas, or about 1/1000th that of the Earth (25,000–65,000 nanoteslas or 0.25–0.65 Gauss). For comparison, a decent fridge magnet has a field that is about 10 million nanoteslas. Consequently, the solar wind is able to interact with the full sunlit disc of the planet—well almost. Above Venus the solar wind energizes the carbon dioxide-rich atmosphere on the sunlit hemisphere and breaks apart some of the atmosphere's constituents. Atomic oxygen and ionized oxygen are directed to the night-side of the planet by thermal winds flowing above the cloud tops. The movement of these ions is analogous to an electrical current and results in the formation of an induced magnetic field (Chap. 5). Once in darkness, the oxygen can recombine with electrons or with carbon monoxide and emit light. Although much of the energy released is in the ultraviolet part of the spectrum, a faint greenish glow might also be apparent when oxygen ions recombine with electrons.

A similar phenomenon occurs on Mars, but at much lower altitudes simply because the wishy-washy Martian atmosphere allows far deeper penetration of the solar wind than happens on Venus. This is due to a more limited braking of incoming particles by the lower density of the gas, but also because the thinner gas is less able to develop its own internal and repulsive magnetic field in response to the solar wind slamming into it.

In some ancient regions of the southern Martian crust there exist broad, linear strips of magnetized crust with a field strength up to 1500 nanotesla (Fig. 6.11). Their origin predates the large Hellas impact basin and likely formed within the first 400 million years of the planet's formation. Only, here, where magnetic fields generated within the crust is sufficiently strong, is the Martian atmosphere afforded a limited degree of protection. However, it is here that magnetic fields funnel charged particles into particular regions of the atmosphere. Thus, the protection that is provided is certainly limited and the net result of these regional fields is that more of the atmosphere is energized than might otherwise be the case were Mars surrounded by a smooth bow-shock like Venus.

In 2004 the ultraviolet spectrometer, onboard Mars Express, discovered ultraviolet emission from localized regions above

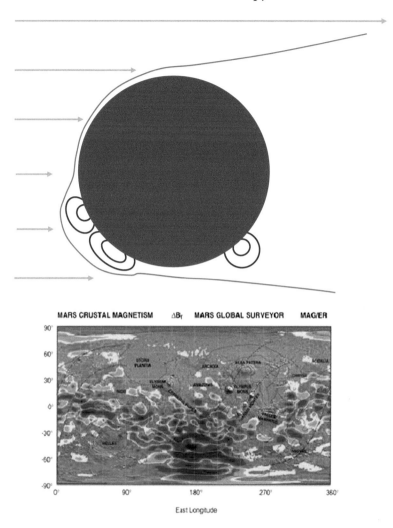

Fig. 6.11 The effect of localized magnetic fields around Mars. Unlike the Earth, Mars has localized fields which partly encase regions of the planet's surface (*purple ovals*). Extending out into space, these fields distort the bubble of space around the planet located within the bow shock (*blue line*). This not only protects parts of the Martian surface from the impact of solar wind particles (*yellow arrows*) and extra-solar cosmic rays, but also directs the solar wind into other area causing aurora and influencing the stripping of gases from the Martian atmosphere. The pattern of magnetic stripes is shown in the graphic from Mars Global Surveyor, *lower* (NASA)

Mars. This was clearly distinct from the diffuse emission as it was localized to one geographical region Jean-Loup Bertaux (Service d'Ae´ronomie du CNRS/IPSL) and colleagues had identified the first true aurora on Mars. Through the channeling of charged particles from the Sun, the localized magnetic fields on Mars were producing aurora over the southern hemisphere of the planet. In essence this had been expected, yet their discovery was important, for it still remains the only example of regional aurora above any planet, where the crust is generating the magnetic field that guides the particle flow.

Eleven years later there was a slightly more surprising discovery by MAVEN. Again, observing in the ultraviolet portion of the spectrum, MAVEN discovered very widespread aurora over the northern hemisphere of Mars. Like the southern aurora these occur at significant depth within the Martian atmosphere. Whereas, terrestrial auroras occur exclusively above 100 km, the Martian auroras manifest more deeply; down to 60–80 km from the Martian surface. Again, this is a consequence of the dilute Martian air, which affords limited protection from energetic solar and cosmic particles.

Mystery Clouds Across the Southern, Morning Terminator

From 1995 through to 2014 there were a number of amateur astronomer (archived), Hubble observations and later MAVEN observations of odd clouds seen near the Martian terminator. All of these were seen above the same geographical region, 195° West and 45° South, near Terra Cimmeria. The problem with these clouds is that they appeared to lie at great height within the Martian atmosphere: up to 250 km above the surface. This is almost twice as high as many auroras seen across the portion of the Martian globe (above) and clearly places them in the Martian ionosphere and exosphere—effectively out into interplanetary space. Each cloud that was seen—and the best observed occurred in March and April 2012 was several hundred kilometers across and lasted for about 10 days. More perplexing the clouds were only seen at the morning terminator but had vanished by the time Mars had rotated bringing the elusive clouds around to the evening side. Inspection of images (Fig. 6.12) of Mars does appear to imply that these clouds

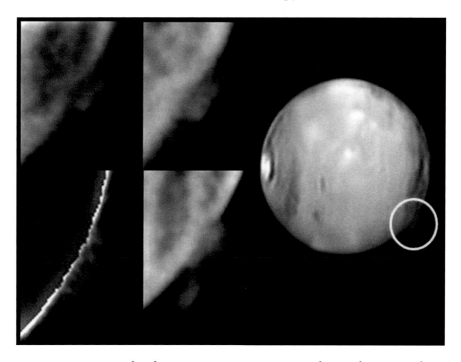

FIG. **6.12** Mystery clouds on Mars. Appearing at only one location, these clouds of what appear to be dust extend across 500–1000 km of Martian terrain and only appear in the morning. They occur at extraordinary heights, greater than aurora. Possible explanations include some sort of unknown and very elevated aurora and dust clouds that have somehow reached the edge of space. Images: Donald Parker (Association of Lunar and Planetary Observers, Pennsylvania)

do extend some way across the day side. This is evident from the parity in the colors of features seen at that latitude on the Martian surface and in the clouds. However, they appear to dissipate during the Martian day. So, what are these mysterious, seasonal and periodic clouds and how do they get so high in the atmosphere?

There are a few possibilities based on their location above the Martian surface. Firstly, they occur where magnetic fields are strongest, so they might be auroras. However, the color is wrong (they look dusty rather than blue–green or red) and they only appear in the early morning sky. Moreover, they would have to over 1000 times brighter than terrestrial aurora to be visible from Earth. Alternatively, they might be dust—the color is right—but how did dust get so high up: wind strength is going to be negligible as the air density is miniscule.

Martian dust is likely ruled out, but another possibility is dust from the small Martian moon, Phobos. The authors of the 2015 *Nature* paper, Agustin Sánchez-Lavega and colleagues suggest that the clouds, if that's what they are, would be best modeled as carbon dioxide or water ice. Getting the ice to this height—100 km and above the surface—is problematic, but not impossible; but why is it only visible across the morning terminator?

The suggestion is that these odd clouds were created in the Martian mesosphere where temperatures can be lower tan –150 °C at night. Here trace amounts of water vapor and/or carbon dioxide might freeze out forming tiny particles (around 100 nanometers (100 billionths of a meter) across that eventually assemble into clouds. The location, nearest where Mars hosts its strongest magnetic fields might be an important factor. The particles will need to be levitated to higher altitudes (and warmer altitudes) where they can be seen above the terminator. As is likely, air density is too low for wind to be a factor. However, particles at the heights the clouds are seen will become charged through the action of charged particles from the Sun and through the action of cosmic rays. Once charged, magnetic fields in the southern hemisphere of Mars could transport them to great heights. Enhanced and magnified images imply some form of filamentary structure to the clouds. This would favor the action of magnetic fields in structuring the gas. However, at present there is no real evidence to support any of the possible mechanisms that levitate clouds to such great altitudes.

If the clouds are ice we can begin to understand why the clouds vanish during the Martian day. As the planet rotates and the clouds are swept into the full glare of the Sun, they will soon sublimate once more and disappear from view. Presumably, much of the material that makes up these clouds will become lost to space as at 250 km they are well within the grasp of the solar wind.[1]

MAVEN has also detected dust at high altitudes. There appears to be rather a lot about the depleted atmosphere of Mars that we do not yet understand.

---

[1] David Andrews of the Swedish Institute of Space Physics suggests that these clouds are levitated by coronal mass ejections. These impact the ionosphere and generate enough ionization to lift dust or ice grains from further down.

Martian Ozone

Like Venus, Mars has ozone within its atmosphere; and like Mars this is produced not by the reaction of abundant oxygen with ultraviolet light, but indirectly from oxygen released when carbon dioxide is split by ultraviolet radiation. On Venus, ozone is largely confined to the night side of the planet at an altitude of about 100 km (Chap. 5), but on Mars a much more complex pattern of ozone production and destruction is found. This difference, once again, relates to the tenuous nature of the Martian atmosphere.

With a low atmospheric density, ultraviolet radiation penetrates deeply and ozone can be produced low down. It is circulation within the atmosphere ultimately determines where the greatest concentrations of this will be found. Nowhere does the concentration of ozone come anywhere close to that in our atmosphere. At its densest it is 300 times less dense than found in our ozone layer and thus similar to that above Venus. This is enough to somewhat limit the ultraviolet radiation that reaches the Martian surface. The greatest concentration of ozone occurs near the South Pole in the winter. Mars has a complex distribution of this gas with height. Not far above the Martian surface is a persistent layer that covers most of the Martian surface. Above this, in the northern spring and summer, lies a second ozone layer in 30–60 km up. Over the southern pole lies the third and most plentiful layer at an altitude of 40–60 km. This third layer never forms near the northern Pole during its winter and more unusually, decreases in altitude with decreasing latitude, being lowest nearest the Martian South Pole.

This pattern of ozone creation and destruction relates to changes in the circulation of the Martian lower atmosphere. Remember, that during the southern summer a large circulatory cell expands until it delivers air from the Martian equator almost the entire way across the southern hemisphere (Fig. 6.5). This sweeps up any ozone and not only delivers it more generally to the rest of the planet, but brings it into contact with hydrogen gas that has been released from the breakdown of water vapor by ultraviolet light. Ozone is then destroyed. During the Martian winter, the developing Ferrel Cell isolates the polar atmosphere from the rest of the planet. It also cools so much that any moisture that is present freezes out as frost. This leaves any free oxygen the chance to reform more ozone. This brings us a gentle reminder about developing an extensive hydrogen economy on Earth. Hydrogen is notoriously difficult to store safely in large quantities. It is highly likely that in our attempts to use hydrogen we will inadvertently release yet another agent that will deplete our ozone.

Martian Methane: Mars not Dead, Yet

Methane is a contentious gas. On Earth, as a greenhouse gas it is primarily produced by anaerobic bacteria in paddy fields or the large colons of mammals, and as such is quite literally the "butt" of jokes. Methane is also stored in large quantities in three terrestrial locations: prehistoric gas fields dating back several tens or hundreds of millions of years; historic to prehistoric methane in permafrost; and finally as frozen, water-caged complexes known as clathrates on the deep ocean floor. In every one of these cases methane has a biological origin, having been produced by anaerobic fermentation. Methane may also be produced through chemical reactions between the mineral olivine, water and carbon dioxide. During the geological process known as serpentinization, olivine takes up water and in the process hydrogen is released that can chemically reduced carbon dioxide to leave methane.

In the coldest reaches of the universe, hydrogen ions that are produced by ultraviolet irradiation of deep, dark nebulae can also initiate a bewildering array of chemical reactions. Amongst the myriad of products is methane in abundance. Comets and meteorites can deliver this to the terrestrial planets, but it is thought that they were probably not born with much of this volatile gas. On a planet like Mars conditions on its surface are strongly oxidizing: that is methane effectively wouldn't stand a chance of surviving long—perhaps a few years at most. When scientists at the Goddard Flight Center discovered methane in the Martian atmosphere in 2003 it caused something of a stir. What was all the more interesting was that the methane seemed to come and go in a seasonal pattern. Enter the Curiosity rover.

During the first year of the rover's operations it found little evidence of methane in the Martian atmosphere. Stringent upper limits were placed on the gas of less than 1 part in 5 billion. For comparison oxygen comprises 0.1 % of the Martian air which is

1 part in a thousand, or a million times denser. That's where the situation remained for the best part of a year, thus flying in the face of the terrestrial measurements. In the winter of 2013–2014 NASA reported that Curiosity had discovered a tenfold increase in atmospheric methane. While not quite a surge, the level had gone up so much that it implied some sort of source on Mars. Moreover, the rapid changes in the concentration clearly implied some sort of local source for the gas, which is then rapidly destroyed. Immediately ruled out were photochemical reactions between water and carbon dioxide in the Martian air.

Meteorite impacts or atmospheric detonations might do the trick but researchers at Imperial College in London ruled out this as their calculations suggested such an effect could not explain the observed variability. At present, researchers are left with two possibilities: Martian microbes (presumably underground) or some form of subterranean volcanic or hydrothermal activity that is producing methane. Sadly, in each case there is no evidence to support either. Subterranean microbes are not impossible—and there are clearly many examples on the Earth, as deep as 3 km underground: we simply can't tell if this is true. The chemical reactions they employ to produce methane are very similar to that involving olivine. Given that serpentinization would be limited to deeper, warmer terrains, discriminating between the two processes will be challenging.

Worse still is our inability to identify the process or processes that is rapidly removing the vented methane. Methane only peaks for a few weeks on Mars implying a very rapid process of elimination. What this is also remains in the realm of speculation and could involve chemical reactions or sequestration, where clathrates or other compounds absorb but do not necessarily destroy the methane.

Experiments done by Viking in 1976 offer a clue—as do measurements made by the Phoenix Lander in 2008. Viking went looking for life, essentially by offering up lunch to any waiting microbe. Glucose was mixed with Martian soil and a rapid chemical reaction ensued. Carbon dioxide and water vapor were released as though the glucose had been respired. However, on repeated exposure nothing else happened. It appeared that the observed reaction was a one of chemical reaction involving materials in the Martian soil, meaning it looks like highly oxidized iron compounds or hydrogen peroxide may be the source of the chemical reactions that ate Viking's meal. The discovery of perchlorate compounds by the Phoenix Lander in 2008 might offer an alternative explanation for the observations. These compounds make very handy components of "caps"—small explosive devices used by children throughout the western world to annoy their parents and bystanders. These can explode on impact with solid surfaces, particularly when mixed with sugar. In the chemical reaction, the perchlorate oxidizes the sugar, surrendering its oxygen atoms to power the reaction. On Mars similar reactions would readily do away with methane and any organism unwitting enough to be near them when it was venting the gas.

Perchlorates are something of a double-edged sword on Mars. Aside from aggressively oxidizing organic compounds, perchlorates—and this is most likely calcium perchlorate on Mars—will help stabilize liquid water. Under standard Martian conditions water is either solid ice (as frost or buried ice) or a gas. Exposure of solid ice results in sublimation—the direct conversion of solid to gaseous water. However, add some perchlorate—and the observations suggest 0.6 % by mass—and water is stable in its liquid form. Now, although such water would be a tad aggressive to terrestrial microbes, it doesn't necessarily mean it would be to Martian microbes.[2] With some irony, perchlorate might make life's existence rather unsavory but might also allow the presence of liquid water that allows it to develop in the first place. If an organism could eke out a living, perchlorate would make a very handy alternative to oxygen gas. Still not convinced, look at life on Earth now. Life on our little blue world's surface is bathed in one of the universes most noxious and reactive gases. Oxygen is a brutal gas to organic compounds; just ask any anaerobic bacterium. Yet life on Earth thrives. A dash of perchlorate and a whiff of peroxide might just be the tonic to perk up a long-frozen Martian microbe. Never doubt the resilience of life to withstand and adapt to harsh conditions.

---

[2] In September 2015 NASA confirmed that liquid water does periodically flow on the surface of Mars and that this is likely stabilized against evaporation by dissolved perchlorates.

Regardless of the source the presence of methane in the Martian atmosphere is almost certainly the most interesting discovery of the last 15 years—even dwarfing the observation of mystery dust clouds. If methane has a biological source or a geological one, it implies Mars is not quite dead yet. The biological significance is obvious, but the geological significance is certainly on a par: the Martian giants, the volcanoes of Tharsis and beyond, might still hold a lit candle to the steadily ticking clock of Martian senescence.

The Lingering Death of Mars

The Martian atmosphere of today is a pale reflection of the one it once had. Mars once held a deeper, denser atmosphere, the means by loss of which remains controversial. Early on Mars likely had an atmosphere similar to the early Earth and Venus. After all, each of these worlds was born from roughly the same materials. Thus we would expect Mars to have had a fairly hot, wet atmosphere for at least the first 50–100 million years, or so. This might well have been similar enough to the Earth's for it to have hosted hot, pressure-cooked oceans much like the Earth and probably Venus. As Mars is roughly 50 % further from the Sun than the Earth, this early wet super-greenhouse would have rapidly broken down with any moisture and much of the carbon dioxide raining out to be sequestered in Mars' early oceans. All of that is fairly uncontroversial. It's what happens next that a mystery.

Measurements by Curiosity reveal that most of the Martian atmosphere has been lost to space: what is left on Mars is a pale and tattered imitation of the planet's former blanket. At some point between 4.4 and 1 billion years ago Mars dried out. The generally accepted but poorly constrained thesis is that the solar wind stripped the Martian atmosphere away leaving the currently desiccated world. The problem is how to do this.

There are several different ways in which gases can escape the atmosphere of a planet. Some are straightforward and relate to the temperature of the gases: such mechanisms are called thermal mechanisms and they are driven by the temperature of the gas and the gravitational pull of the planet as a whole. Other mechanisms involve more complex reactions or physical extractions by the solar wind. These rather more robust pathways are called non-thermal mechanisms and include: stripping by the solar wind; chemical reactions within the gases in the atmosphere; extraction by impacting comets or asteroids; or electrical levitation.

The manner in which mechanisms have the greatest effects depends strongly on the mass of the planet; the mass and temperature of the gases in the atmosphere and the strength of the solar and more broadly stellar winds that impact on it, as well as the chemical composition of the gases in the atmosphere. For Venus the relatively large gravitational pull of the planet, combined with the relatively massive principle gas, carbon dioxide, means that relatively little gas can escape through simple Jean's escape. However, Mars is a different matter altogether. Two papers in the journal *Science* summarized Curiosity's findings. In the first Paul Mahaffy (NASA Goddard) and colleagues described the skewed ration of carbon-12 to carbon-13.

As with oxygen in the water in the Earth's atmosphere, there is a fixed abundance of carbon isotopes in the universe that is set by nuclear reactions in stars and the processes that give rise to the formation and isolation of compounds. Carbon-12 is far more abundant than carbon-13, but because it is lighter, compounds, such as carbon dioxide, which are made from it are easier to lift into space than those containing the heavier isotope. On Mars there is an enrichment of about 5 % carbon-13 relative to the amount on Earth. This suggests that Mars has lost significant amounts of carbon (most likely as carbon monoxide) through a process called sputtering. Here, cosmic rays impact the top of the atmosphere and energize carbon dioxide. This splits the carbon dioxide and in most cases the oxygen atom (or ion) escapes because it is lighter than carbon monoxide. However, in the right circumstances carbon will escape as carbon monoxide ions leaving an uncharged oxygen atom behind.

In the second paper by Mahaffy's colleague, Chris Webster (Jet Propulsion Laboratory) looked at oxygen isotopes. Like carbon, oxygen was strongly enriched in the heavier isotope, again suggesting that Mars had lost a significant amount of oxygen. Comparison of Curiosity's results with those obtained from examining Martian meteorites suggested a rather violent scenario. Pieces of Mars often end up on Earth as meteorites. These will be delivered when large impacts blast rocks into space. The most famous Martian visitor is ALH84001, the meteorite reputed to possibly show signs of Martian life. This rock dates to four billion years ago. Other iron-rich basaltic rocks are called Shergottites and are often younger, dating to a few hundred million years old. Comparisons of ALH84001 and these younger Shergottites implied that Mars had effectively attained its present atmosphere as early as four billion years ago. Curiosity's measurements broadly agree with this but allow for an additional slower loss in the intervening years.

It looks like Mars took on its present aspect a long time ago. How might this have happened? Consider a few factors: firstly Mars has a low mass so it's relatively easy for Mars to shed gas. Secondly, a low mass means that Mars would have cooled down faster and subsequently lost its early volcanism that would otherwise replenish gas. That is not to say the Tharsis volcanoes are utterly dead; but they certainly can't and aren't supplying the atmosphere with gas at a detectable level now. Sulfur dioxide gas—a key gas from volcanoes—is effectively undetectable at Mars, now. Thirdly, Mars receives less radiation from the Sun because it is further away from it than the Earth or Venus.

Picture the scene. Volcanic activity on Mars is waning and becoming episodic. Although there are oceans, Mars shows no evidence for plate tectonics. Those planet-wide stripes are certainly reminiscent of the magnetic striping pattern on the ocean floor but they are not the same and occur in crust far too thick to subduct. As far as we can tell, without subduction there are no plate tectonics. Without plate tectonics carbon dioxide drains out through sequestration into the oceans and ultimately into the Martian crust as carbonate rocks. Simultaneously, without extensive volcanic activity there is insufficient gas being returned to the atmosphere. Like Titan (Chap. 9), with a low gravity the Martian atmosphere is relatively puffy and accessible to the solar wind. Once the geodynamo on Mars fails—apparently within the first few hundred million years, water vapor is attacked by ultraviolet light and cosmic rays and is lost to space.

Once sufficient water vapor is split, and once most of the atmospheric carbon dioxide has either reacted with rocks to form carbonates, or been split high in the atmosphere to liberate oxygen and ultimately carbon monoxide, the temperature of Mars falls.

There is little or no cycling of gases from the crust into the mantle and back again so once these have gone into the crust there is no coming back. As temperatures continue to decline what water vapor that remains first rains then snows out on the Martian surface. By four billion years ago, despite ongoing volcanic activity, the Martian atmosphere has largely been lost. There will be waves where the density of gases periodically increases. These will coincide with larger bursts of volcanism. On Mars once the crust stabilized this was confined to the Tharsis bulge, where an underlying hotspot appears to have fueled volcanic activity in bursts separated by calm intervals lasting 200 million years. These volcanic outbursts could have raised the density of gas high enough to allow liquid water to flow. Although there is good (but still controversial) evidence that much of the Martian northern hemisphere was initially flooded by an ocean, most of the Martian record implies only periodic inundations. These could have been caused by volcanic activity melting glacial deposits, or indirectly thorough volcanic out-gassing raising the pressure and temperature of the Martian atmosphere to the point where surface ice melted. This still remains unclear.

With all of the evidence in hand, what is (or are) the likeliest explanations for the loss of the Martian atmosphere?

Jean's Escape

The simplest to understand, but the least important mechanism for most planets is Jean's Escape. Particles pick up energy from solar and other radiation sources. This is manifest as an increase in the kinetic energy of the particle. As kinetic energy scales with mass, it is the lightest gases, hydrogen and helium that are easiest to accelerate through this mechanism. Heavier gases, such as oxygen (32 g for every 6 hundred billion trillion atoms) or carbon dioxide (44 g per 6 hundred billion trillion molecules) are usually too massive to be moved efficiently by this route. If the gravitational pull is low, because the mass of the planet is low, then the velocity that the particles need to achieve in order to escape is also correspondingly low. Within the Earth's atmosphere oxygen atoms have a mean velocity less than 1.1 km per second—or less than one tenth the escape velocity. Therefore Jeans escape has no measurable impact on the loss of gases similar to or more massive than oxygen on Earth. On Mars the escape velocity is only about 4 km per second and although this is still higher than the mean velocity of oxygen atoms, it is closer. The key term is mean velocity: many oxygen atoms have the mean velocity, but a significant proportion will have velocities that are higher than this and can, therefore, escape easily to space.

Both Venus and the Earth are sufficiently massive that very little of the particles making up their atmospheres are energetic (or hot) enough to escape in their own right. In Jean's Escape (thermal) mechanism gas particles gain sufficient kinetic energy from the Sun that they can escape the planet's gravitational pull. With the exception of hydrogen and helium, most of the particles in our atmosphere and that of Venus are simply too massive and hence too slow moving for this mechanism to really work. For this mechanism to be efficient the gravitational pull of the planet has to be low or the temperature of the gases higher. For Mars the gravitational pull is sufficiently low enough that the Jeans mechanism can work, however, it isn't quite enough to strip Mars down on its own.

Charge Exchange

If an energetic ion trapped within the planet's magnetic field collides with a slow moving neutral atom, the atom can pick up the charge of the ion. This leaves the fast moving ion free of charge but still with enough kinetic energy to escape. This is not an issue for Mars or Venus, which lacks an appreciable field, but it does contribute to the loss of gases above the Earth.

Hydrodynamic Escape

In this situation the atmosphere is heated so strongly by its star (or by tidal forces) that it simply flows off its parent world, much like a wind into space. The high temperatures generate high pressures (Chap. 1) that are sufficient to lift the gas free of its gravitational shackles. Given enough kinetic energy and this gas escapes beyond the gravitational "limit" of the planet and is captured by the solar wind. This process is very evident in a few, unfortunate extrasolar worlds, which lie perilously close to their parent stars. HD 209458b, or GJ 436b, the latter discovered by the now defunct French craft CoRoT, show strong ultraviolet absorption in a large comet-like cloud that extends away from each world. In both cases Hubble's ultraviolet camera was used to examine the cloud of gas. Although GJ 326b is likely to survive its ordeal largely intact and is already six billion years old, the younger HD 209458b is not going to be so fortunate. Losing at least 10,000 metric tons per second (or 10,000 times the rate of loss of gas on Earth, Mars and Venus), HD 209458b is likely to be stripped down to its core before its parent star becomes a red giant (Chap. 10).

Solar Wind Stripping

Like Venus, Mars lacks any appreciable magnetic field, and what field there is localized primarily to regions in the southern hemisphere. Consequently, like Venus the solar wind can directly interact with the top of the Martian atmosphere. Like Venus, it's not entirely a one-way street. Sunlight and energetic particles from the Sun ionize gases at the top of the atmosphere forming an ionosphere. In turn these ionized gases move around under the influence of the solar magnetic field and under the influence of light. As a result a magnetic field is induced within these flowing, ionized gases, which then shield the bulk of the atmosphere from the scouring action of the solar wind. Around this region a bow shock forms that also deflects the bulk motion of the solar wind, before it can whittle away at Mars's tenuous gases. However, as the Martian atmosphere is less dense than that of Venus, the effect is weaker. Moreover, with a weaker gravitational pull, many of the gases that Venus could hold onto, Mars can't and it loses a more substantial portion of its atmosphere per second than either the Earth or Venus.

Like Venus, Mars does suffer from wind stripping. With a weaker gravitational pull and a less dense atmosphere, the Martian air is more prone to stripping than that of either Venus or the Earth. This is particularly true when the solar wind is running at its strongest and the Martian bow shock is shoved closer towards the surface of the planet. At these times, the denser Venusian atmosphere develops a stronger induced field and hence is more thoroughly shielded. However, Mars is unable to pull off this trick as its thin atmosphere cannot respond to the change in the strength of the wind and induce a stronger magnetic shield in the way that the Venusian atmosphere can.

Mars has another chink in its armor. As we've already seen, during the largest Martian dust storms the middle and upper portions of the troposphere warm so much that the entire atmosphere puffs up. This, in turn, exposes more of upper atmosphere to the erosive effects of the solar wind. Mars, it seems, has a rather self-destructive side, for the more atmosphere that is lost, the greater the difference in surface temperature and this leads to greater storms at particular times of the year.

If we look at the Martian atmosphere today, in all solar wind stripping accounts for less than a third the rate of gas loss from Mars through these non-thermal processes. However, MAVEN data suggests that it was this process, and the additional erosive effects of solar storms, that ripped the Martian atmosphere out of Mars' puny gravitational clutches. Today, other non-thermal processes, including stripping by asteroid and comet impacts; and the effects of charging of the atmosphere by solar and cosmic radiation. None of these are relatively important. Mars, thus, lost most of its atmosphere early on to the young Sun's more powerful solar wind. MAVEN data suggests that this happened between 4.2 and 4.0 billion years ago. Meanwhile, today it loses most of its gas through thermal effects: the Jeans Mechanism (described above), with non-thermal mechanisms accounting for the remaining loss. Remember, that in the Jeans' mechanism gas particles have to attain sufficient energy to overcome the gravitational pull of the planet. Very little gas in the atmosphere of Venus comes anywhere close, despite lying being closer to the Sun and its gases being heated more strongly. Mars loses out simply because it is a small planet with a low gravity. The lack of an appreciable magnetic field is simply the icing on the cake. Could you combine Mars with Mercury and you'd get a world with enough gravity (and with a magnetic field) that would be far more habitable in the longer term? I am not aware anyone is planning to maneuver these two worlds together. However, there is a small, but reasonable chance that repeated gravitational pulls from Jupiter might lead Mercury to come off its orbit and collide with Mars in the next few billion years. Earth: watch out!

While Venus loses gas through the charging and repulsion, Mars is actively stripped. Venusian gases also undergo more chemical reactions within their atmosphere than do those of the Earth or Mars. This is because the particles within the Venusian atmosphere are closer to the Sun and, therefore, receive more energy, which in turn encourages chemical reactions amongst them. These reactions can release further energy that helps whisk them out of the atmosphere.

The atmosphere's of Mars, Earth and Venus lose mass at a similar rate. Unfortunately, Mars has less to lose (2.5×10^{16} kg for Mars versus 5.15×10^{18} kg for the Earth) and, was its current rate of loss to be maintained, would bleed dry in less than 2.9 billion years. Compare that to the Earth, where we wouldn't lose our atmosphere for another 591 billion years or so (Chap. 5)—if the Sun didn't get us first. Mars won't be a place to go when the Sun begins to parboil the Earth. However, this figure assumes that the atmosphere isn't replenished from surface or interior stores— which, of course, it is. As the atmosphere bleeds away to space, ice—both water and carbon dioxide, sublimates on the Martian surface and partly compensates for the loss. However, Mars will never approach the conditions it might have had when the solar system was young. The planet has lost too much gas and has dwindling stores from which to replenish this.

There is one final possible mechanism through which Mars lost most of its atmosphere—and might explain why it happened four billion years ago. Enter the joker: the Late Heavy Bombardment. From 4.2 to 3.9 billion years ago, the Earth and Moon appear to have suffered catastrophic impacts, with objects up to 100 km across. These massive objects appear to have begun their life far out near the current orbit of Neptune. During this interval, Jupiter and Saturn appear to have booted Uranus and Neptune out of their earlier and warmer orbits to the outer fringes of the solar system. During these maneuvers, vast quantities of icy debris were scattered in all directions, with a significant amount falling in towards the Sun. Mars would not have escaped this bombardment. In the case of the Earth, our substantial gravitational pull would have held onto most of the impact material—and more importantly much of the gas energized and blasted upwards by the impact.

In such an event, the blast wave expands radially in all directions from the impact point and most of the material that escapes (at least temporarily) into space vents upwards like an enormous rocket exhaust through the hot channel the incoming asteroid or comet has carved through the atmosphere. This forms an enormous mushroom cloud that is centered on the impact point. With a low gravity, not only can more of the impacting object escape into space, but also a larger volume of hot gas can be vented, as well. Unable to hold onto this, such material is captured by the solar wind and vented into interplanetary space. Such impact erosion was likely more significant to early Mars than any gains it would have made from the impactors themselves. However, the precise contribution of atmospheric erosion by impacts is still unclear. All we can currently say with confidence is that Mars was bled dry early on. From four billion years ago, Mars underwent a slower process of attrition that was only delayed by the slow sublimation of its remaining icy inventory and a declining input from volcanic activity. This process appears to have been primarily caused by the Jeans Mechanism and Solar Wind stripping of what little atmosphere, Mars retained.

How then would we sum up the manner in which Venus, Earth and Mars lose atmosphere? On Venus, gas loss is primarily through two non-thermal processes: electric field acceleration and photochemical reactions (Chap. 5). In the former case, electrons are driven off and this creates a strong enough electric field to accelerate ions. In the latter sunlight splits carbon dioxide into carbon monoxide and free oxygen. When these flow to the night side and recombine they release sufficient energy to propel many of the oxygen molecules into space. Mars, by contrast loses its atmosphere to the solar wind, when particles achieve sufficient energy to escape. Thus, Mars is stripped by a combination of thermal and non-thermal processes that lift gases gradually out into space, with a likely additional contribution to atmospheric loss from impacts during the Late Heavy Bombardment.

Despite our perception of a deep, impenetrable atmosphere, the Earth has already lost most of its original bulk. Much hydrogen has escaped through the Jean's mechanism, but most of our atmosphere has been lost to our planet's surface. Vast quantities of carbon dioxide, along with most of the atmosphere's water have condensed onto its surface, or have chemically reacted with it. Such processes do happen on Mars and a substantial body of water remains frozen. Carbon dioxide also freezes out in a seasonal pattern. Venus will have also lost some of its atmosphere to its surface through chemical reactions with erupted rocks. How much this is remains in the realm of mystery—and will stay that way until someone comes up with a credible means of sampling the Venusian surface.

All in all, Mars suffered solely because it has a relatively low mass. For not only is Mars unable to maintain a magnetic field, but it incapable of generating a sufficiently strong gravitational pull to hold its atmosphere in place. With no means of replenishing its mass, Mars bled first to space and then later to its surface. That Mars can maintain any real atmosphere today is a testament to the slow sublimation of what remains of its former riches from its planet's icy surface.

Future Mars

Mars is a desiccated world, but it is not one without potential. Large reserves of water ice and carbon dioxide ice carpet the planet. One of the more interesting discoveries of the last few years has been the discovery of glaciers and glacial deposits buried in dust near the Martian equatorial highlands. Quite how much water remains is unknown but it may be enough to refill most of the Martian northern basin were it to be thawed out.

Many people have hypothesized about the potential to reclaim Mars as a habitable world. In order to realize this dream the pressure within the atmosphere must rise to at least one tenth that at the surface of the Earth. Any less than this and water will not be stable in its liquid form. To put this in perspective, at the tropopause the average pressure is around one fifth that at the Earth's surface. This is still 20 times the pressure currently found at the surface of Mars. So how will we raise the pressure to the point at which water could be stable at reasonable (0–20 °C) temperatures?

In a paper published more than 20 years ago in *Nature*, Christopher McKay and James Kastings outlined what would be the simplest—and fastest—scenario. This was revisited by McKay and Margarita Marinova a decade later, with a consideration of the ethical issue associated with terraforming. In essence terraforming Mars will all be about thawing it out. In order to do this cheaply and sensibly future terraformers would likely flood the lower atmosphere with fluorinated carbon compounds. The simplest and possibly the best would be tetra fluorocarbon (CF_4). Not only is this easy to make—and could be manufactured on Mars—but it is sufficiently stable that the fluorine won't escape its chemical bonds and erode any future, protective ozone layer.

After a few million tons of tetrafluorocarbon are released the Martian greenhouse will strengthen and the temperature will begin to rise. Get it right and in about 100 years the rising temperature will sublimate carbon dioxide and water from the frozen Martian crust. Moreover, the water will react with the copious amount of perchlorates and decompose them. This will release oxygen gas and begin the process of making the planet suitable for microbial and more complex life—notably the construction of a thicker ozone layer.

The process should become self-sustaining with the increasing concentration of greenhouse gases driving higher temperatures, more thawing and thus the release of more greenhouse gas. Although the next steps would take millennia, eventually the balance of energy trapped by carbon dioxide, carbon tetrafluoride and water vapor will match that escaping and hold the temperature at 10–20 °C. Atmospheric pressure will never equal that on the Earth, but it could reach 300–400 millibar, or roughly the pressure found just below the summit of Mount Everest. Humans would need to wear oxygen masks and any large animals would need to be kept under pressurized domes, but otherwise Mars would be somewhat akin to life on the Tibetan plateau. Complex plants could take over the job of creating oxygen and maintain habitability, as well as producing food for the future Martians although the low atmospheric pressure would likely slow their growth.

If we decide to let nature do the work we will have some time to go. For it won't be until the Sun is around 15 % brighter than it is at present that Mars will receive enough radiation to begin to thaw it out. For us, this will be around 1.4–1.6 billion years in the future. If we are lucky, just as the habitable phase of the Earth wanes and finally expires, Mars will blossom. Imagine watching the Earth turn from blue to brown from the greening surface of Mars? However, the word "blossom" should not be taken too literally. At its temperate peak Mars will be about as warm as ice age Earth If one wished to wait until Mars was as warm as the current Earth, we would need to wait another 4.5 billion years. For only at this point would the Sun be as bright to make Mars feel truly like a second home.

How long could Mars remain habitable? With no intervention and no magnetic field to protect its low gravity and, consequently, puffy atmosphere, the same damage will begin to destroy what humans or nature have created within a few hundred million years (at most). Stripping of gases by the solar wind; sputtering by cosmic rays and Jeans Escape will strip the water down to hydrogen and drag it away into space. Carbon dioxide will follow; and there will then begin a protracted phase where Mars returns to its current dry, desertified state. Unless we chose to intervene and continually replenish the planet with water, perhaps from comets, there will be no going back. Mars will be left bone dry and as bereft of air as Mercury by the time the Sun leaves the main sequence. Long before Mars could hope to attain the terrestrial temperatures of today it will be utterly airless and lifeless.

Conclusions

Mars is a complex world. Its atmosphere has undergone a traumatic series of changes since the planet formed 4.5 billion years ago. Soon after this time an extraterrestrial visitor to our Solar System might have seen three, pristine habitable worlds: although free from oxygen, Venus, Earth and Mars would have hosted dense, cool, moist atmospheres with extensive oceans carpeting their surfaces. Although Venus is somewhat problematic in this scenario (Chap. 5), that the solar system could have been habitable in this regard is truly fascinating. The loss of two of these worlds was due to the success of greenhouse gases in one and the failure of them in the other. Venus spawned a moist, then dry super greenhouse and its atmosphere boiled dry, while the atmosphere of Mars was whittled down and eventually froze dry. Only the Earth had a combination of sufficient mass and greenhouse gases to maintain its habitability.

While Venus paints a picture of how the Earth will be, Mars reminds us the planetary ecosystem is fragile and malleable in equal measure. Add extra greenhouse gases to Mars and it will warm and thaw. By the same token adding additional greenhouse gases to the Earth's atmosphere is currently causing it to warm. The fatal fallacy that adding extra greenhouse gases will have no effect is a virus which permeates the internet and leads many to believe carbon dioxide and methane are irrelevant (Chap. 3). Close examination of our terrestrial cousins, Mars and Venus exposes the flaw within this argument. For if it were true all those ideas of terraforming Mars through the addition of more greenhouse gas would clearly come to nothing. Moreover, there would be no link between high density of greenhouse gases and the high temperatures found at Venus.

That the Earth has an average temperature of +15 °C instead of –18 °C, while Venus has a mean temperature of 470 °C instead of –43 °C is a testament to the additive power of greenhouse warming. By trapping much, but not all of the outgoing radiation these gases maintain planetary temperatures. Should we wish to terraform Mars (or even Venus) it will only be through the manipulation of the concentration of these gases that these dreams become possible.

A well-known terrestrial meteorologist is keen to equate the ability to dream with burning fossil fuels. I might agree but not in the way he envisages the power of dreams. Surely, the only way we can ensure our future prosperity is by imagining about how we make it sustainable; not by keeping our fingers crossed and daydreaming our way into an irretrievable situation. Dreaming is all well and good only for as long as you can separate what is achievable and what is merely a wasteful desire. There is a vast world of available energy out there: why restrict ourselves to those forms which are most harmful for our future?

References

1. Fenton, L. K., Geissler, P. E., & Haberle, R. M. (2007). Global warming and climate forcing by recent albedo changes on Mars. *Nature*. http://humbabe.arc.nasa.gov/~fenton/pdf/fenton/nature05718.pdf.

2. Wang, H., Richardson, M. I., John Wilson, R., Ingersoll, A. P., Toigo, A. D., & Zurek, R. W. (2003). Cyclones, tides, and the origin of a cross-equatorial dust storm on Mars. *Geophysical Research Letters, 30*(9), 1488. http://www.gfdl.noaa.gov/bibliography/related_files/hqw0301.pdf.

3. Haberle, R. M., Leovy, C. B., & Pollack, J. B. (1982). Some effects of global dust storms on the atmospheric circulation of Mars. *Icarus, 50*(2–3), 322–367.

4. John Wilson, R. (1997). A general circulation model simulation of the Martian polar warming. *Geophysical Research Letters, 24*(2), 123–126.

5. Bertaux, J.-L., Leblanc, F., Witasse, O., Quemerais, E., Jean, L., Stern, S. A., et al. (2005). Discovery of an aurora on Mars. *Nature, 435*, 790–794.

6. Leshin, L. A., Mahaffy, P. R., Webster, C. R., Cabane, M., Coll, P., Conrad, P. G., et al. (2013). MSL science team. *Science*, 341(6153), 1238937. doi:10.1126/science.1238937.

7. Pamela G. Conrad., Dan Harpold., John J. Jones., Laurie A. Leshin., Heidi Manning., Tobias Owen., Robert O. Pepin., Steven Squyres., Melissa Trainer., MSL Science Team. Science 341(6143), 263–266.

8. Webster, C. R., Mahaffy, P. R., Flesch, G. J., Niles, P. B., Jones, J. H., Leshin, L. A., et al. (2013). Isotope ratios of H, C, and O in CO_2 and H_2O of the Martian atmosphere. *Science, 341*(6143), 260–263.

9. McKay, C. P., & Marinova, M. M. (2001). *The physics, biology, and environmental ethics of making Mars habitable*. http://www.ifa.hawaii.edu/~meech/a281/handouts/McKay_astrobio01.pdf.

7. The Gas Giants

Introduction

Saturn and Jupiter have more in common than they have in isolation, thus Chap. 7 compares and contrasts the two giant masters of the Sun's planetary realm. Despite some issues of scale, the broad brush strokes of each planet are rather similar. Within this broad canvas there are specific differences in the appearance of each world and the behavior of its weather. The underlying reasons such as differences in the amount of sunlight each world receives are understood, but how this manifests as differences in the appearance of each world remains largely in the realm of speculation. As more probes venture into their territory there remains much to be learned and more than enough to discuss, pontificate and speculate over. Welcome to our giants.

The Structure of the Giants

While the interiors of the terrestrial planets can largely be ignored when it comes to weather, primarily locked in a solid crust as they are, the atmospheres of all four of the giants are so tightly and intimately intertwined with their cores that they must be addressed. In particular, the gas giants, Jupiter and Saturn, have atmospheres that emerge from the sea of liquid hydrogen and helium that forms the bulk of their interiors. Additionally, there can be a strong transfer of energy by convection from their liquid centers to the atmosphere above (Fig. 7.1).

Jupiter has an atmosphere that is roughly 5000 km thick, and Saturn's is similar. Of this depth, only around 100 km lies below the cloud tops on Jupiter (Figs. 7.2 and 7.3). At Saturn this figure is somewhat higher at around 320 km. On each world the clouds lie where the pressure is conducive to their formation at

© Springer International Publishing Switzerland 2016
D.S. Stevenson, *The Exo-Weather Report*, Astronomers' Universe,
DOI 10.1007/978-3-319-25679-5_7

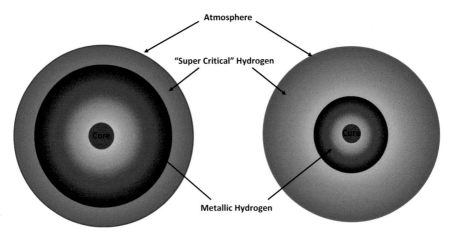

Fig. 7.1 The internal structure of the gas giants. The bulk of their volume (and mass) consists of liquid hydrogen and helium surrounding a possible "rocky-metallic" core. Both planets have diameters in excess of 100,000 km. However, Jupiter as a little over three times the mass of Saturn the degree of internal compression is far higher. Jupiter and Saturn's atmospheres are roughly 1000 km thick, with most of this above the cloud tops. Roughly 100 km beneath the ammonia cloud layer on Jupiter the pressure reaches 10 bars and the atmosphere gives way to an increasingly thick soup of hydrogen. This transition is somewhat deeper on Saturn at around 320 km, due to Saturn's lower gravity. Beneath this hydrogen slowly morphs into a thick fluid. 20,000 km down on Jupiter and 30,000 km down on Saturn the liquid hydrogen becomes metallic and its motion generates each planet's magnetic field. Temperatures reach 35,000–40,000 K at the core of Jupiter and around 11,700 K inside Saturn, where helium rain may explain some of Saturn's extra heat

around 0.1–1 bar for ammonia and up to 10 bars for water vapor (Fig. 7.2). These conditions are reached at slightly deeper levels on Saturn with water vapor condensing below 10 bar and the dominant ammonium hydrogen sulfide layer at around 7 bars of pressure. The differences are a consequence of the lower temperature of the Saturnian atmosphere and the difference in gravity: mighty Jupiter squeezes its layers more tightly than fluffier Saturn due to its larger mass.

At about 10 bars of pressure (equivalent to around 100 m depth in our oceans, or one ninth the pressure at the surface of Venus) molecular hydrogen gradually takes on the appearance and properties of a liquid. This is not an abrupt change but a gradual one. On Jupiter this layer is 20,000 km deep and for the first

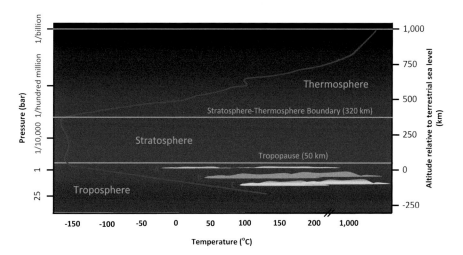

FIG. 7.2 The structure of Jupiter's atmosphere. Clouds occur in three decks in the troposphere at or below 1 bar in pressure. 1 bar corresponds to terrestrial atmospheric pressure at sea level. Cloud deck *A* consists of ammonia ice crystals at a temperature of –150 °C and an altitude of 23 km above "terrestrial sea level". Below this (*B*) are clouds of ammonium hydrogen sulfide at around 1 bar and water (*C*) at –50 km or so. Above the tropopause there exists a stratosphere with broadly constant temperature (*red line*), and above 320 km we reach the thermosphere where charged particles from the Sun cause temperatures to rise rapidly (Chaps. 1, 5 and 6)

10,000 km or so the mix of hydrogen and helium is more gas than liquid. Beneath this you'd be hard-pressed to think of this substance as a gas. On Saturn, the liquid hydrogen layer extends from 320 km down to a depth of 30,000 km, or half the way to Saturn's core. Hydrogen, in this state, is known as a super-critical fluid— something we encountered in Chap. 5. To recap, in this odd state, hydrogen adopts the ability of a liquid to dissolve substances, but retains an ability to effuse through solids like a gas. It is also an excellent conductor of heat.

Within this layer there are some important processes taking place. On Jupiter and Saturn, when temperatures exceed 1000 K (727 °C) and around 40,000 bars of pressure methane begins to transform into other compounds. In the absence of other materials this tends to be longer chain of hydrocarbons and eventually graphite, but in the dense interior of the gas giants the outcome is less certain. The presence of ammonia and hydrogen sulfide as well

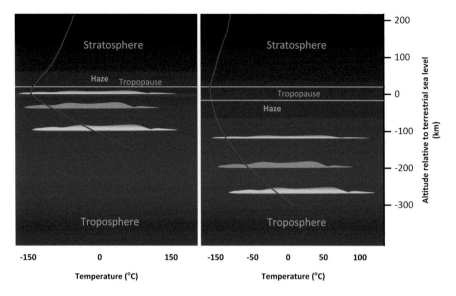

FIG. 7.3 A comparison of the lower atmosphere of Jupiter (*left*) and Saturn (*right*). The cloud layers on Saturn have the same composition as those above Jupiter (Fig. 7.2), but lie at lower altitudes. This is simply because Saturn has a colder atmosphere so the condensation levels are lower down. The lower atmospheric pressure and lower temperature gradient means that the cloud decks are more spread out The upper atmosphere of Saturn is also filled with a haze of carbon-based compounds which is largely absent above Jupiter. This mutes the appearance of the banded cloud structure on Saturn. The tropopause marks the lid of the troposphere as on Earth, while "0 km" marks 1 bar of atmospheric pressure, or sea level on Earth

as water favor the production of more complex compounds, until at even greater temperatures and pressures these break down into others. This means that while the bulk of both Jupiter and Saturn is hydrogen and helium, there is a substantial amount of heavier elements that can't be ignored.

In some speculative models these heavier compounds form a kind of hot "icy" mantle around the core of these planets. Indeed, the presence of a core at all is unclear on Jupiter. Even if it began life with one several tens of times the mass of the Earth, it could well have dissolved by now into the metallic hydrogen ocean that lies beneath 20,000 km inside Jupiter. The presence of a core in Saturn is more likely and measurements by Cassini suggest something with a mass between 9 and 22 times the mass of the Earth may well lie at Saturn's heart.

While Jupiter's metallic hydrogen layer is too hot to allow stratification, or the settling of materials, the layer on Saturn is not. It is suspected that the helium which should compose around 24 % of the material has condensed or is condensing. This generates a rain of helium falling through to the metallic layer below. As this material descends it releases gravitational potential energy. This is realized as heat which warms Saturn's interior and allows this smaller world to mimic Jupiter, releasing more than twice the heat energy the planet receives from the Sun. Jupiter, meanwhile, has remained hot from its formation due partly to the steady release of gravitational potential energy as the planet contracts.

Between the core and the "liquid hydrogen ocean" lies a thick layer of metallic hydrogen. Under a pressure three millions of times that found at terrestrial sea level (3 million bars) the super-critical fluid hydrogen undergoes a more profound rearrangement, forming an ordered structure more akin to a metal. In this state the electrons that are normally bound one per nucleus can break free and become delocalized, making the hydrogen a super-efficient conductor of electricity. Stirred by both convection and the rapid spin of both planets, this metallic stew can generate magnetic fields which enshroud each world. Jupiter's is particularly showy, extending outwards over 700 million kilometers to the orbit of Saturn.

The Color of Giant Planets in the Solar System

On Earth the color of a cloud contains information about its nature. Aside from the ubiquitous whites and greys there are a host of other colors that can give a clue to impending weather. For example Fig. 7.4 shows a small sample of the possible variety of shades caused by the effect of water and dust on the light that ultimately enters our pupils. In the dawn or evening sky a yellow hue usually signifies the presence of abundant water vapor and the likelihood of coming rain. A very pale, turquoise blue between clouds has a similar warning during the middle part of the day. Green is rare, but if a green cumulonimbus thundercloud is approaching you can expect severe hail. A copper tint around the edges of clouds or along part of their base (Fig. 7.4a) tends to indicate electrical charging and hence the threat of lightning. In each case it is the scattering of light around dust or water droplets and ice crystals; or the refraction of light through droplets of water that produces these different colorations. In the atmospheres of the giant planets there exists a bewildering array of additional chemicals that can add their own absorptions to expand this array of colors even further.

The chemical composition of Jupiter and Saturn's atmosphere is very similar. The cloud layers are made of water vapor, overlain by ammonium hydrogen sulfide with wispier ammonia clouds on top (Figs. 7.1, 7.2 and 7.3). On Jupiter the ammonia clouds sit in a hazy layer of hydrocarbon compounds and compounds called "tholins" that are produced through the reaction of ammonia and the abundant methane. Methane itself does not form clouds simply because in the atmospheres of both Jupiter and Saturn it is too warm.

How do scientists know what the clouds are made of, particularly since the water clouds are effectively hidden from view? To some extent the presence is inferred because oxygen is one of the universe's most abundant chemical elements and most of it will be found as water vapor or ice at Jupiter's orbit and beyond. Moreover, the presence of lightning within the atmospheres of both planets' atmospheres suggests that water clouds are present. This assumption is because on Earth water vapor is needed to form the ice that ultimately helps separate charge in thunderclouds, however, ammonia has a chemical structure not too dissimilar from water vapor, with molecules able to separate charges just as water does, so this is a little dubious as evidence.

Fortunately, the Galileo Orbiter probe was able to detect water vapor from measurements made in orbit. Additionally, the violent collision of Comet Shoemaker-Levy 9 with Jupiter in 1994 blasted deeper material upwards, beyond the cloud tops. Spectra taken from Earth and from Galileo indicated that there was water present but that it was at much lower level than was expected. Quite why this is so remains unclear. In the case of the Galileo Orbiter, it appears that there was a bit of bad luck involved in the trajectory as the probe descended through a relatively dry belt with limited cloud cover. In the case of the comet, it's plausible that the comet may have vaporized above the 10 bar level where water clouds would be formed. In this case water clouds were present, but the comet was unable to dredge them up to the "surface" where terrestrial spectroscopes, and those on Galileo, could observe them. Alternatively, our assumptions that Jupiter has a thick layer of water clouds is wrong and this layer is so thin that it rapidly gives way to the sea of hydrogen and helium immediately underneath.

However deficient Jupiter appeared to be in terms of water, the cometary impacts did reveal a host of other interesting chemicals in Jupiter's atmosphere. Amongst these were hydrogen sulfide, ammonia, sulfur, carbon disulphide and it was thought that it was some of these sulfur-containing compounds that gave Jupiter's deeper layers their color. Yet despite many attempts the reddish color of these bands could not be reproduced with sulfur chemistry. Instead these deeper belts and the eye of Jupiter's Great Red Spot are red because of some complex carbon-chemistry involving ultraviolet light. The Great Red Spot is red (or rather a varying shade of pink) because the storm carries ammonia high into the atmosphere (Fig. 7.4).

High at the tropopause, ultraviolet light can react with the ubiquitous methane haze to produce acetylene (ethyne)—the gas used in welding. This is carried downwards into the belts or the eye of the great red spot where it encounters ammonia that has been driven upwards. With a bit of a spark from ultraviolet radiation the two gases combine to produce a pinkish compound. Without the influence of solar ultraviolet light the clouds would mostly be white as they are on Earth. Experiments suggest that were ammonium hydrogen sulfide irradiated, the clouds would appear slightly green. One can think of the pink coloration as a kind of Jovian sunburn. Wherever the atmospheric circulation brings ammonia and acetylene together and are exposed to sufficient ultraviolet radiation they take on the characteristic pink coloration. One can see this process in action with some of Jupiter's new storms that are joining the Great Red Spot (GRS) to pockmark Jupiter's otherwise stripy complexion.

Out by Saturn the colder and lower pressure atmosphere leads to condensation of ammonia and ammonium hydrogen sulfide at deeper levels (1–2 bars instead of 0.5 bars at Jupiter). Consequently, there is a greater burden of haze above the clouds which obscures them, much like smog above a terrestrial city. Saturn, thus, appears rather muted in color with less pronounced belts and zones (Fig. 7.5).

Even further out in the Solar System, by Uranus and beyond, methane clouds contribute white coloration, but it is the deep

Fig. 7.4 The color of clouds. Jupiter's Great Red Spot (*center*) is a distinct pinky-orange color, with distinct differences depending on the depth to which we see the cloud. Is this a true color caused by a chemical substance, or is it caused by something more prosaic, such as scattering of sunlight? An approaching thunder-storm delivers the usual dark grey but also a distinctive copper color indicative of charge separation (**a**, *above*). (**b**, *below*) shows a partial circumzenith arc produced by refraction of sunlight through high, thin ice clouds. (**c**, *above*) shows cirrocumulus clouds demarcating the Polar Front Jet Stream. Scattering of light from the setting Sun gives an orange–pink glow on the horizon. (**d**, *below*) shows a pervasive yellow glow caused by scattering of light on water vapor. *Image credits*: Great Red Spot (NASA/Voyager 2); Remaining photographs by author ((**a**)—Hunstanton, UK; (**b**)—Glasgow, UK; (**c**)—Nottingham, UK and (**d**)—Kyle of Lochalsh, UK)

expanse of methane in general that colors these planets blue or greenish-blue. The red and yellow bands that are visible in the atmospheres of Jupiter and Saturn are simply buried far below the deep ocean of blue methane. Saturn forms a yellow transitionary member, Uranus appears somewhat green, while the transformation is complete at Neptune with only the deep, and oddly terrestrial, blue visible. Why blue? This comes down to Rayleigh scattering.

This process contributes to our reds and yellows of sunset and the blue of our daytime sky, but out at Uranus and Neptune, methane also preferentially scatters blue light. The longer wavelengths are absorbed, meaning Uranus and Neptune are blue for the same reason that Earth has a blue sky during the daytime.

At Jupiter and Saturn the yellowish haze also shares its origin with ultraviolet light. Carl Sagan and Bishun Khare christened the chemicals that were thought to color Saturn and Jupiter's upper atmosphere, tholins. Originally, the term was restricted in use to describe those chemicals thought to provide the haze in the atmosphere of Titan, but the term can be more generally applied to the atmospheres of all cold worlds, as well as the reddened surface of Pluto and other Kuiper Belt objects. The precise composition of these chemicals is less well defined, though they are more than likely produced in a series of complex photochemical reactions between ammonia, methane and ethane. With the aid of cosmic rays, ultraviolet restructures the bonds between these molecules and produces a complex range of much larger nitrogen and carbon-based molecules. These are able to absorb a considerable range of wavelengths, but their low density means that they merely mute the colors of Jupiter's clouds rather than obscure them. At Saturn, with deeper cloud layers and consequently, the greater depth of tholin haze, means that the banded pattern of clouds is blander in its complexion. Above Venus sulfates, principally sulfuric acid (Chap. 5), are the driver of Rayleigh scattering. Within the atmospheres of the Earth and Mars Rayleigh Scattering is predominantly caused by silicate dust (Chap. 6). Scattering is bolstered within the Earth's atmosphere by organic molecules produced by living organisms.

Belts and Braces: The Bands and Storms of Jupiter

In the atmospheres of both gas giants, clouds are organized into colorful bands called belts and zones. The belts are thought to represent areas of down-welling, with the zones as regions of up-welling. Where air rises ammonia is taken up several kilometers above the level of the belts. Here it cools and condenses forming

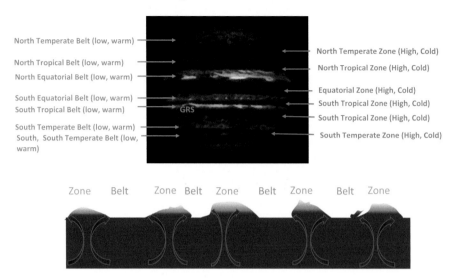

North Temperate Belt (low, warm)

North Tropical Belt (low, warm)

North Equatorial Belt (low, warm)

South Equatorial Belt (low, warm)

South Tropical Belt (low, warm)

South Temperate Belt (low, warm)

South, South Temperate Belt (low, warm)

North Temperate Zone (High, Cold)

North Tropical Zone (High, Cold)

Equatorial Zone (High, Cold)

South Tropical Zone (High, Cold)

South Tropical Zone (High, Cold)

South Temperate Zone (High, Cold)

GRS

Zone   Belt   Zone   Belt   Zone   Belt   Zone   Belt   Zone

Fig. 7.5 The organization of Jupiter's cloud belts and zones. *"Belts"* represent areas where cooler air is descending and warming and any clouds tend to evaporate except at deeper levels. In *"zones"* upwelling air cools and ammonia clouds condense that overly the deeper ammonium hydrogen sulfide layer. You can imagine each zone represents the top of a Hadley cell, while the belts represent terrestrial Horse Latitudes (Chap. 1). The rapid rotation of Jupiter generates the much stronger Coriolis effect that drives the formation of this highly banded atmosphere. Colors in the lower rendering are not representative. Infrared image of Jupiter: NASA/JPL/CALTECH

white stripes across Jupiter's face (Fig. 7.5). In the deeper zones, air over-spilling from the neighboring zones descends; warms and the clouds within it evaporate, exposing the ammonium hydrogen sulfide layer. A contributing factor to the freshness of the zones is their altitude. At Jupiter these zones are more clearly visible because there is less obscuring haze high up. The belts and zones also counter-rotate around Jupiter and Saturn at speed (Fig. 7.6), so as air rises and falls it also moves to the east or west with substantial speed giving an overall corkscrew motion.

What is the underlying mechanism that generates these belts and zones within the giant planet atmospheres, and why do Jupiter and Saturn differ in their organization of these cloud patterns? On the Earth, Venus and Mars the driving cells are the Hadley cells that overlie the equator. These pump energy into the atmosphere

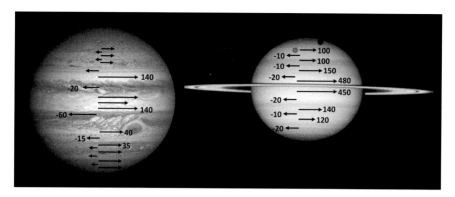

FIG. 7.6 The direction and speed (velocity) of the winds on Jupiter (*left*) and Saturn (*right*). On both planets winds are predominantly eastwards. Air rises in zones and sinks in belts, giving the overall flow a corkscrew-like pattern as the air barrels across the face of each world. Airflow around the Great Red Spot (GRS) is westerly along its northern flank, at around 180 km per hour, and easterly at 140 km per hour on its southern flank. Airflow around the GRS peaks at 120 m per second (432 km per hour), while it is roughly stagnant, with modest upwelling in the heart of the GRS. At Saturn, one of the most interesting atmospheric features is the north polar hexagon. It turns out the 100 m per second westerly jet at the top of the image is key to the formation of the north polar hexagon. Meanwhile, further south, the 20 m per second retrograde (–20) jet drove the great storm of 2011. Images: NASA/HST

from the sun-warmed surface. As air flows north or south, away from the equator, the Coriolis Effect directs the air towards the east (westerly winds). Air moving in the opposite direction towards the equator has less eastward velocity than the ground underneath so appears to curve to the west (as seen in Chap. 1).

North of the Hadley cell in the Martian and terrestrial atmosphere lie the Ferrell cells. Venus has a more subdued equivalent polar collar, which is broadly equivalent to these. Air flow in the Ferrel cell is overspill from the Hadley cells and its motion to the east is driven by air flowing down and around the Hadley cells. If you need an analogy, run a bath and watch the water circulating. A fast cell spins where the water moves outwards away from the tap. Further away, a second cell spins in the opposite direction carrying momentum towards the tail of your bath. On Saturn and Jupiter this effect alone can't explain the observed airflow. For one, equatorial winds on Jupiter and Saturn are westerly—and very strong. Indeed, on both Jupiter and Saturn the predominant wind direction is westerly at the level of the cloud tops. This super-rotation—where air moves more rapidly in the direction of the planet's spin than the underlying planet— is unexpected and was looked at in Chap. 5. It will be looked at again in Chap. 9 (Titan) and 10 (exoplanets). Models by Tapio Schneider and Junjun Liu (Caltech) indicate that superrotation is driven by convection at the equator. Convection from the deep interior generates eddies and Rossby Waves (Chap. 1), which allow momentum to flow into the equatorial regions, accelerating winds into two prograde jets.

There are some differences in direction with latitude and considerable variation in speed (Fig. 7.6). On Jupiter an easterly jet stream is found roughly 15° north and south of the equator, with the same feature found at 40° north and south in the atmosphere of Saturn. Polewards of these latitudes the atmosphere is organized into alternating easterly and westerly bands with the frequency increasing with increasing latitude. As the equatorial air on both worlds is prograde (rotating in the direction of the planet's rotation) unlike the Earth, movement must be driven not by convergence at the equator but divergence. Immediately above the axis separating each Hadley cell there is an easterly jet stream (east to west) on Earth, while both Jupiter and Saturn have a westerly (west to east) jet stream.

There are more bands of circulation within each planet compared to the Earth simply because both Jupiter and Saturn rotate faster. The Coriolis Effect is more than ten times as powerful so winds deflect to the east or west over proportionately less of the planet's face. Remember when looking at Jupiter that each belt or zone is roughly 1–3 times the width of our planet. Thus, the air is still moving a considerable distance to the north or south before it is deflected eastwards or westwards. If Jupiter's rotation slowed to that of Earth (1700 km per hour at the equator) it would take more than 10 Earth days to rotate on its axis whereas now Jupiter accomplishes this feat in less than 10 h (9.6). Saturn is only marginally slower.

One must also remember that the cloud decks form very inconsequential layers within the atmosphere of Jupiter and

Saturn. Jupiter's atmosphere is more than ten times deeper than ours. The cloud decks form a visible shell only one hundredth that depth. Most of Jupiter's atmosphere is the upper stratospheric and thermospheric layers (Figs. 7.1 and 7.2). At present very little is known about these other than their bulk composition (mostly hydrogen) and their temperature.

One thing that sets all the giants apart from their solid, diminutive cousins is the intimate connection between the interior and the atmosphere. Jupiter releases a little more than twice the energy it receives from the Sun and this energy is the principle driver of atmospheric movement. Heat can convect directly from the interior to the top of the cloud deck. Diagrams in most books imply a series of distinct layers, but in actuality the liquid interior gradually morphs from one state to another. The atmosphere is merely the outer veneer where chemicals from within the interior broth can condense against the frigid chill of outer space. Chemicals, such as water vapor will break apart and reform within this stew, depending on the temperature and pressure, but the chemicals are still present, simply mixed within the giant planet's depths. The super-rotating atmosphere can be considered the surface manifestation of the circulation within the planet as a whole, even if they are not moving in exactly the same way. The weather on Jupiter and Saturn is more a product of heat coming from within than heat coming from without. The Sun still plays a role in their weather, but it is likely subordinate to the release of internal heat from the planetary interior.

Great Red Spot and Other Vortices

Perhaps the greatest meteorological icon in the Solar system is Jupiter's Great Red Spot (GRS). This anticyclonic storm measures 12,000 by 14,000 km across, first spied by Galileo, has apparently been raging for over four centuries. Indeed, despite the rather fractious environment in which the storm rotates, mathematical models suggest that this storm should be stable—particularly if it is sustained by devouring smaller storms that wander too close.

Perhaps because the GRS inhabits the highly turbulent environment of Jupiter's stormy equatorial belt, the storm now appears be waning. Just as with the common misunderstanding of the storm's likely nature are misunderstood, the apparent rumors of the spots inevitable demise may be greatly exaggerated.

Before we consider the likely fate of the spot, let's look at how the spot is structured and how it obtains its red color. In 2010 the GRS was imaged in the far infrared portion of the spectrum. This more than any other set of observations gave a clue to the storm's structure and perhaps its origin.

Observations of the GRS and flanking white ovals suggest that these are always cold cored and high in the atmosphere: up to 8 km higher than the surrounding cloud decks. The GRS and white ovals are a few degrees colder than their surroundings suggesting that they are produced by uplift of air. Within this column, like the lower zones, air cools until ammonia condenses and it is this that produces the white coloration. However, the GRS and later the GRS Junior (below), all develop pink hearts. Given what we now know about the chemistry of this pigment, these can only form where air is subsiding and bringing ammonia into contact with acetylene from higher in the atmosphere. Around the flanks of the GRS air is warmer than in the core, again implying subsidence. As air descends it compresses and warms and in both terrestrial and Jovian systems this causes the evaporation of clouds. Far below the level of the visible red eye, lies an area of warmth, 3–4° hotter than the eye wall (Fig. 7.7). Whether this is warmed by convergence or is an underlying plume of warmer material is unclear.

Measurements of air flow imply that air is not spinning within the "eye" of the GRS. Instead it is roughly stagnant, much like the air in the Venusian polar anticyclones in Chap. 5. Vertical motion could be present but has been hard to determine directly. This led researcher Leigh Fletcher to a rather complex model (Fig. 7.7). In 2000 the merger of three small ovals produced a larger oval which gradually reddened and warmed. These eventually formed The Red Spot Junior (or more prosaically "Oval AB"). These observations imply that the white ovals are high in the atmosphere near the tropopause then, when they have generated sufficient vorticity, or spin, they extend downwards into the deeper cloud decks and tap warmer layers. The introduction of acetylene to the ammonia-rich

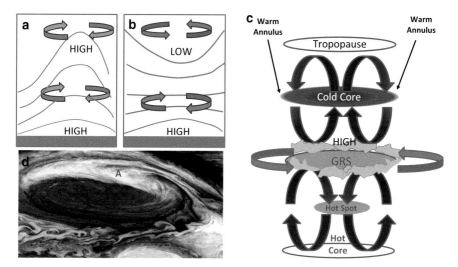

Fig. 7.7 Not all anticyclones are alike. Terrestrial anticyclones are regions of calm, descending and warming air. "Anticyclone" means a region where the air rotates around the core in the opposite (clockwise) direction to that seen in cyclones. Terrestrial anticyclones can have a warm core (**a**) with air circulating in a clockwise direction, throughout, or have a cold core with high pressure at the surface and low pressure aloft (**b**). The Great Red Spot (GRS) appears to be even more complex (**c**). In Leigh Fletcher's model (*right*), pressure is highest at the level of the red cloud deck where air converges from all sides. This air is then sucked slowly upwards and outwards by the storm's rotation. Here, it cools and ammonia clouds condense (see *A* in photo, **d**). Thermal imaging by ground-based infrared telescopes identified a warm ring (annulus) of descending air where clouds evaporate and a much deeper hot spot below the visible red spot. GRS Image Voyager 2/NASA

clouds then sets in motion the reddening of their color. However, this may be misleading.

You can envisage a number of scenarios that fit all of the data together. Storms begin as regions of upwelling from deep down like terrestrial, mid-latitude cyclones (Chap. 1). Warm to begin with, as they deepen they develop a cold core. Ammonia clouds form above and within these domes due to the drop in pressure and temperature. If the storms can draw additional energy from flanking jet streams they continue to intensify.

Consider terrestrial hurricanes, although the analogy is not precise. When the velocity exceeds a certain limit it induces downwelling in its center. In a terrestrial hurricane this is in the center of the eye, which soon transforms from a cold to a warm core through the transfer of heat from the surrounding cumulonimbus towers. In the case of the GRS and Oval AB, the area of maximum down-welling is within the eye wall, while the center of the eye has broadly stagnant air. This may mean that the GRS and hurricanes are fundamentally different or it may simply be a reflection of scale. Remember, each storm is roughly the size of the Earth. If eye-wall winds exceed a critical threshold the storm will open outwards through the effects of inertia (or what we mistakenly call centrifugal force). Winds of several tens to hundreds of meters per second will exert considerable force on their surroundings. The critical event in the lives of these storms appears to be the opening of the "eye". When the threshold is reached air begins to descend from the upper troposphere to the top of the underlying cloud deck. Air is descending along the eye-wall and passively filling the space vacated by the parting cloud wall. Within the eye of the GRS a counter flow becomes established that brings air upwards in the center of the "eye", maintaining the high dome under the tropopause that was observed by Voyager 2, Leigh Fletcher and others (Fig. 7.7).

Further down, beneath the cloud deck, the presence of the warm core implies either a region of hot upwelling or possibly, as Leigh Fletcher, conjectures, another counter flow bringing air downwards where it warms through convergence, much like a terrestrial anticyclone.

The GRS is maintained through its ability to drain energy from surrounding jets, but also by absorbing smaller storms that add spin to its bulk. Red Spot Junior has grown through the same process: merging of smaller, cold-core white ovals increases the vorticity, or spin, to the point that they can generate their own internal circulation. Once it is established the resulting storm is a rather passive feature within the atmosphere. Air barely moves within them and then only in response to the circulation of air around them.

The only question that then remains is what kick starts the white ovals in the first place. At present the only clue is the warm, lower core. There are two possible sources of this warmth. In the first the closest terrestrial analogy is a warm, blocking anticyclone (Fig. 7.7a). In these terrestrial systems, a pool of warm air meanders with a kink in the polar front jet stream (Chap. 1). This has the effect of enhancing the kink further until the Rossby Wave (Chap. 1) expands to breaking point leaving a pool of warmer air cut-off on the cold, Polar side of the jet. This cut-off feature then acts to further block the passage of the jet stream, which then has to flow around it. Such blocks are the most persistent features of terrestrial mid-latitudes. Depending on their initial warmth, they can last for weeks or even months in what is otherwise a very chaotic flow. Such terrestrial blocks can be maintained by the continued reintroduction of more warm air and only fade when their supply wanes causing their internal temperature to fall.

Alternatively, the warmth is associated with a warm-cored thunderstorm complex bringing heat from deeper down. This forms a large storm with a cold cap. This is the white oval stage. These release latent heat as various gases like ammonium hydrogen sulfide, water and ammonia condense, remaining warm-cored deep down, but cold on top. This warm core then blocks or redirects the flanking jets. When the storm is sufficiently deep the core "opens up" revealing its pink heart. Unlike terrestrial low pressure areas, however, these tap their energy from the planet's interior, rather than the Sun. With a warm lower core they can continue as stable, features spinning between the planet's jets. The upper, visible and very cold feature is a passive veneer over an otherwise warm core. This would explain the updraft of cold, dense air within the storm, in apparent defiance of basic physics that would demand cold, dense air sinks. Deciding which, if any, of these hypotheses are correct will await more data that will hopefully come with the arrival of Juno in 2016.

Remarkable though the GRS is, it is certainly no longer as stable a feature as it was once thought. The GRS appears to have been as much as 40,800 km across in 1800, if contemporary paintings are accurate. A grainy black and white photograph taken in 1879 appears to vindicate this, but since at least this time the spot has been on the decline. In 1979 when Voyager 1 passed the spot was 23,200 km across, and today it is approximately 16,400 km in diameter along its longest margin. It would seem that there is a distinct downward trend that has accelerated to over 928 km per year. The cause of the shrinkage is unknown but may relate to the absorption of smaller storms into the GRS. Rather than adding momentum they may be draining momentum from it, so the storm shrinks. Is the storm then going to vanish for good? Before we write the obituary for the spot we should remember that the spot has disappeared before. During the 1680s the GRS vanished but reappeared nearly 20 years later in 1708. The spot may be taking a well deserved rest before emerging reinvigorated in a few decades time.

One final observation: while the Red Spot continues to shrink and become more circular, its color is morphing from pink to orange. Indeed, at the time of writing, it would be more accurate to refer to the GRS as the GOS.

Derechos and the Storm That Ate Itself

Jupiter is pock-marked with multiple storms, but if we cast the Great Red Spot aside, nothing much else springs to mind in terms of intensity or shear grandeur. If instead we turn our attention to Saturn we see storms on a vast scale, both temporary and long-lasting, that put the Great Red Spot to shame. These storms bear a striking resemblance to features we see on Earth only on a far grander scale.

Two Saturnian storms in particular were impressive: the long-lasting polar vortices, one shaped like a hexagon; and the northern spring storm of 2011. The polar hexagon appears to be a long-lasting event with its first detection in August 1981 by Voyager 2. There are some lovely ideas regarding the origin of this storm, but before we look at the storm in more detail it's worth looking back again at the structure of the Saturnian atmosphere (Fig. 7.8). After all, the storm owes its development to the interaction of two layers and the transfer of energy between them.

Like on Jupiter clouds of water lie deep down, just above the liquid hydrogen ocean, at around the 10 bar level. Ammonium hydrogen sulfide lies at 3–6 bar around 100 km higher, while ammonia clouds lie at 0.5–1 bar, just below the tropopause. It is thought that the large Saturnian storms originate in the water layer and presumably tap the internal heat of the liquid interior, just as they do at Jupiter. It is clear that these storms only appear in response to the slow change in seasons at Saturn. Aside from some deep oval storms Saturn was relatively storm free when Voyager

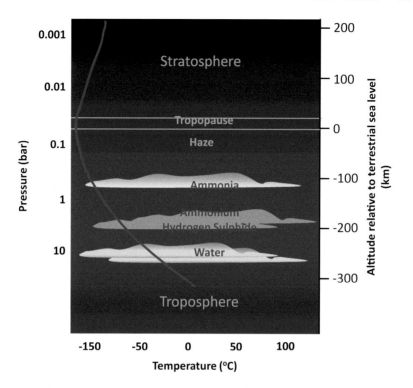

Fɪɢ. **7.8** The structure of Saturn's atmosphere. This is very similar to that of Jupiter, except that it is colder at ay particular altitude relative to Jupiter's. The lower gravity also allows for the clouds to be more spread out. The same bands of cloud are seen here as at Jupiter, although the lower temperature and pressure means that they occupy a greater spread of height than above Jupiter's liquid hydrogen sea. 1 bar is sea level pressure on Earth and occurs just under the Saturnian tropopause (*pale blue line*). Storms likely originate in the deeper layers then punch through and mix with the upper visible ammonia layer

passed. Since the arrival of Cassini, many more larger storms have been spotted and a deeper connection between the atmosphere and the interior has been uncovered.

The 2011–2012 storm was particularly illuminating. This storm appears to have begun as a cluster of small dark storms as early as late 2010, but these initially remained stuck within the deeper water and ammonium hydrogen sulfide layers. In 2015 Cheng Li and Andrew Ingersoll (Caltech) published a model that could reconcile the slow, early development of this storm, as well as the apparent periodicity of storms on Saturn. For observations, tend to suggest Saturn undergoes waves of storm development with a period of 20–30 years. Storms seem to occur during the northern summer; and this was has been the case since 1876 with storms seen in 1903, 1933, 1960 and 1990. The 2010–2011 storm followed the same trajectory as one observed in 1903, but was substantially larger. This time the storm came nearly 9 years ahead of schedule and appeared in the northern spring, rather than the summer. Why this storm chose to arrive early is unclear.

Each of these Earth-sized white storms is situated between the equator and the mid-latitudes in the northern hemisphere. Smaller oval storms are found at all latitudes except in the immediate neighborhood of the poles. Li and Ingersoll propose that Saturn spawns many small storms every year. The problem for these little storms is that they cannot penetrate the full depth of the atmosphere. It is suggested that water vapor at the 10 bar level is the problem. This gas bubbles, as heat is conducted or convected from deeper layers. The high density of the gas then prevents these plumes from penetrating the less dense upper ammonia cloud deck. Li and Ingersoll suggest that without the effective transfer of heat from the water layers far below, the upper ammonia deck becomes so cold and dense that it effectively subducts into the deeper warm layer. As it descends, it shoves the warm air up from below and the formerly deep storm erupts through the ammonia cloud deck. Li further suggests that one reason that such large storms aren't seen on Jupiter is because the planet is intrinsically dry. There is insufficient water to form a layer of air that blocks convection. Air simply rises through the entire depth of the atmosphere on Jupiter, distributing energy and preventing it from building up to the point which it generates large, periodic storms. This would jibe with the observations made by Galileo—and in the wake of the Shoemaker-Levy 9 impacts—both of which implied Jupiter has a surprising lack of water vapor. The arrival of Juno at the Jupiter system in 2016 should help settle that one.

One problem with this model is that it doesn't explain the apparent link to the northern spring and summer when all these storms erupt. I suggest a possible solution in Fig. 7.9. Here, heating of an upper layer that separates the upper cold layer from the deeper warm layer generates convection within it. This only occurs when sunlight heats the layer appropriately in the northern spring and

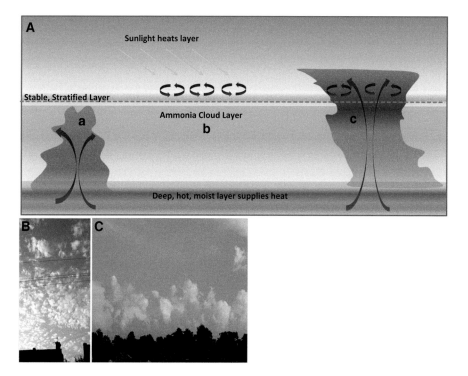

Fig. 7.9 Initiation of the 2011 storm. This possible model has this seasonal storm beginning with deep, moist convection that is trapped below a temperature inversion (**A**, *left*). A combination of heating of the layer from above (*center*) and intense convection below eventually breaks the inversion and allows ammonia-rich convection to rise higher, generating the visible storm. On Earth altocumulus castellanus clouds mark convection within a high layer (**B**). When surface convection breaks through the intervening (**C**) inversion rapid ascent and thunderstorms can ensue. Photos by author

summer. At this stage convection in this layer can link up with deeper convection or destabilize the upper cold layer. Disturbance of the upper layer then initiates subduction of the upper cold air into those below. On Earth analogous situations arise frequently in the spring and summer in the mid-latitudes. Cold air often finds itself trapped above an increasingly hot lower layer. If the middle-layers of the troposphere are heated sufficiently, they spawn further mid-level convection characterized by the appearance of altocumulus castellanus clouds or so-called high base storms (Fig. 7.9). When sufficiently hot and unstable, convection in the lower layer can penetrate any temperature inversion that is present and fuse with the middle layer. On its own this can cause thunderstorms, but throw an upper layer of cold air into the mix and the situation can become explosive. This is a frequent occurrence over the Mid-West of the US in the spring, where the introduction of an advancing cold front, steered by an upper level jet, sets in motion the formation of supercell storms that often spawn tornadoes.

The 55 day-long 2011 storm was unique in that it spread right around the globe, eventually consuming its own tale. Winds on Saturn are predominantly westerly, the 2011 storm erupted north of the equator where there is an easterly jet (Fig. 7.6). Like the terrestrial supercell storms the jet helped steer the storm to the west this jet directed the 2011 storm to the west. On its own this could only have provided enough oomph if it connected more deeply with the storm's base. It is here that we can turn our attention to long-ranging thunderstorms on Earth. These are the infamous derechos. These unique storms provide a clue to the longevity and underlying power of the Saturnian global storm.

Derechos are a world apart from humdrum thunderstorms. These fascinating phenomena are effectively unknown in the United Kingdom but are fairly common in the United States and eastern Europe and Russia. Similar phenomena also occur in Australia, Argentina and South Africa. *Derecho* roughly translates as horizontal wind in Spanish and describes a violent, gale force wind generated from vicious and fast moving thunderstorms. Derechos require a lot of energy in the atmosphere, both to form and then to propagate—advance at speed away from the originating disturbance.

The basic set up for these monster storms is the formation of a violent thunderstorm in very unstable, hot, energy-rich air. The initial thunderstorm can be really rather small and isolated, but given the right forward push, perhaps because the storm has become entwined in an upper level jet-stream, and the storm will become self-sustaining.

Most thunderstorms have a life measured in minutes, with individual cells lasting less than 35 min on the whole. Storms die when rain chills the air within the storm and generates strong downdrafts. Normally, these fan outwards in all directions and choke off the supply of warm air needed to keep the storm alive. Derechos circumnavigate this problem by moving quickly forward, through the warm air, so that the axis of the storm is tilted backwards. As their rain-cooled wall of air descends it drives upwards an accompanying wall of warm, moist air at its leading edge and since most of this downdraft is behind the leading edge of the updraft the storm can continue to advance without cutting off its supply of warm air. This pattern also allows the rapid formation of further storm cells along its leading edge (Figs. 7.10 and 7.11). In turn these replace the trailing, rain-cooled, storm cells at the rear and a wave is launched through the warm, humid air mass. In front of the wave, warm air is blasted upwards by the advancing wall of cold, rain cooled air at the rear. As long as there is a

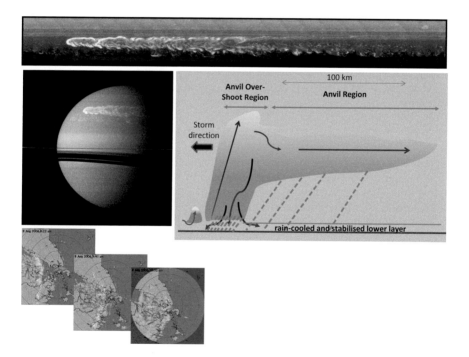

FIG. 7.10 The planet-wide springtime storm on Saturn in 2011–2012. The storm emerged as a cluster of small dark storms before morphing into an enormous, white monstrosity which encircled the planet. The storm bears a striking resemblance to advancing squall lines, such as Malaysia and Indonesia's "Sumatras" (radar image lower left of figure): These arise as smaller clusters of storms over the north coast of Sumatra during the summer monsoon, before fusing into advancing lines of storms. Images: Saturn—NASA/JPL/Cassini; Sumatra satellite images (*left*) http://www. nea.gov.sg/training-knowledge-hub/weather-climate/sumatras

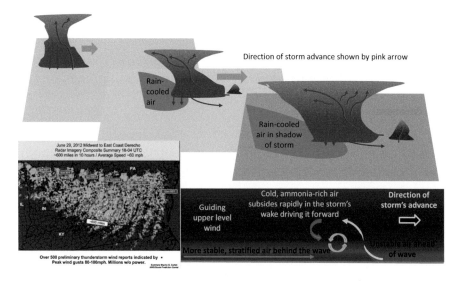

FIG. 7.11 "Derechos", "Sumatras" and the Saturn Springtime Storm are all examples of advancing waves. These storm systems are initiated in unstable air and if conditions are right will advance in the general direction of the air flow, swallowing unstable air along their leading edge, uplifting, cooling and releasing its energy. The June 29th 2012 derecho, in the US, began as a cluster of storms that rapidly organized into a wave. This swept across nearly half of the northern US from the Great Lakes to the east coast. It might have gone further if the cool Atlantic waters hadn't been in the way (lower left radar image). Similarly, the Saturnian storm began as a cluster of dark storms and could have gone on indefinitely had it not run into its tail and run out of unstable air. Derecho radar image courtesy NOAA

supply of warm unstable air in front of the wave—and as long as the storm cell can maintain its forward momentum—the Derecho continues. In its wake the warm, unstable air has been replaced by a deeper, colder and much more stable air mass, generated by the advancing storm. The atmosphere then becomes more stable, overall with warm air aloft, and cooler, drier air below.

A couple of Derechos are notable for their violence, longevity and for their degree of analysis and are worth describing, here, for comparison. On July 5th, 2002, a derecho struck eastern Finland generating winds of hurricane strength, which in turn produced over 400 storm-related wind damage events. The Finish derecho was the most northerly event of its kind ever observed. Like many derechos in the States, the storm originated when a cold front stalled over the western portion of the country.

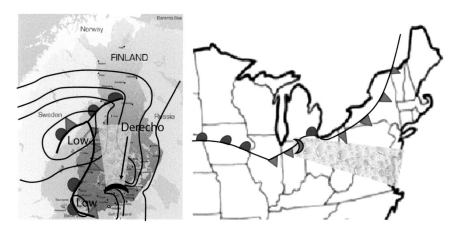

FIG. 7.12 The 2002 Finish derecho and 2012 US derecho. Both storm systems formed ahead of stalled cold-fronts: the Finish storm to the north of the Bay of Finland and the US storm to the SE of Lake Michigan. However, the Finish storm moved perpendicular to the prevailing wind, driven by upper level flows of air. The US storm moved more conventionally along the direction of the wind-field from NW to SE. The Finish storm damaged a track 450 km long, while the US system affected a track over 1100 km in length. Both storms died out when the supply of suitably unstable air was cut-off. The derecho tracks are shown as *light shaded areas*. Underlying US and Finish maps courtesy of http://www.proteckmachinery.com/stats.php?p=us-map-blank-vector and http://www.vector-eps.com/finland-vector-map/ respectively

Here, things diverged from the usual scenario that afflicts the States. On July 4th the unstable, warm and humid southerly flow of air was moving out of Germany and Poland towards the Arctic Circle. Over Lithuania a severe thunderstorm developed a classical comma-shaped (bow-echo) appearance that is typical of derechos (Fig. 7.12). This storm generated winds of over 100 miles per hour before dying out over the Gulf of Finland.

The following day this feature had moved north before merging with the system's warm front over northern Lapland. In its wake a narrow ridge of higher pressure poking into the warm sector over southern Finland into which air was descending from above. A belt of thunderstorms developed ahead of this region and soon these became organized into a classical comma (or bow-echo) shape and advanced quickly towards the northeast. The trajectory followed that of the Lithuanian storm system taken the day before. You can see, here, the link to the Saturnian storm. Descending air kick starts uplift in the air along its forward edge. This generates further storm cells along the forward edge of the downdraft.

Over the next several hours the Finnish system surged northwards, scooping up warm, unstable air in its path and replacing it with drier air descending in its wake. The storm crossed the length of Finland over the ensuing 12 h, leaving a trail of damage 450 km long.

The Finnish storm was severe in its own right, but the US storm of 2012 was even more remarkable. Spawned along the southern margin of another stalled cold front, this system developed during the tail end of the June 2012 heat-wave that brought temperatures into the 40s Celsius across much of the mid-west. Like the Finish system, this derecho began as a cluster of thunderstorms. These formed to the south of the Great Lakes along a wave in the otherwise stationary cold front. Given the vast amount of heat energy that was available, the cluster of storms soon organized themselves into another bow-shaped structure that was driven aloft by northwesterly winds.

What was extraordinary about this storm was its longevity. Powered by the high temperatures, high humidity and the strong upper level winds, the derecho crossed ten states from Indiana to Washington D.C. and finally out into the North Atlantic, moving at an average speed of 100 km per hour. Thirteen people lost their lives—primarily killed by trees felled by winds up to 130 km per hour (80 mph). One might imagine that such a violent storm would have at least drained the heat from the baking North American interior: there was no such relief, once the clouds cleared, the sun continued to bake down—the causative frontal boundary waving back northwards once more. The storm had crippled many power lines, leaving four million people without power. This was a critical effect adding insult to injury: with no power, cooling systems in buildings were rendered useless. Consequently, a further 34 people died as a result of heatstroke, or from its related effects.

Derechos are fairly well understood and relatively common in the US, with 50 or more occurring every year. In most cases storms begin on the northern side of an east–west trending frontal boundary—along its upper surface. As these develop they migrate southwards into the deeper, warm air and begin to draw energy from it. In this location, the storms soon organize into the classical bow-shaped structure that drives eastwards with the prevailing winds.

Although the mechanism underpinning their formation is distinct, derechos share many common features with other line squalls—organized systems of rapidly advancing thunderstorms. Amongst these are the tropical Sumatras (Fig. 7.11). Like the derechos of the mid-latitudes, these begin as a cluster of thunderstorms over northern Sumatra (Indonesia). These storms form during the northern summer monsoon as southwesterly winds draw very humid warm air northeastwards towards the west coast of Malaysia. The storms usually form in the late afternoon before crossing over the Straits of Melaka during the first half of the night. At night time the waters retain a lot of heat and readily give this up to the overlying air, causing the storms to intensify. After a short time the storm systems impact the west coast of Malaysia and Singapore. Here, deprived of further moisture, the Sumatras decay. Before they expire, the storms can generate winds up to 100 km per hour, along with 1–2 h of torrential rain, thunder and lightning.

The link to the Saturnian springtime storm can only be inferred (Figs. 7.9 and 7.10), but the appearance and propagation of the storm to encompass the planet are very similar to the propagation of line squalls on Earth. Saturn's springtime storm emerged at 37° N, with Cassini detecting lightning ahead to the appearance of the white cloud top. The storm tracked the trajectory of a seasonal easterly jet, and propagated faster to the west than the jet could have carried it. That these white storms are seasonal implies the Sun helps in part determine their development. Measurements show that these storms originate well below the level that the Sun's energy can reach—at a depth where pressures are 10–12 times that on the surface of the Earth. Here water and ammonia clouds condense and form a deep layer that rises and falls by convection.

Inevitably, one wonders how the Sun can control the formation of a storm that lies outside its reach. The answer might lie in the way the Sun controls the top of Saturn's atmosphere. As Saturn moves round its orbit and enters springtime once per 29.5 years, the top of the northern hemisphere's atmosphere warms and the weak westerly jet stream emerges. As on Earth this might mix colder air down into Saturn's deeper layers and break up any stable layer lying on top of the deeper clouds. Once this process has begun, storms that are normally trapped deeper down in the atmosphere can begin punching upwards, much as thunderstorms do on the Earth, once overlying stable layers have been breached. In the

US the development of supercell storms—and, in a slightly different manner, Derecho-bearing storms—accompany the breaching of stable layers above the embryonic storm cells. The storms remain small and isolated until the upper stable layer is degraded. At this point convection becomes explosive and a major storm cell is initiated. On Earth the Sun provides all of the energy to carry out this process, while on Saturn the Sun may merely breaks the seal and allows Saturn's hot interior to do the rest. Another link between terrestrial line-storms, such as derechos and Sumatras, is the alignment of the winds at different levels. On Earth where the upper winds blow in the opposite direction to the movement of the storm, any instability is sheared off and storms tend to die away. Align the two movements and storms can propagate aggressively. The upper wind helps get them going, much like giving a car a nudge on a steep hill, when the break is released.

What of Saturn's other storms? The white ovals are essentially the same as those on Jupiter with cold cores that project a few kilometers above the surrounding belts and zones. Unlike Jupiter, but continuing a trend that extends out to Neptune, the belt of equatorial westerlies is broader than it is in Jupiter. In general, the wind flow pattern is much simpler with a broad westerly belt overlying the equator and tropics. Only north of 30° N and 30° S do the westerlies give way to easterlies. Polewards of 50° or so from the equator narrower belts of easterlies and westerlies are found (Fig. 7.6). Within the southern hemisphere, one rather interesting and lightning-generating storm has been spotted. This "Dragon Storm" seems to be a semi-permanent feature, much like Jupiter's Great Red Spot. Although the storm appeared in July 2004 and vanished a few weeks later, the location seems to match a long-term region of storm generation and provides a further clue to the working of Saturn and Jupiter's atmospheres.

The Dragon Storm was illuminating for three reasons. Firstly, the storm generated bursts of radio waves in precisely the same fashion as terrestrial thunderstorms. Although lightning was not directly seen, the storm is believed to have generated a considerable battery of electrical discharges. Secondly, its location where multiple other convective storms have formed suggests that there is an underlying plume that regularly generates visible atmospheric storms. In this regard, it should originate with rising warm gas from the deeper interior. Infrared measurements of the GRS also indicate it has a deep warm core. Thus, both the GRS and the Dragon appear to have a warm heart. Within the Earth's atmosphere the largest storms have warm cores—as do the strongest anticyclones (above). The key is the pressure warm air creates. By pushing on its surroundings it creates pressure that can divert airflow around them. Both the GRS and perhaps the Dragon Storm appear to have this pushiness.

Finally, the Dragon Storm presented atmospheric scientists with a view of how such storms pump energy into the surrounding atmosphere. Unlike the GRS of Jupiter, in March 2004 a precursor to the Dragon Storm was seen shedding smaller, dark "stormlets". Some of these peeled away from the main storm area and migrated with the surrounding jet streams until they were eventually consumed. Others merged to produce larger storms that eventually morphed into larger, white ovals. This suggests that at least at Saturn, warm-cored deep storms eventually drive the formation of much higher, cold features. Again, this implies that perhaps features like the GRS are driven first by the generation of a warm core deep down, which then diverts the jet stream.

Aside from illuminating how storms develop and contribute to the overall running of Saturn's atmosphere, the Dragon Storm presented an odd puzzle. Radio signals, associated with lightning, were only detected when the storm was below the night–day terminator. As soon as the storm rose over the horizon Cassini saw the radio signals dropped off. This was repeated over many rotations of Saturn over several weeks. Some researchers concluded that the source of the lightning was related to the visible storm but that they were separated by such a great distance that the lightning source was in view while the cloud was not. Perhaps instead the emissions are beamed, much like a laser, in the direction of Cassini (this would be coincidental); or perhaps illumination changes the structure of the cloud so that it no longer produces sufficient charge separation.

One of the intriguing general features of the gas and ice giant atmospheres is the steady increase in mean wind speed with increasing distance from the Sun. This is also reflected in the increasing simplicity of the airflow. One suggestion is that nearer to the Sun solar radiation helps generate more turbulence and such turbulent motion robs the atmosphere as a whole of its energy. This leaves less to power the jet streams. Perhaps this is weakly analogous to the situation on Earth. When the winds blow most directly from west to east, they also blow strongest. When the winds become, as meteorologists term, more meridional and less zonal, with a stronger north-south component, the general speed decreases. At Uranus and Saturn (Chap. 8) there are effectively five belts over the entire globe. A large easterly belt overlies the equator, with two broad westerly belts extending to each pole and a relatively quiet polar region with slower westerly winds. As more energy is available to blow the zonal winds, Saturn has stronger winds (around 500 m per second, or 1800 km per hour), and these only accelerate as we move further out from the Sun.

The Polar Hexagons

Discovered by Voyager 1 in 1980 was a large hexagonal cloud feature that appeared to rotate with Saturn. Superficially, the structure appears utterly shocking: this is an apparently perfectly, hexagonal feature, thousands of kilometers side-on-side that sits above Saturn's North Pole. Infrared observations show that the hexagon is rather deep; extending to several bars of pressure—essentially to the base of the atmosphere. From the perspective of Cassini or Hubble each edge seems perfectly straight with perfect 120° angles. However, scale and distance are important; let's look at this storm and reconsider its geometry. A side length of 13,800 km is wider than the diameter of the Earth. Each corner is roughly 500 km across, meaning what appears to be a sharp turn is actually rather smooth.

Even without the sharp turns, how then do we recreate a hexagonal feature? Recall Chap. 1's atmospheric features called Rossby Waves. Due to the Coriolis Effect winds rotate counterclockwise around the poles (west to east). Because air is trying to move north and south to even out pressure, the atmosphere organizes itself into a series of waves. Now, on Earth surface features, such as the Rocky Mountains, tend to cause buckles and kinks that alter the flow, but on Saturn there is no surface to impede movement. There is still a north–south pressure difference because the atmosphere is warmer across the more intensely illuminated latitudes, and because the spinning motion of the planet influences convection from the deep interior. As Fig. 7.6 indicates, at 78° N where the hexagon is found there is a sharp change in wind velocity with latitude from 100 m per second east to near stationary or slightly to the west. This abrupt change is all that it takes to generate the waves. Indeed, it is the absence of such an abrupt change in wind velocity near the South Pole (and in the atmosphere of Jupiter) that prevents a similar hexagon from forming here.

Put this together and the result is Rossby waves within Saturn's atmosphere. In the atmosphere of a planet with a sizable spin, like Saturn, the atmosphere will naturally set up jet streams that rotate around the poles in a series of these Rossby waves. Within the core of this six-wave pattern is the polar vortex that is largely cut-off from the rest of the atmosphere.

The Rossby Waves model has been successfully recreated in the laboratory. Ana Aguiar and colleagues (Oxford University) continued their research using a large tank of fluid, in which they could alter the velocity at different distances from the tank's center. The research clearly showed that where there was a suitably steep gradient in the velocity of the winds, a polygonal pattern rapidly becomes established. In experiments shapes with up to 8 sides were appeared, with six sided polygons the most common. In experiments where tanks of fluid were used, rotating the middle of the tank faster than the fluid on the outside resulted in the formation of turbulence in a narrow zone between the periphery and the center. The polygonal shape arises when the turbulence begins to organize into a fixed number of smaller vortices that form the motion of the fluid on the slower outside of the tank. As the vortices mature they naturally space themselves out and then organize the rest of the flow around them. Ultimately, each vortex forms a corner of the polygon. If this sounds abstract, remember that on the Earth the jet stream will naturally generate identical waves with sides thousands of kilometers long (Chap. 1). "Abrupt" changes in direction aren't visible merely because Earth is a smaller world where the bends in the jet take up a proportionately larger length of the overall jet.

What of the southern vortex? Cassini images are telling. Here, the circumpolar winds organize white oval storms into a broad belt around what is clearly an eye-wall. Earlier we used this term for the wall of clouds around Jupiter's GRS, but strictly speaking this is more of a gap in the main cloud decks. Saturn's southern vortex has, instead, a clearly defined eye wall: a region of clouds that rises above the main deck (Fig. 7.10). Within this wall the level of the cloud deck bottoms out. Inside this lower shield of cloud rests a smaller polygonal feature analogous at least in terms of shape to the northern hexagon. Quite why the two hemispheres show distinct patterns is unknown but conceivably might be due to the season. It could be that the southern storm and northern storm alternate structures depending on the relative amount of solar heating. That won't be clear until Saturn has completed one of its 29.5 Earth-year long orbits.

Speaking of heating, oddly enough if one was planning a balmy vacation in Saturn's cloudy atmosphere, the southern polar region would be the place to be as it is a good 63 °C warmer than anywhere else on the planet. While the cloud tops drop out at –185 °C on average, the South Pole is almost as warm as a Martian night at –122 °C. Certainly not tropical by any standards, it is still significantly warmer than one would expect for distant Saturn. Quite why the South Pole is so much warmer than the rest of the planet is mysterious. Solar heating is unlikely, as this temperature is far above those reached at Jupiter, and right now, the south pole remains warm, while the Sun is heating the northern hemisphere most strongly. Instead this extra heat is almost certainly coming from Saturn's interior. If so the difference in each vortex might reflect differences in the delivery of heat to the atmosphere. Saturn might always have a "warm" south pole regardless of the season.

Ice Fall from the Rings

While the South Pole is the place to be for an artificial "summer" on Saturn, the place to go for snow lies at the equator if you are prepared to go high enough up. Saturn's rings orbit the planet at a distance of 7000 km from the cloud tops. Although they seem permanent, various processes, from collisions to the action of Saturn's magnetic field, are constantly whittling away at them. While shepherd moons do their best to keep the rings in shape, material will naturally spread outwards and inwards from the main body of the rings. The icy material that falls inwards is captured by Saturn.

These icy debris drizzle down into Saturn's upper atmosphere. Although this persistent snowy fall is miniscule and evaporates long before it reaches Saturn's cloud-tops, it does contribute water vapor to the upper atmosphere of Saturn.

If you want proper snow on Saturn then look just beneath the tropopause. Here, ammonia ice crystals barrel around the planet in the most severe blizzard you can imagine. 100 km lower and this blizzard morphs into an ammonia rain, as temperatures climb. Jupiter has much the same, but here the snow has considerably less far to fall before it encounters the ammonium hydrogen sulfide layer and chemically merges into it.

Io's Electrifying Connection to Jupiter

The giant planets all generate strong magnetic fields that actively deflect the solar wind (Fig. 7.13). Jupiter's field is roughly 14 times stronger than that of the Earth, while Saturn's is just approximately half the strength of our planet's. The Jovian field—its magnetosphere—completely encases the Galilean satellites, with Io, Europa and Ganymede buried deep within it. Callisto, lies nearly two million kilometers from Jupiter and is effectively separate from the remaining, inner three moons. As each lies within Jupiter's magnetic field they are subject to a constant pummeling from charged particles that are trapped within the field. In turn each moon sheds copious charged particles into the magnetosphere. Most of these remain trapped along the orbit of the satellite, but as the moons sweep through the cloud of ions, some are driven out of the ring and along the magnetic field line towards Jupiter (Fig. 7.14).

Io adds another dimension to this relationship. Io is also tortured by strong gravitational forces that keep its interior molten. Consequently, Io is racked by violent volcanic activity. Blasting hundreds of kilometers above the satellite's surface, these eruptions contribute a wealth of sulfur, oxygen, sodium and chlorine ions at a rate of approximately 1000 kg (1 metric ton) per second. This is approximately the same rate of loss as the Earth, Venus and Mars lose from their atmospheres. Like those oxygen and hydrogen ions shed from the outer Galilean satellites, this broth of charged particles is swept in a pair of arcing columns of ions along Jupiter's magnetic field lines (Fig. 7.15).

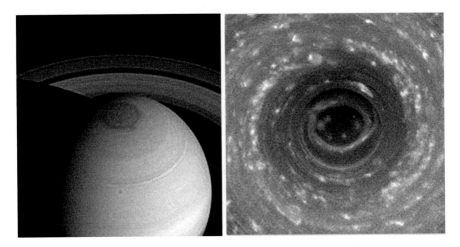

FIG. 7.13 The Polar vortex storms of Saturn. The North Polar storm (*left*) is clearly hexagonal with edges over 13,800 km long and rotates with the same period as Saturn as a whole (10 h 39 min). Within the hexagon there is a more circular elevated storm region, separated from the walls of the hexagon by a lower "moat". Over the south pole there is another storm with broadly circular eye walls. However, superficial inspection show that within this structure is an inner vortex which is roughly hexagonal with roughly straight walls (*right*). Images by Cassini (ESA/JPL/NASA)

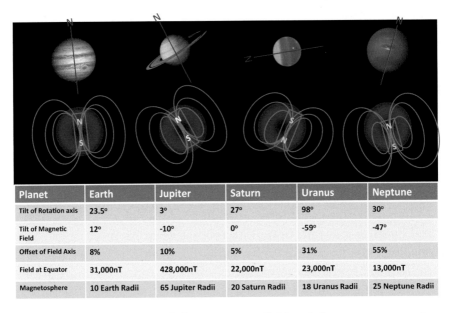

Planet	Earth	Jupiter	Saturn	Uranus	Neptune
Tilt of Rotation axis	23.5°	3°	27°	98°	30°
Tilt of Magnetic Field	12°	-10°	0°	-59°	-47°
Offset of Field Axis	8%	10%	5%	31%	55%
Field at Equator	31,000nT	428,000nT	22,000nT	23,000nT	13,000nT
Magnetosphere	10 Earth Radii	65 Jupiter Radii	20 Saturn Radii	18 Uranus Radii	25 Neptune Radii

FIG. 7.14 A comparison of the magnetic fields of the giant outer planets. Aside from Jupiter's, which dwarfs all the others, most have field strengths comparable with the Earth. Jupiter and Saturn generate theirs in their core, while Uranus and Neptune generate theirs further out. Image credits: Jupiter (HST/NASA); Saturn Cassini/NASA; Uranus: HST/NASA; Neptune Voyager 2/NASA

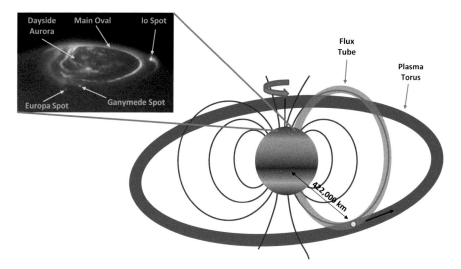

FIG. 7.15 The complex interaction of Jupiter's large Galilean moons with its atmosphere and magnetosphere. Jupiter's powerful magnetic field sweeps ions across the surface of its moons and also traps ions released by these satellites. Such ions are directed into two flows, one following the orbit of the satellite through the planet's magnetosphere, and the other flowing along the planet's magnetic field lines to the cloud tops. Within this second flow, known as a flux tube. A giant electrical circuit links the moons to the planet. Io has a particularly aggressive interaction, as its volcanoes release as much mass each second as the Earth loses to the solar wind. Io's flux tube thus carries a whopping 5 million Amp current. With 5 trillion Joules of energy transferred each second. As the flux tubes from Io, Europa and Ganymede reach Jupiter they generate intense foci of aurora (*inset* photo, NASA)

As these particles move along the lines they accelerate until they smash into the thermosphere of Jupiter. Here, ionization causes permanent aurora focused into spots and arcs (Fig. 7.15). Within Io's flux tube runs a 1 trillion ampere current that flows back and forth along the flux tube as though it was a giant copper wire. More than 2 trillion joules of energy is transferred every second. In many ways the flux tube forms a far more visible version of the circuit that links the terrestrial ionosphere (and the thunderstorms below) to the magnetic field. Were you able to hover in the vicinity of where the flux tubes connect with Jupiter, as well as impressive aurora, beneath this shimmering blue canvas would flicker and flare hundreds of lightning bolts as Io's flux tube probed down towards the metallic hydrogen mantle. The presence

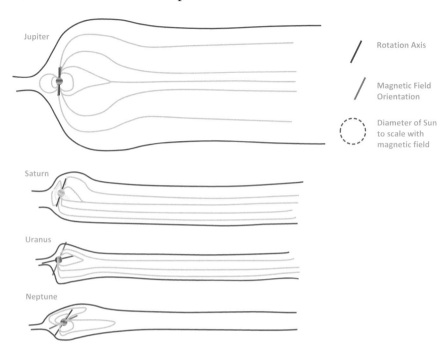

Fig. 7.16 The scale of the magnetic fields around the giant planets. Jupiter's dwarfs the others and stretches 700 million kilometers to Saturn. Those of Uranus and Neptune are steeply inclined to the oncoming solar wind and whip the field around in different directions as the planet rotates. Planetary field lines are in *yellow* while the solar field lines are in *red*. The *spin axes* and *magnetic field axes* are indicated above along with a comparison to the diameter of the Sun

of Jupiter's Galilean satellites adds far more conducting material than is available from the exosphere and solar wind; while the rapidly rotating powerful magnetic field acts as a super-charged dynamo driving the flow of charge into and out of Jupiter's upper atmosphere. Out at Saturn, Titan also sheds particles into the magnetosphere of Saturn, though this is not enough to power similar, punctuated auroral displays as are seen at Jupiter.

The flux tube connection points are embedded in more general auroral emission (Fig. 7.14). The day lit side of the polar region glows permanently with a constant barrage of charged particles arriving from the Sun that are grabbed by Jupiter's powerful magnetic field and like the Earth, some aurora are focused into a large ring centered around the magnetic pole. Here, the interaction of the magnetic field lines with the atmosphere generates another circum-polar current in Jupiter's ionosphere.

Filled with ions from its satellites and from the solar wind, Jupiter's magnetosphere is a powerful source of radio emission in the 0.6–30 MHz range. However, Jupiter also generates maser emission as so-called Alfvén waves carry pockets of ionized gas into Jupiter's Polar Regions. As this radio emission is beamed it is not always detectable from the Earth. When it is, it emphasizes Jupiter's regal status by exceeding radio emission from the Sun. One must remember that with the exception of sunspots, Jupiter generates the strongest magnetic field of any object in the Solar System. Jupiter really is the King.

Is Mighty Jupiter Losing Any Atmosphere?

Previous chapters showed that Venus, Earth and Mars are shedding their atmospheres to space at a rate of about one metric ton per second. This is mostly through mechanisms involving ionization of their gases and the subsequent loss of them to their magnetic fields or the solar wind. Mars also loses gases directly as their kinetic energy can overcome the gravitational attraction of the planet. By contrast, Jupiter's strong gravitational field and its relatively cold upper atmosphere guarantees that most of its atmosphere will remain bound for hundreds of trillions of years. Some gas is shed through its magnetic field. Like Venus, particles become charged high in its upper atmosphere and here, where the gravitational pull is weakest, these particles can become embroiled in the strong magnetic field. From here, they can be accelerated into the magnetosphere and hence slowly out into space. A small amount might find its way to Saturn, where the immense field of Jupiter occasionally sweeps past. Jupiter and Saturn thus form a frigid pairing, much like Venus and the Earth (Chap. 5).

However, the rate of loss is trivial. Even compared to Venus, where trillions of years would be needed to bleed the atmosphere dry, Jupiter and Saturn will remain impressive, if frigid, gas giants until the distant dark future of the universe. The only moment of risk for these impressive worlds will come during the few hundred million years that the Sun is a red giant, when cloud top temperatures will rise above 100 °C at Jupiter and well above 0 °C at Saturn. This will be enough to drive enhanced mass loss, but still nothing near enough to drain these vast worlds of much of their substance.

Conclusions

Jupiter and Saturn are the true giants of the solar system. Each manifests weather on a scale dwarfing the small terrestrial planets in their entirety, but when one looks closely at their super-charged phenomena, one returns to familiar territory. Saturn's mysterious hexagon is a grossly expanded manifestation of our polar front jet stream; the Great Red Spot shares some aspects of its structure with warm cored high or low pressure areas; Saturn's northern spring time storm of 2011 behaved like a derecho (or tropical Sumatra). Weather, it seems is the same the universe over, with each planet's atmosphere flowing to the same underlying laws. This is not surprising in itself: what is amazing is how these laws play out and how scale can make such an impressive difference at least to our eyes.

This is a theme that will continue as we head outwards to the ice giants, Uranus and Neptune, and to their icy satellites. Beyond our cosmic shore we will encounter Rossby waves and circular storms once more, forming a unified picture of weather across the universe.

References

1. Sánchez-Lavega, I. A., del Río-Gaztelurrutia, T., Hueso, R., Gómez-Forrellad, J. M., Sanz-Requena, J. F., Legarreta, J., et al. (2011). Deep winds beneath Saturn's upper clouds from a seasonal long-lived planetary-scale storm. *Nature, 475*, 71–74.
2. Li, C., & Ingersoll, A. P. (2015). Moist convection in hydrogen atmospheres and the frequency of Saturn's giant storms. *Nature Geoscience, 8*, 398–403.
3. Transient internally driven aurora at Jupiter discovered by Hisaki and the Hubble Space Telescope. (2015). *Geophysical Research Letters. 42*(628), 1662–1668.
4. Barbosa Aguiar, A. C., Read, P. L., Wordsworth, R. D., Salter, T., & Hiro Yamazaki, Y. (2010). A laboratory model of Saturn's North Polar Hexagon. *Icarus, 206*(2), 755–763.
5. Emily Lakdawalla. (2010). *Saturn's hexagon recreated in the laboratory*. Planetary.org.
6. Godfrey, D. A. (1988). A hexagonal feature around Saturn's North Pole. *Icarus, 76*(2), 335.
7. Sánchez-Lavega, A., Lecacheux, J., Colas, F., & Laques, P. (1993). Ground-based observations of Saturn's north polar SPOT and hexagon. *Science, 260*(5106), 329–32.
8. Sayanagi, K. M., Dyudina, U. A., Ewald, S. P., Fischer, G., Ingersoll, A. P., Kurth, W. S., et al. (2013). Dynamics of Saturn's great storm of 2010–2011 from Cassini ISS and RPWS. *Icarus, 223*(1), 460–478.

8. The Ice Giants

Introduction

Superficially the ice giants, Uranus and Neptune, are simply smaller cousins of the gas giants, Jupiter and Saturn, but it doesn't take much examination to realize that this analogy is only skin deep. Aside from their smaller size and much lower tropospheric temperature, the interiors of these worlds are composed chiefly of the ices ammonia and water. Hydrogen makes up nearly 83 % by mass of their atmosphere, with helium second at 15 %, but this is a relatively thin veneer over a thick sea of liquid dominated by water and ammonia. This structural difference has fairly profound consequences for the internal workings of each world, including how its atmosphere and magnetosphere behave. Their relatively low mass compared to Jupiter and Saturn means that their upper atmospheres extend for thousands of kilometers. Most surprisingly, that of Uranus extends out past its rings and affects their behavior as well.

The outer two giants also play host to the solar systems strongest winds. Yet, despite many similarities, there are perplexing differences between them. Neptune, like Jupiter and Saturn, releases more energy than it receives from the Sun, while Uranus appears dead, releasing only what it receives. Quite why these two worlds have evolved differently remains unclear. Uranus had also appeared utterly bland when Voyager 2 swept past it in 1986, until in recent years Uranus has had something of a meteorological renaissance on showing an atmosphere that appears as dynamic and interesting as our solar system's other pale blue dot, Neptune. Yet many gaps remain in our understanding and will likely do so for decades to come unless more missions are launched their way. This chapter is a story of what is known and what remains to be discovered.

© Springer International Publishing Switzerland 2016
D.S. Stevenson, *The Exo-Weather Report*, Astronomers' Universe,
DOI 10.1007/978-3-319-25679-5_8

Inside the Ice Giants

Both ice giants have a very similar structure. The visible atmosphere extends fully 1000 km below the cloud tops. Until Yohai Kaspi (Weizmann Institute) and co-workers completed studies of the gravity fields of both planets this layer was poorly understood. Using various observations of the wind-driven cloud patterns and shape of the planet it was possible to work out the distribution of mass and density for both worlds. In each case Kaspi concluded that the "weather" on both worlds was confined to a layer above 2000 bars pressure on Uranus and 4000 bars pressure in the more massive Neptune. In terms of depth, this meant that each planet had winds confined to approximately 1100 km from the visible surface. Those mighty jets, which are seen pushing the weather on Uranus and Neptune around, were fairly shallow in nature encompassing less than 1 % of the planets mass. Kaspi's team obtained the same result as that of earlier modeling of the jet stream, which also implied that weather was limited to the outer skin of these large planets. Given the apparent differences in terms of the heat that these worlds emit and their appearance, such a strong physical connection might seem surprising. The bottom line is that they each are very similar in mass and density, and probably formed at similar positions in the solar system. This is most likely roughly where Saturn is found today.

A different set of models were produced by Jonathan Aurnou and Moritz Heimpel. In these the circulation was made to extend deeply towards the core. When this manipulation was carried out the number of jet streams that was seen in the lower atmosphere decreased from 20 to around 3–5. This is clearly at odds with observations of Jupiter and Saturn, which both have around 10–20 jet streams. It also implies that they have shallow circulations. Indeed, most models of giant planet atmospheres suggest that the lower boundary for circulation occurs at the depth where the gases become cool enough for silicates and metal oxides to condense. This occurs at temperatures of around 2300 °C, which, in Uranus and Neptune is broadly at the depths Kaspi concluded—around 1100 km below the visible "surface".

Beneath the meteorological domain at a little over 1100 km down inside both Uranus and Neptune, the atmosphere gives way to a thick layer dominated by hydrogen and helium gas. This is broadly analogous to the liquid hydrogen and helium layer seen in Jupiter and Saturn, except that it is at a lower, overall density. As we continue deeper at a depth of 5000 km the hydrogen–helium soup gives way to what might loosely be called a deep, hot ocean. The interior of both worlds is never dense enough to make hydrogen become metallic. Instead this hot icy ocean forms the lowermost layer above the planetary core. Like the liquid hydrogen layer inside Jupiter and Saturn, this material gradually morphs from gas to liquid state and then, possibly to a plastic-like semi-solid layer nearest to the core. Dominated by water and ammonia, this extends down to a depth of roughly 20,000 km, where it should be compressed further into a type of solid ice. Pressure in this layer rises steadily from 300,000 bars to 6 million bars at the core-mantle boundary. It is possible that somewhere within this deep partly liquid shell, methane is compressed and transformed into diamond—at least in the more massive Neptune, giving Neptune a diamond rain onto a jewel-encrusted core. At the heart of each world is thought to lie a small hot, rocky-metallic core with a mass perhaps a half that of the Earth. The core probably simmers at around 5000 °C in Uranus and somewhat hotter at 7000 °C in Neptune.

Given their similar mass and history, why does Uranus appears so much colder than Neptune from the vantage of its cloud tops and why does Uranus radiate so much less energy than its near identical twin? Two possible answers emerge: either Uranus experienced a catastrophic collision which stirred up its interior and allowed heat to escape, or Uranus's interior is structured in such a way that heat is prevented from escaping. The latter hypothesis seems more credible as you would imagine that a catastrophic impact would release more heat into the interior, causing warming rather than cooling. If this "insulating blanket" is a seasonal feature associated with layering in the atmosphere, it could well alter as Uranus moves from one season to the next.

What of the diamond layer? Aside from the jewel-encrusted possibility of diamond rain, at the pressures and temperatures concerned, the diamonds that fall will most likely melt into a rather interesting diamond sea that overlies the small, rocky core.

The Structure of Uranus and Neptune's Atmospheres

Uranus has an atmosphere structured much like that of Saturn and Jupiter. However, the lower gravitational field means that this is much more extended. It is also much colder, lying almost twice as far from the Sun as Saturn. This means that the layers of cloud we have encountered before (ammonia, ammonium hydrogen sulfide and water) lie at much greater depth and are all but invisible beneath a thick layer of hydrocarbon haze at 130–260 km above the 1 bar level. Figure 8.1 shows this in more detail.

Despite the bland-seeming appearance Voyager 2 encountered, which might suggest a permanent and pervasive fog, the atmosphere of Uranus is otherwise really clear. Indeed, Uranus has one of the clearest atmospheres in the solar system. Only Neptune exceeds its clarity. Like Jupiter and Saturn before it, and Neptune subsequently, the atmosphere of Uranus can be divided into broad layers analogous to those above the Earth. Extending downwards from 56 km at the 1 bar level to 300 km below is the troposphere. The atmosphere is at its most dynamic in the troposphere with extensive cloud decks and very strong winds powering westward flow around the tropical latitudes and eastward flow nearer the poles.

Winds exceed 240 m per second, or the best part of 870 km per hour, near the mid-latitudes on Uranus. On Neptune the pattern of wind strength with latitude is broadly reversed, but the general pattern of direction is maintained. The same equatorial easterlies and mid-latitude westerlies are found, but on Neptune the winds are strongest in the easterly jet along the equator, with winds in excess of 400 m per second or a nifty 2,100 km per hour. These are the fastest winds in the solar system—a good three times faster than winds on Jupiter and more than four times faster than the fastest tornadic winds on Earth (Fig. 8.2).

Uranus has the coldest tropopause of any planet in the solar system, with temperatures barely above –220 °C. Oddly, the coldest region of the atmosphere currently lies 25° south rather than at the dark southern pole, but this regime is changing as Uranus moves from northern summer to autumn. Uranus, like Jupiter and Saturn has an atmosphere dominated by hydrogen and helium.

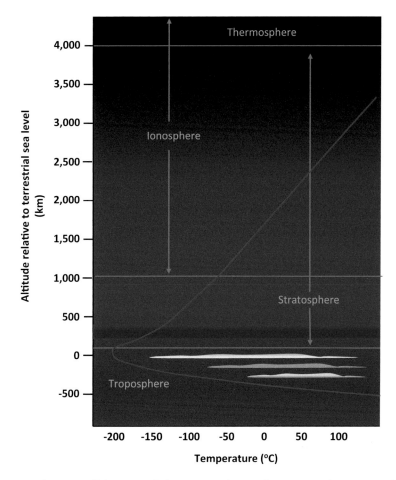

FIG. 8.1 The overall layout of the atmosphere of Uranus. The atmosphere is very extended compared to that of the Earth (and much colder overall). Extending out to around 10,000 km is an ionosphere, which may, like the Earth, have different layers embedded within it. This overlaps a thermosphere and stratosphere with temperatures that steadily fall from around 523–573 °C at 4000 km up to –220 °C at the tropopause which is situated at 56 km above the 1 bar level (equivalent to sea level on Earth). It is currently unknown why the upper atmosphere is this hot. The stratosphere has a hydrocarbon haze layer at an altitude of 130–260 km above the 1 bar level. Within the troposphere lie four cloud decks. The coldest and highest is composed of methane at 1–2 bar; ammonia or hydrogen sulfide clouds at 3–10 bar; ammonium hydrogen sulfide clouds at 20–40 bar and finally water clouds deep down at 50–300 bar. Thus Uranus (and Neptune) have cloud decks much like Jupiter and Saturn, but these are found at greater depths and pressures where temperatures are sufficient to allow their condensation

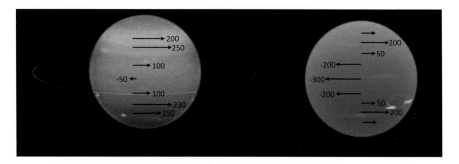

FIG. 8.2 The circulation at the cloud-tops of Uranus (*left*) and Neptune (*right*). Both planets show three jet streams in their atmospheres, but the overall strength is different. While Uranus has only a weak easterly jet stream above its equator, with winds of 180 km per hour, Neptune's winds scream around its equator at well over 1000 km per hour (*left*). Both planets have strong westerly jet streams at mid-latitudes with Uranus having its peak winds in its northerly (Sun-facing) jet. The southern jet is around 20 m per second slower

Unlike the former two worlds, the amount of helium is broadly the same as found in the Sun, indicating that little if any helium has settled into the planet's deep interior. The remainder of the atmosphere (around 2 %) is methane. Rayleigh scattering by this compound gives Uranus its deep aquamarine color and Neptune its vivid blue. Above the methane clouds the atmosphere has far less methane, most of it having condensed out at lower levels.

The upper troposphere and lower stratosphere of both Uranus and Neptune hosts layers of hydrocarbon haze produced, as with Jupiter and Saturn, when ultraviolet light and cosmic rays impact and energize molecules of methane. The broadest of these bands lies in the stratosphere between 160 and 320 km above the nominal 1 bar level. Overall, it is the narrow distribution of these haze layers that ensures that the remainder of the atmosphere is clear (Fig. 8.3).

On Uranus, the methane cloud deck was found to lie at 1.2–1.3 bars with deeper ammonia, ammonium hydrogen sulfide and water clouds implied from spectral measurements, and through implication from the general similar (overall) composition of Uranus and Neptune to Jupiter and Saturn. Water clouds are thought to lie at 50–300 bars, ammonium hydrogen sulfide at 6–10 bars, although this is uncertain (Fig. 8.4).

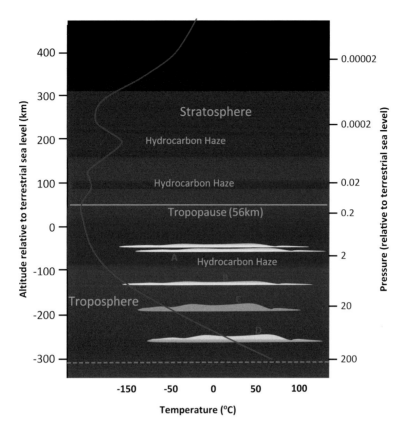

FIG. 8.3 Zoom-in on the lowermost stratosphere and troposphere of Uranus. Temperatures rise across the tropopause as we descend through the troposphere, the base of which lies at around 320 km below the 1 bar level (*dashed blue line*). Various layers of hydrocarbon haze fill the upper troposphere and lower stratosphere. These hazes consist of ethane and acetylene mixed with more abundant methane. Methane clouds (*A*) occur within 50 km of the 1 bar level, with ammonia (or hydrogen sulfide) (*B*), ammonium hydrogen sulfide (*C*) and water clouds (*D*) at greater depths (Fig. 8.1). Temperatures rise from around –220 °C at the tropopause to over 200 °C at the base of the troposphere. More limited heat flow from the Uranian interior means that convection, and hence clouds, are more limited in extent in the atmosphere of Uranus than the other giant outer planets

As well as bright clouds, which appear to becoming more common, Neptune and Uranus also share a propensity for dark spots, the most famous was the anticyclonic Great Dark Spot that was visible when Voyager 2 visited in 1989, but subsequently shrank and faded.

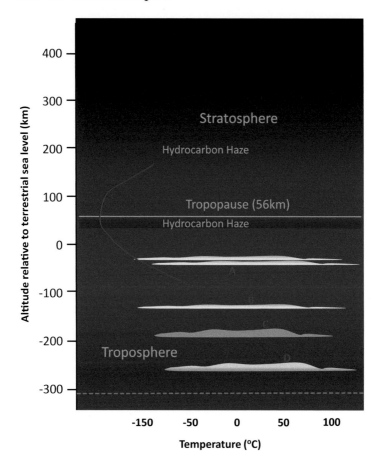

FIG. 8.4 The structure of Neptune's lower atmosphere. Like Uranus the temperatures at the tropopause are extremely low at around –220 °C. Convective clouds of methane (*A*) are visible in Voyager 2 and Hubble Images. The deepest layers, visible in the larger storms are dark but the composition of this layer is unknown. It is likely that like Uranus, underneath the visible methane clouds lie deeper layers of ammonia (*B*), ammonium hydrogen sulfide (*C*) and water vapor (*D*). These are too deep to currently observe directly. Like Uranus various layers of methane and other hydrocarbons form hazes at higher altitudes

Lawrence Sromovsky and Patrick Fry used HST and Keck images to analyze cloud movements in the atmosphere of Uranus. There appeared to be no overall change in wind speeds from 1986 to 2004. However, they did find a wealth of new information. Storms came in a wide variety of forms: some lasting hours; while another, dubbed S34, appeared to be a long-term feature throughout the period. S34 also appeared to oscillate in latitude from 32 to 36.5° S suggesting that it was entrained in a Rossby Wave (Chap. 1) as it pummeled around Uranus.

Above the tropopause is a stratosphere with steadily rising temperatures (Figs. 8.1 and 8.3). Heating occurs by conduction from the layer above and through the action of cosmic rays and ultraviolet radiation from the Sun. Although not detected thus far it would seem likely that atmospheric gravity waves exchange energy between the different atmospheric layers, as they do on Venus (Chap. 5) and other worlds. We'll say more on these phenomena in Chap. 10.

Finally, above 4000 km is the thermosphere. At 850° K (577 °C) this is by far the warmest in the outer solar system, bar Jupiter. At present the reason for this intense heat is unclear, but it may relate to the shape and activity of Uranus's magnetic field. Overlapping the stratosphere and thermosphere is an ionosphere where energetic radiation has stripped atoms and molecules of one or more of their electrons, leaving them positively charged (Chaps. 5, 6 and 7). It is through such action that particles can attain enough energy to escape both Uranus and Neptune.

Uranus has an odd thermosphere in other regards. For one, it is vast: stretching 50,000 km, or several times Uranus's radius, deep into space, it is more like the vast coma of a comet rather than an atmospheric layer. The thermosphere encompasses the planet's dark rings and exerts drag on them. Ultimately this causes particles from the rings to rain into the upper atmosphere. The thermosphere will also bombard the rings with energetic particles and could well contribute to chemical reactions that drive their darkening. Uranus's extensive thermosphere also necessitated NASA carrying out course corrections for Voyager 2, which plowed through it en route to Neptune. The underlying cause of this puffed up thermosphere is unclear but it may relate to heating from charged particles that are trapped within and energized by the planet's peculiar and ever-changing magnetic field (Figs. 8.6 and 8.7).

Neptune does not share such an extensive or warm thermosphere, ensuring that it hugs more tightly to its cloud tops and experiences a lower rate of loss to interplanetary space. Neptune's somewhat higher gravitational pull also contributes to this effect. However, where Neptune holds a cooler thermosphere, Voyager 2 detected something odd in its stratosphere. Here, temperatures rise to over 480 °C over the equator, while further north and south over the mid latitudes they remain resiliently below zero. At either Pole the air appears to warm again, however, not to the heights seen over the equator. Why is the atmosphere so warm here? The suggestion is that over the mid-latitudes, where winds blow from west to east, air is rising and cooling. High in the stratosphere this reverses and arcs back, mostly towards the equator but also, to a lesser extent, towards the Poles. As this air descends—and one must remember that it is moving at hundreds of meters per second to the east as well as downwards under gravity—it compresses and warms up. This makes the equatorial regions of Neptune particularly warm, high above the frigid, tropospheric cloud layer. At present, with fairly limited data on Neptune available, this remains conjecture.

Finally, clouds on Neptune are more apparent in part because it has more dispersed cloud decks. In many images methane cirrus casts particularly striking shadows on the lower, hazier layers. These filamentous clouds are situated at altitudes 50–100 km above the deeper tropospheric haze layer. Belts are tens of kilometers to a couple of hundred kilometers wide and thousands of kilometers long, blown across Neptune's face by the powerful tropospheric winds.

Seasons

Despite these similarities there is one obvious difference between Uranus and Neptune: the angle of tilt. Uranus lies tilted at an angle of 97.7°, while Neptune has a much more modest tilt of 29°. This has obvious implications for its climate, seasons and for their wild magnetic fields.

When Voyager 2 arrived Uranus had its south pole tilted towards the Sun, while Neptune was experiencing its southern spring. Over the last 30 years Neptune has been growing steadily brighter at near infrared wavelengths (Fig. 8.5). This appears to be the result of an increase in the abundance of methane clouds high in Neptune's atmosphere. Sunlight, striking the upper troposphere, is warming it as spring advances. This allows methane, which is normally frozen into ice, to sublimate (turn directly to gas).

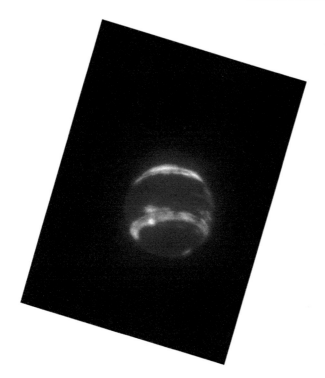

FIG. 8.5 Neptune in the infrared. This doesn't show heat so much as where the high, cold methane clouds are formed. Methane clouds appear to be confined to the temperature latitudes where the westerly jet streams are found. This forms two nearly continuous bands. Nearer the equator methane clouds are much patchier. Since 1980 Neptune has been growing brighter at near infrared wavelengths because the cover of methane clouds has been increasing in the southern hemisphere. This probably relates to the change in season

Gaseous methane then leaks out into the stratosphere where it influences the chemical make-up of Neptune's upper atmosphere.

This pattern should continue for another 30 years or so while the southern hemisphere moves into summer and the upper atmosphere is warmed more thoroughly by the distant Sun. Meanwhile, the southern hemisphere of Uranus has now moved into its autumn, while the northern hemisphere appears to be perking up.

Where would you find the warmest conditions along the cloud tops in the troposphere of Uranus? You might expect that after 20 years the southern hemisphere would be the warmest, but on average the equator is the sultriest place to go. By contrast, the warmest place on Neptune's cloud tops is found across the South Pole where the Sun has warmed the region for 40 years. Quite why each planet shows such contrasting responses to heating is, as with so much of these worlds, unclear at present. Regardless, it further adds to the idea that there is something odd about the way Uranus releases heat from its interior and transfers this (and energy from the Sun) through its atmosphere.

Dark Spots

Both Uranus and Neptune display deep, dark spots. Perhaps the most iconic feature of Neptune was its rather gothic take on Jupiter's Great Red Spot (GRS). First observed by Voyager 2, this dark storm gradually migrated towards Neptune's equator then petered out in the early 1990s. Although this anticyclonic feature didn't persist, it has been followed by many more similar features on Neptune and on Uranus in the last few decades. The Great Dark Spot appeared at the same latitude as the Great Red Spot on Jupiter and the "Storm Alley" on Saturn. Perhaps, just perhaps, there is a common structural feature of the atmospheres of these planets that preferentially creates storms at this latitude. It could, however, just be coincidence. Clearly, some atmospheric modeling is needed.

In 2006 the Hubble Space Telescope spotted the first Uranian dark spot north of the equator. Overall, Uranus is now showing a much more turbulent atmosphere, rich in methane clouds, as its seasons move on with the unrelenting advance of Uranus along its 84 year-long orbit. It seems Voyager 2 arrived when Uranus was at its least interesting.

Each storm reveals a window into the deeper interior of these planets. However, while air may well be descending in the heart of these anticyclonic storms, the presence of high, methane clouds around their periphery suggests that, like the GRS, air may well be rising at higher altitudes within the troposphere (Chap. 7). Despite differences in longevity, these anticyclonic storms may share a common structure and perhaps a common origin at deeper levels within each planet's lower atmosphere[1].

---

[1] The L-class red dwarf W1906+40 also hosts a long lived dark spot with a size similar to Jupiter's red one.

Such anticyclonic storms indicate volatility in the atmospheres of these planets where otherwise east-west, west–east winds dominate. This brings them somewhat closer to Jupiter and Saturn, both of which have atmospheres studded with similar vortices (Chap. 7). Given what we know about the Earth we might expect circular storms to develop at the interfaces of the westerly and easterly jet streams, where air is forced to change velocity abruptly. Such dramatic differences in air flow would be expected to set up waves in the atmosphere which would favor the development of storm systems. Why these features appear to be more persistent on Jupiter and not Uranus or Neptune may simply be a reflection of scale. Winds on Uranus and Neptune are far faster than those on Jupiter and Saturn and this might shear storms apart. It might be that there is a deeper connection between the Great Red Spot and Jupiter's interior than the dark spots share with the icy mantles of their worlds. It is remarkable that, despite a large difference in the amount of heat escaping Uranus and Neptune, they look so very similar otherwise. Indeed, their similarities seem to grow as both worlds move coincidently towards their respective equinoxes.

The Twisted Tale of Ammonium Metal

Deep within the interior of Uranus and Neptune, water and ammonia dominate. These two chemicals can react with one another to form ammonium hydroxide. This chemical breaks into two ions, ammonium and hydroxide in solution. These make the entire icy layer strongly conducting, a process exacerbated by the effect pressure on ammonium ions. While both Uranus and Neptune are too insubstantial to form metallic hydrogen, they can compress ammonium ions so that they behave in much the same way. In this hot, watery soup electrons are free to wander around between the ions that make up its chemical structure. At very high pressure it is thought that water also behaves in an odd way, with the oxygen atoms forming a solid crystalline lattice through which the hydrogen ions and electrons wander. Both of these processes make the entire interior layer of ice highly conducting. The observed fields are off-set from the center of each world, implying that they are produced in relatively thin shells within the ice layer. This, in turn implies that either only a small fraction of the "icy mantle" is in a liquid form, or that only a small fraction has ammonium and water in the appropriate form to conduct electricity. Whatever the mechanism, each ice giant is able to stir this region to produce a large circulating magnetic field that wraps around each world.

The magnetosphere is generated by the spin of Uranus and Neptune. As Uranus spins once every 16 h, its rapid rotation drives the movement of ions through the planets' interiors. In turn, this creates a powerful electrical and magnetic field that envelopes the planet. This off-center generator has some odd effects. As each planet spins rather than the field moving uniformly with the planet, the field whips around at uncomfortable angles as though the planet was holding its field generator at arm's length. The effect is less pronounced at Uranus but has quite marked effects out at Neptune. Uranus, with its strongly tilted axis, has its magnetic field broadly aligned with its spin but lies on its side relative to the plane of the solar system. By contrast, Neptune sits more or less upright, with its magnetic field highly inclined to its spin axis (Figs. 8.6 and 8.7). In both instances this means that the magnetic field points towards and away from the Sun.

Consequently, as both Uranus and Neptune rotate, their magnetic fields are contorted into cork-screw shapes that whip out behind each planet in the direction of the solar wind. What is even weirder is the strong variation of the magnetic field with latitude; something we do not experience much on Earth. Because the field is not located in the geometric center of Uranus, the field you would experience at Uranus's cloud tops would vary from less than a quarter that of the Earth (0.1 Gauss) in the southern hemisphere and up to twice as strong as ours in the northern hemisphere (1.3 Gauss).

Neptune's field is very similar as a result of the greater off-set of the magnetic engine from the center than is found in Uranus (Fig. 8.6). Both Uranus and Neptune also display two different kinds of magnetic field with one embedded within the other. The dominant field is a strong dipole field—an Earth-like field emerging from the north magnetic pole and flowing towards the south in a donut-like configuration. Both worlds also show an underlying quadrupole field with other blebs and folds punctuating the dipole

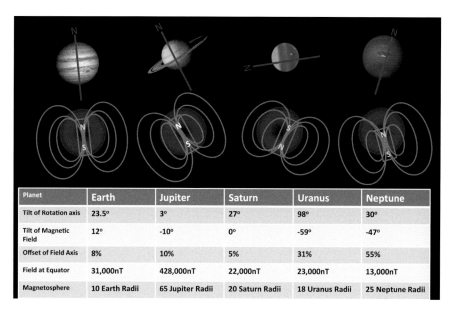

Planet	Earth	Jupiter	Saturn	Uranus	Neptune
Tilt of Rotation axis	23.5°	3°	27°	98°	30°
Tilt of Magnetic Field	12°	-10°	0°	-59°	-47°
Offset of Field Axis	8%	10%	5%	31%	55%
Field at Equator	31,000nT	428,000nT	22,000nT	23,000nT	13,000nT
Magnetosphere	10 Earth Radii	65 Jupiter Radii	20 Saturn Radii	18 Uranus Radii	25 Neptune Radii

FIG. 8.6 A comparison of the magnetic fields of the giant outer planets. Aside from Jupiter's, which dwarfs all the others, most have field strengths comparable with the Earth. Jupiter and Saturn generate theirs in their core, while Uranus and Neptune generate theirs further out. Image credits: Jupiter (HST/NASA); Saturn Cassini/NASA; Uranus: HST/NASA; Neptune Voyager 2/NASA

field. This means that there is substantial variation in the strength of the magnetic field an observer would experience at Neptune just as there is at Uranus. Like Uranus the field strength varies, but in this case this variation is a temporal one that is associated with Neptune's 16.1 h rotation. Thus while there is a varying magnetic field strength with latitude on Uranus, this would be fixed from hour to hour. Meanwhile at Neptune the field changes rapidly every hour. Imagine you traveled to Neptune only to find that your compass steadily pointed to a new north over the course of a day.

At 23 Uranian radii beyond the center, the magnetic field forms a bow shock that deflects the bulk of the solar wind. With a weaker solar wind at Neptune's orbit Neptune is able to stem the flow at 35 Neptunian radii. Within the magnetosphere, charged particles fill much of the space, as they do around the other giants. While Jupiter and Saturn gets most of their charged particles from the solar wind or from the satellites embedded within it, Uranus

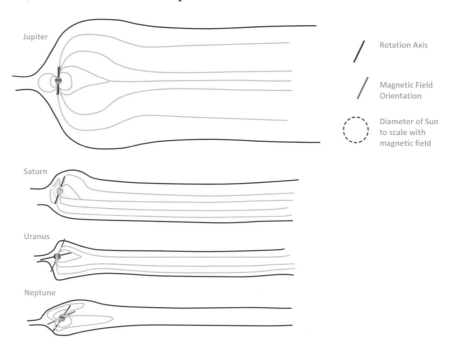

Fig. 8.7 The scale of the magnetic fields around the giant planets. Jupiter's dwarfs the others and stretches 700 million kilometers to Saturn. Those of Uranus and Neptune are steeply inclined to the oncoming solar wind and whip the field around in different directions as the planet rotates. Planetary field lines are in *yellow* while the solar field lines are in *red*. The *spin axes* and *magnetic field axes* are indicated above along with a comparison to the diameter of the Sun

appears to populate its field with charged particles from its highly extended ionosphere. This is supplemented by material blasted from the surfaces of its icy satellites and its rings, as well as a little material from the solar wind. Uranus's magnetic field has wide swathes carved into it as its moons sweep through the magnetized sea of particles. In turn, these particles appear to be responsible for darkening (or space weathering) the surfaces of the icy moons.

Do Neptune and Uranus experience aurora? In 2011 the Hubble Space Telescope detected aurora in spots over the Uranian equator. Why here? Again, this is thanks to the oddly aligned magnetic field. At particular locations across the equator the magnetic field lines slice through the cloud tops, through the planet and back out towards the Southern Pole. This forces particles to slam into Uranus's atmosphere at distinctly odd locations. By contrast, in an archived image of Uranus, taken in 1998, there is also evidence of a ring of aurora near the planet's Northern Pole. A weaker region of emission also appears in the southern hemisphere. Similarly, an observer on Neptune can experience aurora at various latitudes and longitudes depending on the angle of the fast rotating magnetic field compared with the observer, which, as we've seen, alters throughout the Neptunian day.

Aurorae are far less intense at Neptune than they are on Earth. While terrestrial aurorae generate over 100 billion Watts of power, those at Neptune are wimpy 50 million Watts (50 MW) affairs. Compare this to the output of the Three Gorges hydroelectric plant in China, which produces over 22,000 MW. You could power 450 Neptune's worth of aurora with that monster.

Conclusions

Uranus and Neptune form an interesting pairing of icy worlds in the Solar System's deep, dark outskirts. Receiving as little as 1/400th and 1/900th the solar energy we receive, respectively, each world is a domain of ice. Most of this ice is hidden and is distinctly warm. While Neptune releases over twice the energy it receives from the Sun, Uranus remains frigid. Their cyan atmospheres are oddly Earth-like in appearance yet mask a deep, frigid sea of gases.

At present relatively little is known about each world. This wall of ignorance is perpetuated by the enormous orbital periods of each planet. Uranus takes the greatest part of a human lifetime to complete one orbit, while Neptune takes twice this long. Consequently, it will take several generations of humans to observe the full seasonal range of each world. That said it is already apparent that each planet displays marked contrasts from summer to autumn or winter to spring. Uranus, with its wild tilt, will undoubtedly experience the greatest variation. By the time a Uranian year is complete, which of our models will be vindicated and which will find themselves consigned to the trash can? Already we are playing catch up to a world in transition. Hopefully, NASA or another agency will have sent probes back to one or both of these worlds long before their birthday comes round once more.

References

1. Kaspi, Y., Showman, A. P., Hubbard, W. B., Aharonson, O., & Helled, R. (2013). Atmospheric confinement of jet streams on Uranus and Neptune. *Nature, 497*, 344–347.
2. Lawrence Sromovsky & Patrick Fry. (2015). *Dynamics of cloud features on Uranus.* http://arxiv.org/pdf/1503.03714v1.pdf.
3. Hubbard, W. B., Nellis, W. J., Mitchell, A. C., Holmes, N. C., Limaye, S. S., & McCandles, P. C. (2013). Interior structure of Neptune: Comparison with Uranus. *Science, 253*, 648–651.
4. Lamy, L., Prangé, R., Hansen, K. C., Clarke, J. T., Zarka, P., Cecconi, B., et al. (2012). Earth-based detection of Uranus' aurorae. *Geophysical Research Letters 39*(7). http://hubblesite.org/pubinfo/pdf/2012/21/pdf.pdf.
5. Aurnou, J. M. & Heimpel, M. H. (2004). Zonal jets in rotating convection with mixed mechanical boundary conditions. Icarus. *169*, 492–498.

9. Ice Dwarves: Titan, Triton and Pluto

Introduction

Titan, Triton and Pluto form a triad of similar icy bodies in the outer solar system. Each has a diameter around 2500–3500 km across with an interior dominated by ice and rock. Each plays host to an atmosphere dominated by nitrogen. Here, however, the similarity ends. The atmospheres of Triton and Pluto are tenuous affairs, a thin veil of gases that loosely cling to the surface of each world, while Titan hosts an atmosphere thicker than the Earth's. Titan's atmosphere has been rich enough to bequeath the diminutive world with a rich "hydrological cycle" based on liquid methane and ethane. Consequently, Titan displays many of the complex eroded landforms found on Earth.

Triton and Pluto, while far colder than Titan, still host dynamic though tenuous atmospheres that circulate in a manner that would be familiar to us. Both worlds show some aeolian features that are found on the Earth, Mars and Venus, though in smaller quantities. Indeed, as data continues to pour in from the New Horizons probe, the surface and atmosphere of Pluto appear to be far more dynamic than was presupposed on the basis of what we knew of Triton. This chapter, therefore, represents a snapshot of what is otherwise a rapidly evolving picture of weather in our outer solar system.

Titan

Introduction

Titan stands alone in the solar system as the only natural satellite that hosts a significant atmosphere. Triton shares an atmosphere dominated by nitrogen, making Triton to Titan as Mars is to the

© Springer International Publishing Switzerland 2016
D.S. Stevenson, *The Exo-Weather Report*, Astronomers' Universe,
DOI 10.1007/978-3-319-25679-5_9

Earth: a pale reflection of this cloaked world with only the barest wisp of gas. Titan's atmosphere, predominantly nitrogen in content (98.5 %), also contains an abundance of different hydrocarbons with the majority of this being methane. These proportions change as we descend from the stratosphere towards the surface because methane condenses out as clouds at low altitudes. Were one at the surface, nearly 5 % of the atmosphere by mass would be methane. Around 0.1–0.2 % is free hydrogen, but this is mostly found higher up. The hydrogen comes from the breakdown of methane through the action of ultraviolet light. Near the surface Titan's atmosphere is denser than that at the surface of the Earth.

General Structure of Titan's Atmosphere

Titan has an atmosphere layered much like that of Earth (Fig. 9.1). The lowest level is the troposphere where most of the weather action occurs, but unlike our atmosphere where the two retain fairly distinct identities, Titan's troposphere is coupled very strongly to the stratosphere above. Recent evidence from Cassini indicates that this coupling extends all the way through to the thermosphere where a vast circulation extends from near the top of the atmosphere to its base hundreds of kilometers below.

The atmosphere is cold throughout, with temperatures lowest around 50 km above the surface where the troposphere gives way to the stratosphere. The lowest 30 km contains most of the planet's true clouds. These are mostly made of methane with an admixture of more complex hydrocarbons. As in the Earth's atmosphere the stratosphere shows steadily rising temperatures caused by the absorption of ultraviolet light by complex chemicals. While ozone has this effect in our atmosphere, a complex brew of hydrocarbons causes the effect above Titan. These compounds form hazes that fill the space from 100 to 210 km above Titan's frigid surface.

Titan's mesosphere shows cooling but never gets as cold as the troposphere thanks to heat is conducted downwards from the upper thermosphere. Once again, hazes form distinct layers around 300 and 450–500 km up, while in the winter hemisphere the polar hood begins its life this high up before extending downwards into the troposphere.

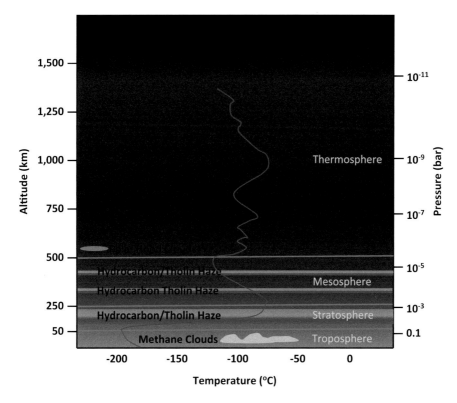

FIG. 9.1 The general structure of Titan's frigid atmosphere. Temperatures fall from around –180 °C at the surface to –200 °C near the tropopause. Like the Earth the atmosphere is divided into four principle layers, troposphere, stratosphere, with rising temperatures, mesosphere, with lowering temperatures and a (relatively) warm thermosphere with rising temperatures. Methane clouds are confined to the troposphere, while layers of tholins, ethane, other hydro-carbons and cyanide hazes fill layers within the stratosphere and mesosphere. Methane clouds form two distinct decks: one at 20–30 km and a lower deck at 8–15 km. Polar Hood clouds can form as high as 300–600 km above the winter Pole

Titan possesses a very extended atmosphere, thanks to its low gravity. The upper atmosphere consists of a thermosphere extending above 520 km, and this strongly overlaps with an ionosphere—the region where much of the gases are ionized. Ionized layers are found as low as 63 km in the mesosphere, but the main region lies above 1400 km, giving Titan an atmosphere more than quarter the width of the moon.

The Weather Report for Titan

What's the weather like on Titan? For starters it is extremely cold, with average temperatures around –180 °C. It's also fairly cloudy, and though clouds as they would be recognizable on the surface of the Earth are relatively rare, there is a persistent high level haze. This is produced when a combination of ultraviolet light from the Sun and cosmic rays energize methane and nitrogen gas. These create a sea of complex hydrocarbons and nitrogen-containing organic compounds that block out much of what is already a distant and dim Sun. That is not to say Titan doesn't have its fair share of clouds, only that they are more diffuse in nature and often localized to specific regions at certain times of the Titan year.

Titan's surface temperature is far lower than one would naturally expect for an object at Saturn's distance. This is the result of two opposing factors: the methane that is present acts as a greenhouse gas, while the low temperature means that much of it is frozen out as ice particles high in Titan's atmosphere. This, coupled to the pervasive haze, blocks much of the sunlight that strikes the satellite. As a result Titan is a good 10–20° or so colder than it would otherwise be. Titan is thus a fairly good model of Earth, where aerosols produced by the burning of fossil fuels impact on our surface temperatures and climate.

What of the circulation? Titan turns out to be an interesting hybrid of Mars, Earth and Venus. Above 20 km the winds are organized into broad westerlies that circulate around the entire globe, much as happens in the stratosphere of the Earth. However, unlike the Earth's stratospheric winds, these move faster than the underlying surface: they super-rotate. We've already encountered super-rotation in Chaps. 5 (Venus) and 7 (Jupiter and Saturn). In each case winds are moving considerably faster than the rotating globe, beneath. The difference with Titan is that the winds on Venus super-rotate from east to west, while Titan's winds super-rotate with the moon's rotation from west to east.

Figure 9.2, shows some that we would expect to find in the atmosphere of Titan. At altitudes of 20–30 km the winds change direction with the planetary Hadley cell (or cells) broken into two layers. Again, this pattern is seen on Venus, but here an upper cell carries heat from the day to the night side, while the lower

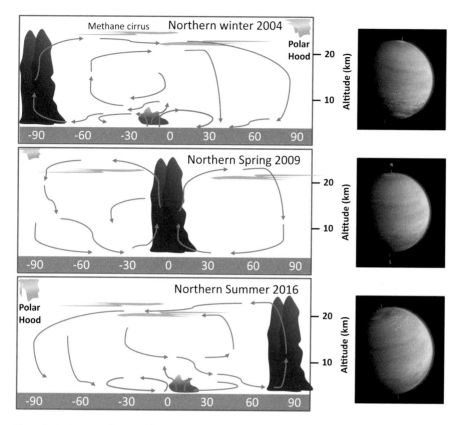

FIG. 9.2 General circulation in Titan's troposphere. Titan has seasons similar to the Earth as it orbits Saturn's tilted equator. Seasons last several years as Saturn completes its 29.5 year orbit of the Sun. The middle and upper troposphere is dominated by a single Hadley Cell (Chap. 1) much like Venus. This takes warmer air from the summer hemi-sphere to winter hemisphere. In the spring the atmosphere splits into two Hadley Cells centered on the equator. The lower atmosphere contains one or more additional cells where heating is strongest. Images: Cassini/ ESA/ NASA

cells carry heat from equator to pole. On Titan the two cells are organized broadly from the moon's warmest to coldest regions. Interaction between these two layers of cells generates turbulence and the temperature differences help form banks of cirrus and cirrostratus clouds. Unlike their terrestrial analogues, these are made of frozen methane crystals. As the air is expected to be saturated with methane these crystals can grow until they are heavy enough to fall towards the surface.

Methane snow does not have the same crystal structure as typical terrestrial snow. Water vapor tends to form hexagonal structures that are dictated by the formation of weak chemical bonds between water molecules. Methane cannot form such bonds (Fig. 9.4) so the crystals adopt a different shape. Experiments suggest that methane snow would be more like the rare terrestrial diamond dust. Diamond dust is occasionally seen on Earth, mostly within the dry continental interiors, during the deepest, coldest parts of winter, when the air becomes super-chilled and ice forms directly in the cold, dry air. On Titan, the air in the region methane snow forms is also likely to be far colder than the temperatures normally required to form methane crystals. This super-cooled methane probably produces eight-sided octahedra or even fourteen-sided cuboctahedra diamond dust that drifts lazily down under Titan's low gravity to the cloud decks below. In amongst these odd methane crystals will be ethane hexagonal crystals, which although structured like water refract the light much more strongly.

As these crystals descend into warmer air the crystals melt forming a fine and very frigid drizzle. By the time these droplets penetrate the lower cloud decks at 8–16 km up, they have absorbed a considerable amount of nitrogen. Although much of this subsequently diffuses back out before the drops hit the surface, it does alter the density of the drops and reduces the speed that they ultimately hit the ground. What would Titan rain be like? In short: rather odd to our eyes. Methane is less dense than water so this, alone, will make it fall more slowly than terrestrial rain. Factor in Titan's lower gravity (roughly one third that on the surface of the Earth) and denser atmosphere, which provides greater air resistance and would marvel at the spectacle of large rain drops drifting lazily downwards. These would have little capacity to erode surfaces as they do on Earth, but they could collect on the already sodden surface and pour downwards under gravity. Erosion could occur here, though this mechanism does not appear sufficient to produce Titan's widespread surface features.

Not a fan of rain? What about snow? In 2006 Cassini discovered a 150 km long, 1.5 km (4800 ft) high mountain range that appeared to be capped with methane snow. The lower clouds are likely most similar to terrestrial stratus or stratocumulus clouds:

shallow in extent with only limited vertical movement of material upwards or downwards. Like a dull winter's day on Earth, a thick bank of stratus delivers a cold and wet countenance to Titan. It is this widespread drizzle that maintains Titan's boggy surface, which Huygens detected when it landed in 2005.

Despite the likely frequent nature of this slight rain on Titan there is clear evidence that Titan's atmosphere must deliver a stronger punch at least occasionally. When Cassini-Huygens arrived at Saturn in 2004 Titan was still in the grip of its northern winter (Fig. 9.2). There was little evidence of the large, convective clouds that would have appeared necessary to spawn more significant precipitation. However, the surface of Titan shows clear and abundant evidence for large flows of what is likely methane. There exist countless dry river channels, much like a terrestrial desert. These cannot have been eroded by fluids deposited as a steady drizzle. Instead, these require a strong burst of heavy rainfall and this in turn requires strong convection and deep clouds. Computer modeling had shown that such convection was possible on Titan where topography or solar heating drove uplift. On Titan the main driver will be solar heating. With a surface saturated in liquid hydrocarbons the atmosphere is fully saturated with methane within a few kilometers of the surface. Under these conditions only a slight increase in temperature (less than 1 °C) is enough to drive the formation of storm clouds over 20 km high. As with water on Earth, when the gaseous methane cools and condenses, it releases latent heat which encourages further uplift. Therefore, given the right push at the surface, methane storm clouds can rise through half the height of the troposphere and generate clouds massive enough to produce substantial rain fall.

Where would we find these flash floods? Figure 9.2 provides some answers. Topography can provide uplift and there are small pockets of convection likely in the mid-latitudes, the most likely locations for rain storms are where the air is rising most strongly. These are the Polar Regions in the summer hemisphere or across the equator during the spring. Indeed, the simplest analogy between Titan and the Earth are the tropics. On Titan the area of greatest uplift and precipitation is the Inter-tropical Convergence Zone (or ITCZ for short: Chap. 1). On Earth this forms the heart of two broadly symmetrical Hadley cells that span 30° north to

30° south. While this belt of convergence moves seasonally north and south with the overhead Sun, it is confined to the mid-riff of Earth. On Titan, with very slow rotation (15 Earth-days, 22 h), the Hadley cells are global in extent and run from Pole to Pole. Thus as the seasons progress, the region of strongest convection moves from the South Pole to North Pole and back again, once every 29.5 Earth-year long Saturnian year. At present (late 2015), Titan is moving into its northern summer and the band of strongest precipitation will move with it, reaching the northern pole in 2016. These storms are thus expected to be the agents that create the seas of Titan, which are most prevalent within the Polar Regions.

Microbursts and Aeolian Features on Titan

Figure 9.3 shows the mean wind directions on Titan from season to season. Winds blow weakly from east towards the west across the equator during the spring season with virtually no wind during the northern winter or summer. When Huygens arrived and descended immediately south of the equator, it encountered first

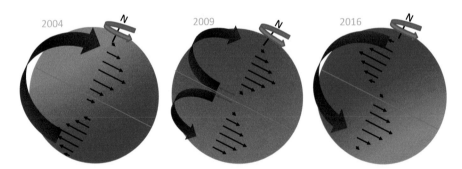

FIG. 9.3 Average wind directions at the surface and at height in Titan's troposphere in different seasons. *Large arrows* show the Hadley cells that carry warm air from the areas that are most strongly heated to the poles or from the summer to the winter hemisphere predominantly at altitudes less than 20 km. The more regional circulations indicated in Fig. 9.2 are not shown. Of particular significance are the equatorial easterly winds that kick in only during the equinoxes. Also not shown are the general movements of air in the upper troposphere (above 20 km), which appear to flow from west to east during the northern winter and summer. The upper cloud deck at 20–30 km appears to separate these two belts of wind. 2004 was the northern winter; 2009, the northern spring; and 2016 will be the start of the northern summer

a strong (120 m per second or 432 km per hour) westerly storm. These winds decreased with height in some surprising ways. Below 7 km—well into the troposphere—the probe encountered two changes in wind velocity. Immediately beneath seven kilometers, the winds changed from westerly to easterly with the speed dropping to around 20 km per hour. Just before touch-down, Huygens detected a change in wind direction back to a westerly breeze, at only 0.3 m per second or roughly 1 km per hour. Tetsuya Tokano (University of Cologne) interpreted this change in winds as caused by the probe descending through the Hadley Cell. Between 7 and 1 km up Huygens sliced through the upper part of the cell; while in the lowest 700 m, Huygens encounters the cross-equatorial flow from north to south that was westerly. It is above 7 km that super-rotating winds are encountered.

Winds don't always follow this pattern on Titan. As with Venus, there is a strong tidal influence on wind direction. Although Titan experiences no significant tidal pull from the Sun, it does get pulled to and fro by Saturn and this effect alters from day to day and season to season as Titan's orbit is elliptical. As Titan orbits Saturn a "tidal wave" moves across the surface. Like terrestrial ocean tides, this atmospheric tide can periodically reverse the average wind direction from west to east by a few kilometers per hour. Compare this to the Earth where atmospheric tides are primarily caused by heating by the Sun. Such tides migrate across the sky in a variety of different waves as the sun heats the surface and greenhouse gases within the atmosphere. The large sun-driven bulge moves generally westward through the day, with smaller bulges moving in different directions depending on what kind of surface lies underneath. By contrast—and this may be surprising— the Moon, which drives ocean tides, has very little effect on the atmosphere. At Titan, the impact arises because Saturn is so much larger than Earth and the Sun far more distant, reversing their contributions to Titan's tides.

Tides are not the only factor that appears to drastically alter Titan's winds either. Across the central 30° or so, north and south from Titan's equator are large dune fields that cover an area roughly equivalent to the continental United States. These peter out moving into the mid-latitudes but can be up to 300 m high. The odd thing is, these all point from east to west and equatorial winds on Titan do not generally persist in this direction, nor are they apparently strong enough to drive the formation of dunes in the first place.

Titan's dunes are either made entirely from rather sticky but frozen hydrocarbon globules, or water ice covered in a frozen, yet sticky, hydrocarbon gloop. These are not expected to move at the sluggish wind speeds experienced at Titan's surface. How, then do Titan's winds move its sand into such large dunes? The suggestion is that during the equinox, strong storm systems develop over the equator. As these rise over 5 km in height they are directed eastwards by the super-rotating upper winds. The super-rotating winds appear to be the source of the dune orientation. Moving against the prevailing surface winds, convective storms travel from west to east. As they move these storms generate powerful downbursts of wind along their leading edges. Much like thunderstorms on Earth (or indeed the Derechos that were encountered in Chap. 7), these eastward-moving downbursts can exceed 20 m per second and, over desert regions, kick up enough sand to generate a dust storm.

The key idea is that these storms have an initially high base which allows them to move with the prevailing, super-rotating westerly winds at height, rather than those weak easterly winds at the surface. This allows the storm to transfer momentum from the upper troposphere (and stratosphere) to the surface of Titan and kick up otherwise static sand that lies across much of Titan's equator.

Titan's Missing Vital Spark

You might imagine that an atmosphere that hosts very active storm cells, more than 20 km high would also play host to some significant lightning. However, on repeated passes Cassini failed to find the characteristic whistler-mode radio noise produced by electrical discharges. The Earth, Venus, Jupiter and Saturn all display these characteristic radio bursts (Chaps. 4 and 5) that signify a discharge of electrons through the planetary atmosphere. Titan does not. Cassini swung past Titan on seventy-two occasions and detected copious radio emission from Saturn. However, when Titan swung in front of Saturn and eclipsed it, the radio receivers fell silent: there was not a single lightning detection from Titan. The inescapable conclusion was that Titan does not host thunderstorms. Its methane cumulonimbus clouds produce precipitation but nothing else. Why should this be? Figure 9.4 suggests one answer to do with the chemical structure of methane.

Because methane can neither readily separate charges nor dissolve substances that are available, charged clouds made from it will have a fairly uniform charge throughout. The only exception would be if ultraviolet light or cosmic rays were able to split the molecules of methane up into charged fragments. As this does not happen in the atmosphere of Titan, its clouds cannot build up the

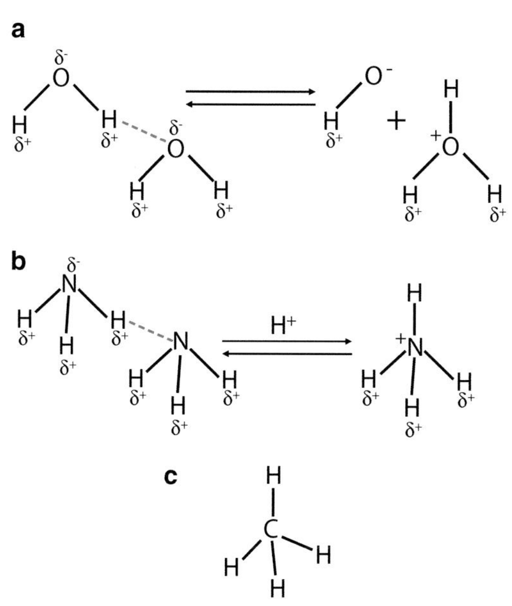

FIG. 9.4 Methane versus the polar covalent molecules ammonia and water. The atoms in water (**a**) and ammonia (**b**) bear small charges (δ^+ or δ^-). These help ammonia and water molecules cohere to one another and adhere to surfaces. These molecules can also gain hydrogen ions from other molecules or in the case of water directly through interactions with other water molecules (**a**). Both these processes allow ammonia and water to separate charges in clouds, most commonly when water or ammonia freeze. Methane (**c**), on the other hand cannot indulge in these sorts of reactions as its atoms have approximately equal charges: the molecules are non-polar) and it cannot dissolve charged substances such as salts in atmospheric dust, easily. Therefore methane clouds might look spectacular, but will never play host to lightning

kinds of charges needed to initiate lightning. Sadly, all of those artistic renderings showing lightning flickering through a sullen sky will need to go into the trash. Lightning remains the preserve of our planets, not their moons.

The "Methanological" Cycle

Methane is a problem for Titan. At Titan's distance from the Sun, methane should be able to react with itself and other chemicals under the influence of the Sun's ultraviolet radiation, leading to its breakdown. The rich stew of chemicals in Titan's atmosphere pays testament to this, but there is still an abundance of the chemical in Titan's air, leading to the belief is that methane can circulate from Titan's interior to its atmosphere.

Under Titan's thick and very rigid water and ammonia-ice crust, there is believed to lie a relatively balmy ocean. The word "relatively" is important here. Hovering at somewhere between –50 and –90 °C this ocean is kept fluid through a mixture of weak tidal heating from Saturn and a small, residual amount of radioactive heating, left over from the satellite's formation. This heating is not enough to thaw pure water, with a sizable dash of ammonia, this ocean can remain liquid down to –97 °C. Although the precise route is unknown, it is thought some of the surface features on Titan resemble cryovolcanoes. These plus cracks in the icy shell are sufficient to allow methane and other gases to leave the ocean and enter the atmosphere, where they can replace the methane that is destroyed through the action of ultraviolet radiation. The remaining methane circulates between the atmosphere and the surface much as water does on Earth.

Rat Poison

What does Titan have in common with a ubiquitous poison found on Earth, and what does the detection of this volatile and highly noxious gas tell us about Titan's atmospheric circulation?

In 2009 Titan's southern hemisphere was heading into autumn. By 2012 the upper atmosphere was cooling dramatically and a cap of cloud began to form over the South Pole (Figs. 9.2 and 9.5).

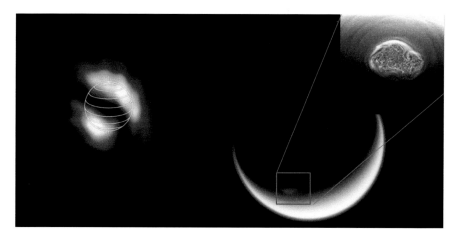

Fɪɢ. **9.5** Hydrogen cyanide and the southern polar vortex of Titan. The polar vortex stands out above the rim of the moon in the Cassini/JPL/ NASA image at center. A color zoomed image of the vortex is shown at top right. Meanwhile, at left is an ALMA (Atacama Large Millimeter/ submillimeter Array) image taken in the microwave portion of the spectrum, which shows the distribution of hydrogen cyanide. This is centered on each pole but skewed in an east-west direction by the circulation of air hundreds of kilometers above the surface of the satellite.. However, instead of winds smearing out the concentrations, there are still high concentrations near each Pole. ALMA image credit: NRAO/AUI/NSF; Martin Cordiner et al./NASA

This happened far earlier than predicted by climate models. Within it various gases began to condense. Amongst these was the highly toxic compound hydrogen cyanide (Fig. 9.5). Cyanide is perhaps best known as the lethal agent of the gas chambers, or as rat poison; but it also adds that acerbic bite to cherry stones, apple pips and banana skins. Within Titan's atmosphere, hydrogen cyanide is produced when methane is split apart by ultraviolet light then partly recombines with nitrogen. At high altitudes the air is very rarefied and will only permit condensation when it becomes saturated. This requires extreme cold. In 2009 the South Polar Region was anomalously warm and there seemed no chance that clouds could ever form here. Within 3 years, at an altitude of 300 km—a distance that would put terrestrial gases into outer space—a cap cloud began to form over the top of the vortex, far above the tropopause, in the heart of the mesosphere (Fig. 9.1).

The picture that emerges is that from 2009 onwards the upper atmosphere began to cool very rapidly, in contradiction of the models that held this off until 2012–2013. Air in this region dropped below –145 °C and hydrogen cyanide began to freeze out forming micrometer (millionth of a meter) sized ice grains. Measurements by Cassini had shown warming at greater depths, around 100 km up. This suggests that air cools rapidly once the Sun sets on the polar region. This air begins to descend slowly from above under its own weight, but warms through compression as it falls. In turn, this caused brief warming in the stratosphere, before these gases, too, cool in pitch blackness. The cyanide and other cyanide-related gases such as acetonitrile, enhance the cooling of the air, overall by releasing latent heat when they re-sublimate and form ice grains. This, in turn, accelerates the downward motion of the air above the Polar Region, helping drive the change in the circulation of the air far below.

What is amazing is that these observations suggest that the entire atmosphere to an altitude of 600 km above the ground, or more than one tenth the diameter of the satellite, is involved in the global changeover in circulation. This is well into the thermosphere (Fig. 9.1). If we compare Titan with the Earth, although there are certainly seasonal changes in the mesosphere and thermosphere with season, these are not thought to directly change circulation within the troposphere, where most of our weather action occurs. Titan's atmosphere, therefore, demonstrates a degree of connectedness not apparent in the atmospheres of the gas giants and the terrestrial planets, with the likely exception of Venus (Chap. 5).

The Loss of Titan's Atmosphere

Before we leave Titan for the solar system's outermost realm, it's worth looking at just how stable Titan's atmosphere is. After all, Titan is a fairly low gravity world, with a mass and diameter roughly twice that of our Moon but less than one fifth the mass of Mars. Its atmosphere extends over 1000 km above its surface. Just compare its troposphere to the Earth's: Titan's is roughly 56 km thick or eight times the average depth of ours. Although Titan's atmosphere squeezes in somewhat more mass (1.19 times the mass) over a smaller surface area, the greater thickness of its atmosphere is a consequence of its low gravity, which cannot compact it down towards the surface as strongly as the Earth can to its own. Titan's low mass and its extended atmosphere should, therefore, make it vulnerable to escape, if we apply the same rules as we did to the other planets.

As with the other planets, there are two main routes for escape of gases: thermal (or Jean's escape) and several non-thermal mechanisms. Titan's low mass should be sufficient for hydrogen and methane to escape through the Jean's mechanism, but heavier gases, such as nitrogen could be maintained. Despite this measurements of the ratio of different isotopes of nitrogen (nitrogen-14 and nitrogen-15) imply that rather a lot of atmospheric nitrogen has been lost. Being lighter, nitrogen-14 should escape more readily than nitrogen-15. As nitrogen is found as N_2 (diatomic nitrogen), molecules can be purely ^{14}N or ^{15}N or consist of one atom of each form. Nitrogen-14 is by far the more abundant and this abundance is set by nuclear reactions in stars. Nitrogen can escape as single atoms (or ions) from the top of the atmosphere when molecules are struck by ultraviolet radiation or cosmic rays. (See Chap. 5 for a comparison with Venus.) Any charged nitrogen ion can then be picked up by Saturn's magnetic field and whisked off into space. Exchanges of electrical charge between particles can also give nitrogen atoms enough energy to escape Titan's gravitational hold. Examination of the ratio of these two isotopes of nitrogen reveals that most of Titan's original atmosphere must already have been lost—yet Titan retains a deep atmosphere.

It is possible that Titan lost a lot of its early atmosphere within the first 50 million years of the satellite's formation. Stronger heating from a young, hot Saturn and more energetic radiation from a youthful Sun, may have whittled its atmosphere down through the process of hydrodynamic escape whereby heating of the atmosphere is strong enough that it creates sufficient pressure to blow much of the atmosphere off into space (described in Chaps. 5 and 6). Hydrodynamic escape works best on light gases such as hydrogen, but as these blow off into space they can carry heavier gases, such as nitrogen, with them. This is known as hydrodynamic drag. Hydrodynamic escape might have been accelerated by heat released from the nascent Saturn or from tidal heating caused by changes to the orbit of Titan around

Saturn, or simply heat released from the formation of Titan, itself. Hydrodynamic escape would have left a greater proportion of the heavier isotope of nitrogen. A lack of significant argon points to the present atmosphere having come from the photolysis (light-splitting) of ammonia that was accreted as ammonia. This process would make nitrogen, but also remove much of it. As the ammonia was split the hydrogen that was also liberated would have escaped Titan's low gravity to space. This process has the potential to drag nitrogen atoms with it – and these are, preferentially, the lighter nitrogen-14 atoms, which are easier to lift. The argon data also points to the Saturnian system being rather warm during its formation. This could be because of heat radiating from Saturn as it coalesced or because, as some models of the developing solar system suggest, young Saturn wandered around: at one point Saturn came well within the region now occupied by the asteroid belt. This would have greatly modified the materials available to form its moons.

Current measurements suggest Titan is still losing rather a lot of its atmosphere to space or to its surface. Using Cassini data, Titan is shedding around seven metric tons of hydrocarbons per day into interplanetary space through the process of electric force field acceleration that we encountered in Chap. 5.[1] While much of the atmosphere appears to have been lost early on through hydrodynamic escape (Chaps. 5 and 6), nowadays this process is subordinate to the electric force field route. To recap, electrons are energized by solar energy and are pulled away by Saturn's magnetic field. This leaves the upper atmosphere positively charged, which then causes it to puff up to the point that it either blows out into Saturn's magnetosphere or is pulled into it by the departing cloud of electrons. This process occurs most strongly above Titan's Polar Regions, earning the name a polar wind. Now, this data only relates to hydrocarbons: hydrogen is also being lost when it is split from methane by sunlight and then drifts off into

---

[1] Getting a precise figure here has proved very frustrating with numbers varying from 0.4 kg per second (34 metric tons per day), —to as much as 300 metric tons per day. My calculations, based on a different published unit (amu), were closer to, but a lot higher than, the UCL figure... I went with the UCL figure, for the described mechanism, in the end and remain vague on the total loss. https://www.ucl.ac.uk/mathematical-physical-sciences/maps-news-publication/maps1535

space and much of this loss takes nitrogen with it. Methane reacts to form other compounds under the influence of ultraviolet light. These rain out onto Titan's surface and are ultimately lost, as well. The total loss is on the order of several tens of metric tons per day—but again getting a consistent figure has proven challenging. Using a ball-park figure of 30–40 metric tons per day, that's probably reasonable and within an order of magnitude of the current rate of loss of gases at Mars (approximately 100 metric tons). With an atmospheric mass marginally higher than the Earth there is a chance Titan can expect to hold onto its atmosphere until the Sun becomes a red giant.

While the atmosphere as a whole seems secure for billions of years to come, the same cannot be said to be true for one of its components: methane. Methane is not stable even in the extreme distant cold of Titan. Ultraviolet light is continuously depleting it by converting it into more complex hydrocarbons, which then rain out onto the moon's surface—or escape to space. Calculations by Christopher Sotin (University of Nantes) and collaborators suggest that the methane was probably released a few hundred million years ago; possibly as a result of a giant impact or perhaps cryovolcanism. Within a few tens of millions of years hence, this will all have been destroyed through the action of ultraviolet light. Titan will run dry and its current active geology might freeze out with it, leaving a dry and desert-world: our ancient Earth analogy will become a frigid mirror of Mars, instead. The current wealth of atmospheric phenomena thus appears to be a temporary blip in the evolution of Titan's otherwise dry atmosphere.

It appears therefore that Titan shed much of its early atmosphere in a dense wind. At some critical point the rate of mass loss slowed to its present rate, probably as Saturn and Titan cooled down and the orbit of Titan stabilized. Convenient though this is, it still begs the question if Titan has already lost a substantial amount of atmosphere, why then, is its atmosphere still dense, while the more massive Mars struggles to retain a dribble of gases despite a similar rate of loss? Moreover, why do Jupiter's large satellites have no meaningful atmosphere at all? Jupiter plays host to two satellites that are more massive than Titan: Callisto and Ganymede, yet neither of these has an atmosphere other than a sparse cloud of oxygen atoms and ions that are kicked off their icy surfaces. There are a few possible reasons. For one, Ganymede and Callisto were born in an intrinsically warmer part of the solar system and, therefore, may not have had access to the same amount of volatile gases as Titan. Secondly, heat from proto-Jupiter would have been significantly greater than that released from Saturn. Running from Io outwards, there is a clear trend in the abundance of light, volatile gases: Io is essentially dry; Europa has a thin, icy shell, Ganymede is a massive combination of metal, ice and rock; while outermost Callisto is mostly ice. This implies Jupiter was hot and radiated enough heat to drive volatile materials away. This would have included those gases needed to form an atmosphere, such as ammonia. Thirdly, Jupiter's greater mass would have meant that impacts from asteroids and comets would have had more energy (and potentially a greater frequency) and thus been able to blast any youthful atmosphere off into space. Mars has lost its atmosphere through an unfortunate combination of low mass and its proximity to the Sun (Chap. 6). It's low mass and low atmospheric density means that it simply cannot effectively hold onto what gases it has (Chap. 6). Venus's higher gravity is more than sufficient to retain a thick veneer of gas (Chap. 5).

All of these effects would have meant Ganymede remained largely airless while less massive Titan was able to maintain a rich, hazy firmament. There is one more factor worth considering: magnetic fields. Thinking back to Venus there is a perception that a planet must have a magnetic field if it is to retain an atmosphere against its star's stellar wind. This is a misnomer, as Venus readily proves. In the case of Jupiter and Saturn magnetic fields may have had contrary effects. Ganymede has its own, relatively strong field but this is embedded within Jupiter's enormous magnetized blanket. Titan is also within the field of Saturn, but it lies further out and Saturn's field is around 25–30 times weaker than Jupiter's. In Jupiter's system the magnetic field energizes particles that have been trapped from the solar wind. These are accelerated to enormous energies before they slam into the surfaces of the Galilean worlds. The effect of these is profound. Secondary particles are then blasted off the surfaces of these icy worlds and driven into Jupiter's magnetic field, each forming a torus around the giant planet. The distress of these satellites is manifest as bright auroral spots in Jupiter's atmosphere (Chap. 7). There is clearly sufficient energy in Jupiter's magnetic field to bulldoze any atmosphere that these satellites may once have had, into deep space. Under the auspices of Saturn's field Titan, by contrast, enjoys some protection from the solar wind. Yes, there is some erosion of its atmosphere from charged particles, but this effect is far less than experienced by Jupiter's satellites. Thus, a magnetic field may not be the protective, nourishing blanket it is always assumed to be. Venus does just fine without one, while Ganymede may have suffered because of one. Mars didn't lose its atmosphere so much for the lack of one, but for its proximity to the Sun and its low mass.

Does Titan retain an atmosphere simply because it is further from the Sun than Ganymede? In part yes. But it seems it had a higher starting point: it was simply born with more volatile materials. Its nitrogen almost certainly came from ammonia. This was vented into the atmosphere—as it may still be now—through cryovolcanism. Even though much of this appears to have been lost to space, it is replenished through the release of more ammonia from an underlying ocean. Ammonia seeps out through cracks or is vented by volcanic activity. In the atmosphere this is split to hydrogen and nitrogen. Some of this is lost to space along with most of the hydrogen, enough is retained to maintain a voluminous atmosphere. Whether this can be maintained for the entire main sequence lifespan of the Sun is unclear. Either way when the Sun does expand to become a red giant, whatever Titan has left will be rapidly removed and ultimately, see its rocky core exposed.

Triton

Introduction

Twenty AU further from the Sun than Titan lies Triton, an odd, massive satellite with a tenuous atmosphere and an orbit that runs retrograde around Neptune. Triton almost certainly began its life further out in the Edgeworth–Kuiper Belt and was captured; perhaps as a binary object, much like Pluto-Charon, before one of the pair was ejected. During this time Triton was undoubtedly tidally heated allowing it to manifest some of the activity we see on its surface today.

Triton is 2,700 km across—nearly identical to, but slightly larger than Pluto. Like Titan and our Moon, Triton is tidally locked to Neptune. However, because it orbits Neptune in a retrograde orbit (one that has an opposite direction to Neptune's spin), tidal bulges on Neptune are gradually pulling the satellite down. In around 3.5–4 billion years Triton will cross Neptune's gravitational Roche Lobe and be tidally shattered. Here, Neptune's tidal pull will act more strongly on one side than the other until Triton comes apart, much like desiccated spaghetti. Before it dies, tidal heating will once again rejuvenate this icy world and perhaps give it an atmosphere as dense as Titan's is today. For now, Triton resides in the deepest freeze the solar system can provide.

Triton's Atmosphere

Triton's atmosphere in some regards resembles a frigid version of that of Mars. However, the surface pressure is 10,000 times lower at one hundred thousandth that found at the surface of our world (around 0.017 millibar). Unlike Mars, the atmosphere is dominated by nitrogen, rather than carbon dioxide: for, at the temperatures found at Triton's surface, carbon dioxide would be as hard as steel. Indeed, fully 99.99 % of the atmosphere is molecular nitrogen (N_2) with carbon monoxide and methane making up a fraction of one percent. Of these, measurements made from the Earth in 2010, suggest that carbon monoxide is more common.

At −237.6 °C, the surface of Triton is too cold for many gases to remain in their free-wheeling forms; and the only reason we see nitrogen, methane and carbon monoxide is that they sublimate from their frozen form under the action of sunlight and cosmic rays. Remember, that on Earth ice will sublimate at temperatures lower than freezing: you don't have to be hot to "boil". The constant action of sunlight (and cosmic rays) provides just enough energy to drive some of Triton's icy surface into the surrounding space where it forms the barest of atmospheres. Remember the term temperature is effectively an average measurement of particle energy. So even at −237 °C some nitrogen molecules, for example, will always have a "temperature" greater than that registered by a thermometer.

Compared with Triton, Pluto is positively balmy with temperatures soaring to –229 °C: this is still below the freezing point of nitrogen (–210 °C); carbon monoxide (–205 °C) and methane (–182 °C), but high enough that we also expected to find a reasonable atmosphere above Pluto.

Triton's atmosphere can be broken into three layers, based on measurements from Voyager 2, terrestrial measurements and from circulation models. The lowest 8 km is largely clear and still air. Triton sports geysers on its surface—something we will return to shortly. These rise vertically for 8 km and clearly indicate that there is no meaningful wind on Triton. There is an expectation of two general and very light flows of air: one is a tidal flow from the far to near side of Triton, with respect to Neptune as Triton completes its 141 h orbit, and the other is a sublimation and condensation flow that runs from the day side to the night side of the satellite. These are clearly so slight that neither seems to deflect the steady, upward flow of gases from Triton's surface. Above 8 km the jets of dusty gas are strongly deflected to the east, indicating a light westerly wind begins abruptly at this point.

More general measurements indicate that above the majority of Triton's surface there is a general easterly flow (prograde with respect to its orbit around Neptune). The term jet stream has been loosely applied. For in Triton's equivalent atmospheric flows, air moves at a very leisurely 3–4 m per second (or less than 20 km per hour). The fastest flows are in the equatorial easterly jet (Figs. 9.5 and 9.6), with more sluggish eastward movements further north and south. These jets mark the base of Triton's next atmospheric layer: its thermosphere. Here temperatures rise, but only to around –178 °C. Nowhere on Triton (or above it) is warm.

The lowest 8 km can be loosely thought of as a troposphere, although there is no pronounced rise in temperature as we move into the thermosphere, above. Unlike on every other world we have encountered so far, the troposphere is not bounded by a temperature inversion (the tropopause), simply an increase in wind shear. Triton is, therefore, unique in this regard. The thermosphere is lightly warmed by the steady impact of charged particles from the Sun and from Neptune's magnetosphere, which whips across it every 16.1 h (Chap. 8).

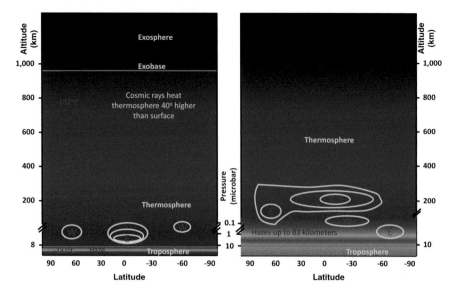

FIG. 9.6 A comparison of the atmospheres of "sister worlds" Triton (*left*) and Pluto (*right*). Although superficially similar, there are differences. Triton's troposphere is the lowest 8 km, above which wind sheer increases because of weak easterly or westerly jet streams. Pluto, on the other hand has a troposphere capped by a strong temperature inversion. The troposphere on both worlds has virtually no wind due to low temperature gradients and friction. Pluto's exobase (where the thermosphere ends) may be higher than 15,400 km from the surface—high enough that Charon may orbit within it

High above Triton's surface the atmosphere is complete when the thermosphere gives way to the exosphere at 950 km. At this point molecules, atoms and ions from the tenuous atmosphere below can escape into Neptune's magnetosphere. There is no aurora marking the upper boundary, just the whisper of gases making their way into the void.

What's the Weather Like?

Obviously it is cold, with largely clear skies that gives a virtually unimpeded view of the cosmos beyond. There are no clouds to speak of, only the barest of hazes that permeates most of the troposphere (Figs. 9.5 and 9.6). These hazes are thought to be composed of nitriles—compounds we encountered at Titan. These are compounds of carbon and nitrogen that are produced when

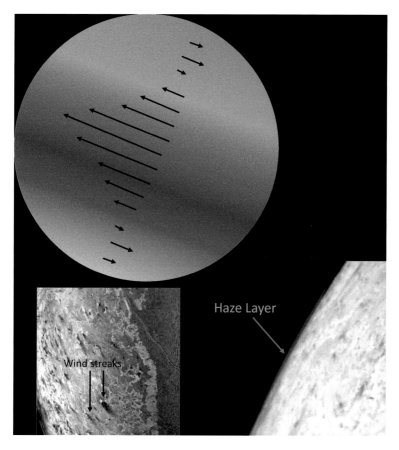

FIG. 9.7 Global winds on Triton above 8 km. Thankfully, with such a tenuous atmosphere winds are very light and follow a simple pattern. There is virtually no wind at the surface because the air is so thin and the temperature gradient across the surface, and vertically, is limited. However, above 8 km winds increase, with speeds up to 15 km per hour at an altitude of 10 km. These blow from an easterly direction across the equatorial and mid-latitude regions. This is evidenced by plumes of dark material that streak to the west at 53° and 57° south (see *inset* image). Over the more polar regions there are gentle westerly winds blowing at roughly 7 km per hour. Given that at this altitude Triton's atmospheric pressure is roughly one millionth that at the Earth's surface, you wouldn't feel this gentle breeze blowing at all. Pluto's atmosphere is warmer but expected to be similar. Triton image: Voyager 2/NASA

ultraviolet light and cosmic rays causes the trace methane and nitrogen to react. Elsewhere there are small, localized clouds of nitrogen ice between 1 and 3 km above its surface (Fig. 9.7).

Aside from this the only real excitement comes from occasional eruptions as nitrogen gas from under Triton's surface. The origin of these is unclear, but it is believed to result from a surface greenhouse effect. Here nitrogen is largely frozen into a broad series of solidified pools or seas. Sunlight can penetrate these and warm them from the base upwards. This, with limited geothermal heat, causes the nitrogen to melt. Normally, the nitrogen would simply sublimate, but if melting occurs at sufficient depth, pressure can rise until the warming solid turns into a liquid. Continued warming raises the pressure until some of this evaporates and is explosively released through the icy shell. Jetting through the lower atmosphere, these nitrogen geysers will shower out a curtain of nitrogen snow, downwind of the jet. Entrained within these jets are any other solid materials and these produce the dark streaks that have proved so useful in tracking Triton's winds.

At present this is all we can say about Triton. The only real data was recorded during the Voyager 2 flyby or through stellar occultation, where Triton passes in from of a distant star. Then, Earth-bound scientists can grab a quick look at a portion of Triton's atmosphere as it interferes with the light from the distant orb. More complete data will require another fly by, or perhaps the insertion of a Neptunian orbiter, forming a more frigid analogue of Cassini. As there are no plans to do this at the time of writing, the forecasters will have to content themselves with the wispy data that exists, or resort to models.

Pluto

Introduction

New Horizon's visit to Triton's near-twin, Pluto, has provided a lot of clues about these distant, icy worlds. It is to Pluto and Charon that we now turn to provide an unexpectedly lively end to this chapter. Both worlds are small, icy orbs that pinwheel around their common center of gravity. Both are likely to have formed in a manner analogous to the Earth and Moon, with a giant impact generating a large debris cloud around the former Pluto. However, the original Pluto was shattered in the impact and the debris scattered with enough momentum to allow the formation of a binary dwarf planet. Although Pluto has the lion's share of the pair's mass, Charon remains the most massive satellite, relative to its parent world, of any partnership in the solar system.

Both worlds have a center of mass lying between them that means that they are tidally locked to one another. Initially, this arrangement would have ensured that a substantial amount of heating would have occurred within each early on. However, 4.5 billion years later, most of this energy should have escaped such small worlds. A small amount of radiogenic heat: energy released by the slow decay of long-lived radioactive isotopes of elements such as potassium and uranium. Rough calculations suggest that around 2.4 milliwatts per meter squared (or around 1/40th of that at the surface of the Earth) will currently be available and this might be enough to maintain a partly liquid interior. Now, by liquid we are not talking terrestrial peridotite—the rock dominating the Earth's upper mantle. Rather, we are talking nitrogen or ammonia ices; materials that will melt under very low temperatures. Even water ice is as hard as steel at the temperatures of Pluto. However, because Pluto may retain a partly molten (or at least soft ice) interior it will be able to convect and transport gases into the atmosphere above, much as volcanoes do on Earth. Indeed, the first images of Pluto appear to confirm this.

In terms of seasons, Pluto is an extreme version of Mars. The dwarf planet and Charon are tilted over at 119.5° with the North Pole pointing downwards (as defined by the direction of spin). And while the Martian orbit varies from 1.4 to 1.7 AU, that of Pluto changes radically from 30 to 50 AU. At its closest approach to the Sun in 1990, the Sun shone down most strongly over the Plutonian equator. With Pluto making a slow retreat to its most distant position (aphelion) 99 years from now, the southern hemisphere is pointing more strongly away from the Sun and cooling down. Within the next 20 years the southern polar cap should become so cold that most of the neighboring atmosphere will condense out onto it. Meanwhile, albeit in a less extreme manner, the northern cap will experience its summer and begin to thaw out. As this will occur when Pluto is furthest from the Sun, the release of gas from sublimation will still not compensate for the rate of loss onto the dwarf's surface. It will only be when the southern cap experiences spring once more (after the year 2114) that the overall mass of the atmosphere will begin to rise again. Here, the combination of greater insolation from the tilt and Pluto's closer position to the Sun ensure the planet can begin to warm up and, at least partially, thaw out.

Pluto and Triton as Non-identical Twins

Before we delve into the small amount of data that currently exists for Pluto (soon to increase further from New Horizons) let's look again at Pluto and Triton together. Pluto has a diameter roughly the same as that of Triton (2370 versus 2700 km, respectively) and both were expected to host similar geological features when New Horizons arrived at Pluto in the summer of 2015. Despite areas of similarly young terrain, Pluto was still a considerable surprise to observers. Not only did its surface play host to some utterly unique features, such as the vast, flat Sputnik Planum (Figs. 9.8 and 9.10), it also had a dynamic multi-layered atmosphere (Fig. 9.8) that defied both expectation and observations made from the Earth. At the time of writing much of this information remains preliminary but

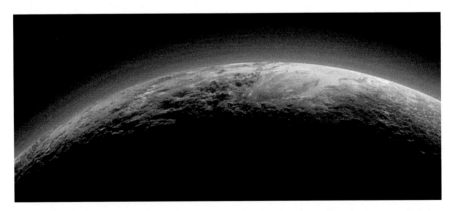

FIG. 9.8 A view of Pluto from New Horizons, 11,000 km from its surface. The edge of Sputnik Planum is visible as the flat area near the center of the image. Norgay Montes is the highest peak to its west. The mountains appear oddly flattened which might indicate erosion, subsidence or perhaps cloud cover? Ice appears to be flowing onto Sputnik Planum from the mountains. Extending over 100 km above the surface are layers of hazes. These appear to reach down into the valleys nestling within the mountains where they form fog-like banks lit by crepuscular rays. New Horizons has confirmed that these layers are produced by atmospheric gravity waves (Chap. 10) as air flows over the mountains

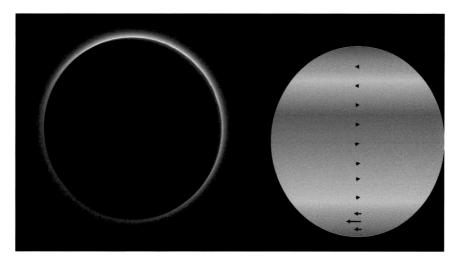

Fig. 9.9 The Solar System's Final Pale Blue Dot. Hazes extend over 80 km from the surface—much higher than expected. Sunlight refracts through the atmosphere, over the upper right edge, partly illuminating the night side and giving rise to twilight. Haze layers are far more complex than was suspected from Earth-bound observations. *Right*: Predicted wind speeds 94 km above the surface—and above the haze layers. Little atmospheric motion is predicted in the lowest 50 km except the barest of breezes caused by the slow sublimation and re-sublimation of surface frosts as sun rises and sets. Pluto image: New Horizons/JPL/NASA

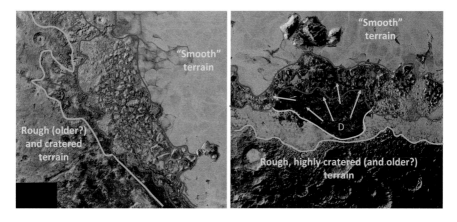

Fig. 9.10 The Surface of Pluto is dynamic. The large and relatively flat area is called Sputnik Planum and appears to be a frozen sea, possibly of nitrogen ice. This abuts a region of more highly cratered, and presumably older terrain along its southern edge (*left*). In a close-up (*right*) of part of this margin there is folded darker terrain with "ripples" (**D**) aligned broadly parallel to the edge of hummocky terrain (*outlined in red*). This latter terrain lies all along the margin of the two different regions. On Earth Hummocky terrain is formed during large landslips, such as that at Mt St Helen's during its 1980 eruption. This suggests the margins of the older terrain are foundering into the flooded basin (*red arrows*). Original, unmodified, images: New Horizons, NASA. Image modifications by author

what is known and conjectured is discussed below. Like Triton, Pluto shows considerable evidence of resurfacing and this may relate in each case (at least in part, or initially) to energy released from tidal interactions with its neighbor: Neptune in the case of Triton; Charon in the case of Pluto. Both worlds also have sufficient internal reserves of radioactive materials to keep them just about ticking.

Looking first at the surface, superficially Pluto is rather like Mars, with some important differences that are only apparent on closer inspection. For one, the pink color of Pluto is reminiscent of the rusty color of Mars. However, while Mars has a fairly old, cratered terrain covered in iron oxides and iron-rich basalts, Pluto has a predominantly water (and other)-ice covered surface, colored pink by compounds suspected to be organic tholins. Spectra appear to suggest that most of the methane[2] ice is concentrated in Sputnik planum and other low-lying regions, while highland areas appear to be water ice, but largely covered in other material.

One of the first surprises was that the atmospheric pressure was far lower than was measured from Earth. Whereas terrestrial measurement, using radio waves and observations of stellar occultation suggested a pressure of 22 microbars (22 millionths the pressure at the Earth's surface), measurements by New Horizons were around 5 microbars at the surface. Why the difference? One suggestion is that the terrestrial measurements are taken for greater altitudes (around 50–75 km) than those made by the passing probe. The latter were made directly by observing how radio waves that were sent from Earth refracted as they passed through the tenuous Plutonian atmosphere to reach the New Horizons probe. These measurements were made at or just above the dwarf planet's surface. However, this discrepancy remains only partly explained at the time of writing .

One of the most intriguing terrains is Sputnik Planum (Fig. 9.10), a vast frozen and very flat and broadly circular basin with an extension that makes the whole feature look rather like a heart. Its broadly pink coloration only added to the analogy. The south western edge of the basin is semi-circular, implying that it is an impact feature that has been flooded by some "lava". New

---

[2] At the time of going to press, it is thought that this methane layer is only a thin covering over what is mostly convecting nitrogen ice.

Horizons has determined that the ice in the basin is methane—at least on its surface. Elsewhere the crust in the highland regions is probably water ice but it is covered almost completely by another material that is unidentified at the time of writing. However, water ice is visible along some of the cracks in Viking Terra. The basin edge also appears to have a broadly consistent geology along its entire edge. This is where my interpretation comes in. The elevated highland region (largely) uniformly gives way to a lighter colored smoother area than then morphs into a hummocky region, adjoining the edge of the flat basin. Some of the nearby, smaller craters are also flooded, perhaps to the same depth (or altitude) as the neighboring basin. This may imply a "water-table" exists across the region with whatever frozen liquid that is present lying at a consistent depth across and through the entire terrain.

So, what of the hummocky terrain: how might one interpret that? One simple explanation that may or may not be true, but that is consistent with terrestrial geology is that the basin margins have either been undermined by liquid and collapsed directly, forming the hummocks, or the material is utterly saturated at a particular depth and this has weakened it so that it has then collapsed. Close examination of the region shows lakes embedded between some blocks implying the region is thoroughly infiltrated with liquids at least at depth.

On Earth you see the same hummocky terrain in places such as land adjoining Mt Shasta in California or Mt St Helens in Washington State, as well as numerous others. Formerly mysterious in origin, nowadays, hummocky terrain is thought to form during massive collapses of mountainous structures. Nearest the source the landscape may be relatively smooth, while at different distances away the landscape is pockmarked by hillocks embedded in a flatter surround. These hillocks are large blocks of the former mountain that have slid downwards and become jumbled. In the case of Pluto there are two other suggestive features of regional collapse: one, an area suggested to be dunes, while the other is simply the pervasive nature of the hummocky terrain that flanks the entire high basin margin and is consistent with collapse (Fig. 9.10).

The supposed "dune field" may be a misnomer. The "dunes" are aligned with the crater wall and the area broadens towards the basin and more distant hummocky terrain. The pattern is again reminiscent of collapse but in this case a more sluggish one where the land has crept downwards and fanned out as it came to rest. In this interpretation the "dunes" are simply folds in the material caused by its movement over the underlying landscape. Such folds form on terrestrial hillsides where the soil is moist and slips over the underlying bedrock. This suggests collapse is ongoing but perhaps now is more restrained in scope than the events that formed the hummocky terrain. The hummocky terrain and "dune-field" would happen if the subsurface was strongly undermined by fluids. On Earth, this is almost always water; on Pluto it is likely to be nitrogen. As these infiltrate the fractured bedrock lining the impact basin, the walls become unstable and collapse catastrophically into the basin floor. Either simultaneous with this, or subsequently, the soft methane and nitrogen ice then flows around the margins of the blocks, softening their edges.

Within the basin the flat icy terrain has an unusual blobby-kind of appearance. It's possible that this pattern reflects regions of upwelling and down-welling within the icy material (or in liquid or slush) underlying the icy surface.[3] Thus, Sputnik Planum may be similar to the frozen surface of the Arctic Ocean. However, unlike the Arctic Ocean, the depth is likely relatively shallow and the contents are probably at least partly frozen to the base allowing the pattern of upwelling or down-welling to become apparent (Fig. 9.10).

Why is a frozen (or partly frozen) ocean under Sputnik Planum important for Pluto's weather? It suggests ubiquitous fluids, and at the pressures within Pluto's atmosphere, nitrogen, carbon monoxide and methane can vaporize forming the observed atmosphere (Figs. 9.8 and 9.9). With an atmosphere, no matter how thin, we have weather. What weather there is probably reminiscent to that found in Antarctica's driest valleys. Here, katabatic winds and sunlight cause ice (in this case water ice) to sublimate. Now free within the atmosphere, the prevailing westerly winds carry the moisture to neighboring areas, where it either condenses as frost or re-joins the general mêlée as atmospheric moisture or ice grains. On Pluto the ice is nitrogen, with some methane and carbon

---

[3] At the time of going to press this has been confirmed and convection within a layer 3 km thick generates the observed pattern in Sputnik Planum.

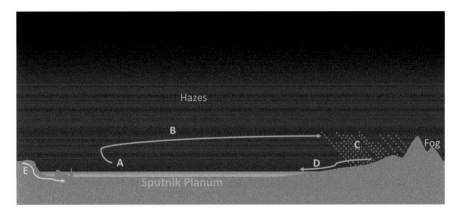

FIG. 9.11 The "methane/nitrogen-cycle" on Pluto. Not to be confused with the biological nitrogen cycle on Earth, here nitrogen and methane gas, liquid and solid cycle from atmosphere to surface and back again in complex ways. At Sputnik Planum (A) nitrogen and methane ponds in a large icy layer, in which heat from the interior or solar energy may liquefy underneath. Nitrogen and methane sublimate from here and enters the atmosphere, along with carbon monoxide. Light, prevailing westerly winds at this latitude (Fig. 9.8) blow these gases towards the 3,000 m high mountains (B). Here, some condense as freezing fog and frost, or perhaps on occasion when Pluto is further from the Sun, form clouds, precipitating snow (C). Such snow either compacts or partly melts and flows back to Sputnik Planum (D) in glacier-like flows. These may be liquid deeper down where geothermal heat melts the ice without sublimation. At E liquid nitrogen and methane underlying the western basin rim and destabilizes it leading to collapses (Fig. 9.9)

monoxide. It too sublimates under the action of sunlight. Carried by the prevailing, light westerly winds, this gas ultimately reaches the highlands to the east of Sputnik Planum. Here, the gases re-sublimate as frosts or perhaps (at least on occasion) precipitate as nitrogen and/or methane snow. After building to sufficient depth, these ices either flow directly back into the basin or partly melt at their base, with a mixture of ice and bedrock liquid nitrogen and methane (held as a liquid under pressure) flowing back into Sputnik Planum (Fig. 9.11). Some of these flows strongly resemble terrestrial water ice glaciers, suggesting a cycling of methane ice, perhaps as snow, from basin to mountains and back again. In the basin such liquids form the flat layer, which may undermine the western wall of the basin leading to its periodic collapse.

Pluto, overall, should be rather similar in terms of weather to Triton—at least for part of its 248 Earth-year-long year, since when it is closest to the Sun, it receives roughly the same amount of solar energy as Triton. However, Pluto has a more elongated orbit around the Sun, which ultimately takes it far beyond the orbit of Neptune. 120 or so years from now temperatures should be low enough for the entire atmosphere to cool to the point at which it collapses onto the dwarf's surface. Some 200 years hence, the process will repeat, with the Sun vaporizing enough of the dwarf's surface to regenerate an atmosphere.

Initially, this "first atmosphere" might be dotted with clouds of nitrogen, methane or carbon monoxide ice. These might erupt from Pluto's frozen surface, much like the geysers seen on Triton. As more ice sublimates and pressures rise, solar ultraviolet light and cosmic rays will be able to act on these gases and produce complex tholins. These materials and related nitriles will then form hazes at different layers within the atmosphere. Some of these precipitate out in a manner analogous to terrestrial frost, coloring the landscape pink. As the density of methane and carbon monoxide rises, their property as greenhouse gases will come into play, trapping more outgoing radiation and raising the temperature of Pluto even higher. This should reach equilibrium a few years into the thaw with a balance between trapping of energy, warming and further sublimation of ices.

Even then, the low gravity of Pluto is insufficient to retain most of its newly constructed atmosphere. As pressure and temperature rise, more and more of the atmosphere reaches the escape velocity of the dwarf world. New Horizons detected a cometary tail of material stretching down (solar) wind of the world, suggesting that a combination of Jeans escape and solar wind stripping is sufficient to remove much of the atmosphere that is vented from Pluto's warmed surface. Interestingly, some of this material appears to be raining out on its sister-world, Charon. Charon wears a red cap over its northern Pole. It is suggested that this was formed (and continues to form) when gases from Pluto that are rich in tholins, spew across the face of Charon. Only at the poles are conditions cold enough for the tholins and other materials to condense out forming a thin veneer, much like the icy cap of Earth or Mars.

Conclusions

Pluto, Triton and Titan form an intriguing triad of dwarf planets. While we think of Titan (quite rightly) as Saturn's greatest moon, it has rather a lot in common with further out Triton and Pluto. All three worlds have nitrogen-dominated atmospheres, the only difference being the temperatures involved, which then causes substantial differences in pressure. If one could shift Titan out to where Triton and Pluto reside, in the twilight limb of our solar system, most of its methane and nitrogen atmosphere would rain then snow out onto its surface, leaving a wispy ghost of its former self. Conversely, move the frigid pair inwards and much of their bulk would vaporize forming (at least briefly) a thick atmosphere. Briefly, because their low gravity—less than half that of Titan—would be unable to retain such a thick firmament for long. Most would escape through the Jeans mechanism—and ultimately a much denser hydrodynamic flow (Chaps. 5 and 6) over the course of a few hundred million years. What would remain would be an icy lump more akin to Europa or Callisto, than the effervescent, but still alive worlds we see today.

Ultimately, none of these worlds will survive in their present form. None retain gases strongly and all will be depleted in mass before the Sun leaves the main sequence. Their end will come when the Sun becomes a red giant. Then, temperatures will rise far beyond the melting point of water at Saturn and easily above temperatures sufficient to sublimate gases out by Neptune. By the time the Sun ceases to be a red giant only their rocky cores will remain. Triton has an even grizzlier fate in store. In a few billion years, its orbit will decay and send it crashing towards Neptune. Triton's loss will be Neptune's gain: a new set of icy rings that will slowly disperse under the increasing glare of the dying Sun.

References

1. Tokano, T. (2009). The dynamics of Titan's troposphere. *Philosophical Transactions A, 367,* 1889. Retrieved from http://rsta.royalsocietypublishing.org/content/roypta/367/1889/633.full.pdf.
2. Burr, D. M, Bridges, N. T., Marshall, J. R., Smith, J. K., White, B. R. & Emery, J. P. (2014). Higher-than-predicted saltation threshold wind speeds on Titan. *Nature, 517,* 60–63.

3. Teanby, N. A., Irwin, P. G. J., Nixon, C. A., de Kok, R., Vinatier, S., Coustenis, A., Sefton-Nash, E., Calcutt, S. B.& Flasar, M. F. (2012). Active upper-atmosphere chemistry and dynamics from polar circulation reversal on Titan. *Nature, 491*, 732–735.

4. Fischer, G., & Gurnett, D. A. (2011). The search for Titan lightning radio emissions. *Geophysical Research Letters, 38*, 8. Retrieved from http://onlinelibrary.wiley.com/doi/10.1029/2011GL047316/pdf .

5. de Kok, R. J., Teanby, N. A., Maltagliati, L., Irwin, P. G. J. & Vinatier, S. (2014). HCN ice in Titan's high-altitude southern polar cloud. *Nature, 514*, 65–67.

6. Tokano, T., McKay, C. P., Neubauer, F. M., Atreya, S. K Ferri, F., Fulchignoni, M & Niemann, H. B. (2006). Methane drizzle on Titan. *Nature, 442*, 432–435.

7. Hueso, T. R. & Sánchez-Lavega, A. (2006). Methane storms on Saturn's moon. *Nature 442*, 428–431.

8. Lorenz, R. D., Wall, S., Radebaugh, J. Boubin, G., Reffet, E., Janssen, M. et al. (2006). The sand seas of Titan: Cassini RADAR observations of longitudinal dunes. *West Science, 312* no. 5774, 724–727.

9. Charnay, B., Barth, E., Rafkin, S., Narteau, C., Lebonnois, S., Rodriguez, S. et al. (2015). Methane storms as a driver of Titan's dune orientation. *Nature Geoscience 8*, 362–366.

10. Zaluchaa, A. M., Michaelsa, T. I. (2013). A 3D general circulation model for Pluto and Triton with fixed volatile abundance and simplified surface forcing. Retrieved from http://arxiv.org/pdf/1211.0009v2.pdf.

11. Schneider, T., Graves, S. D. B., Schaller, E. l., & Brown, M. E. (2012). Polar methane accumulation and rainstorms on Titan from simulations of the methane cycle *Nature, 481*, 58–61.

12. Erwin, J. T., Tucker, O. J., & Johnson R. E. (2012). Hybrid fluid/kinetic modeling of Pluto's escaping atmosphere. Retrieved from http://arxiv.org/pdf/1211.3994v2.pdf.

13. Gilliam, A. E., Lerman, A. & Wunsc, J. (2015). Evolution of titan's atmosphere in relation to its surface and interior. Retrieved from http://www.hou.usra.edu/meetings/abscicon2015/pdf/7772.pdf.

14. Sotin, C., Lawrence, K. J., Reinhardt, B., Barnes, J. W., Brown, R. H., Hayes, A. G. et al. (2012). Observations of Titan's northern lakes at 5 microns: Implications for the organic cycle and geology. *Icarus 221*, 768–786. Retrieved from http://c3po.barnesos.net/publications/papers/2012.11.Icarus.Sotin.Northern.Lakes.pdf.

15. Johnson, R.E., Tucker, O. J., Michael, M., Sittler, E. C., Smith, H. T., Young, D. T. et al. Mass loss processes in Titan's upper atmosphere. Retrieved from http://people.virginia.edu/~rej/papers09/TitanChap15MassLossJohnson09.pdf.

10. Tales of Other Worlds

Introduction

Since 1992 astronomers and planetary scientists have been able to feast upon a bewildering variety of planets that lie outside our parochial system of eight, and a bit, worlds. We now have planetary systems orbiting white dwarfs, red giants, neutron stars, as well as a bewildering array of more mundane main sequence stars.

Kepler doubled the size of the planetary menagerie in the few years it was running at top speed. Even now, with reduced stability, astronomers have managed to keep it throwing out new planetary landscapes, by using the solar wind to keep it pointing in a fixed direction

In turn this array of planets has begun to allow planetary scientists to begin testing ideas about how atmospheres work—often in extreme environments. This chapter explores what is known and what is yet only imagined. It will be apparent where the now routine topics such as Monsoons, Rossby Waves and the ubiquitous jet stream guide our thoughts as they guide our weather. It will also become apparent where there are some surprising contradictions in models that affect how habitable a planet might be. In turn this reflects on how we view the running and habitability of our world.

Exoplanets by the Bucket

Exoplanets fall into a number of different categories based upon their size and location. In general we can recognize planets with masses in the range of several Jupiters, through Neptune-like planets to Earth and Mars-sized orbs. Many of these worlds are locked in star-grazing orbits that toast their surfaces and atmospheres to hundreds or thousands of degrees Celsius.

In general giant planets are confined to more massive stars with only one example known to orbit the universe's most

© Springer International Publishing Switzerland 2016   363
D.S. Stevenson, *The Exo-Weather Report*, Astronomers' Universe,
DOI 10.1007/978-3-319-25679-5_10

abundant stars, the red dwarfs. Red dwarf stars seem to favor planets with masses a few times that of the Earth. Such super-Earths, or super-terrans, may be largely rocky worlds or ones whose entire surface is submerged in a global oceans or a deep atmosphere. At present we cannot discern their true nature, but the presence of such worlds within the habitable zones of their stars suggests that the universe may favor life on such worlds.

One of the first discoveries, and the first planet known to orbit a Sun-like star, was 51 Pegasus b. This by and large set the tone for many of the subsequent finds: a Jupiter-like world in a tight, hot orbit around its star. Such planets are clearly unknown in our solar system and suggested that our orderly view of the universe needed some refining. Instead of planets forming in situ they are prone to migration inwards or outwards. Indeed, this theme was subsequently applied to our seemingly orderly solar system: the so-called "grand track" hypothesis, where Jupiter and Saturn have migrated great distances during their formative years, and consequently grossly modified the layout and subsequent evolution of the solar system as a whole.

At the time of writing over 1890 worlds are known, with over 4690 other worlds listed as candidates. Of these, over 470 fall into multi-planet systems: planetary systems with more than one planet orbiting its star (or in a few occasions, stars).[1] These fall into some broad and interesting categories. Of those that have a known mass (and these were primarily found by radial velocity or transit-timing methods), the majority are Jupiter-class worlds. However, when we look at planetary radius—where planets are found by transits of their host stars, the vast majority are so-called super-Earths (or super-terrans) with radii (and probably masses) a few times that of our world. Next most common are Neptune-class (or ice giant) worlds.

The difference in numbers of each type of planet that is found is down to the manner of how the planets are detected. The radial velocity method is best-suited to find massive planets in short period orbits—so-called hot Jupiters and hot-Neptunes. While the transit and transit-timing methods only require that the planet crosses the face of its star. The latter method, therefore has less

---

[1] Updated lists can be found at: http://exoplanetarchive.ipac.caltech.edu/docs/counts_detail.html.

bias but is clearly limited to the few percent of star that have orbits suitably aligned with ours so that we can observe a transit.

Amongst these largely biased searches there are still some apparent trends. Red dwarfs rarely host Jupiter-class worlds, favoring super-terrans or warm to hot-Neptune class worlds. This is probably a function of the mass of the star and planet forming cloud, which is typically around one tenth that thought to have spawned our solar system. As the mass of the star increases so does the likelihood of hosting planets: only 3.5 % of red dwarfs host planets while 14 % of A-class, Sirius-like, stars do. Moreover, planets forming around stars with low metal contents also tend to have fewer massive Jupiter-class planets.

Of the planets we are interested in throughout this chapter there are a few types that stand out. First are the ubiquitous hot Jupiter and Neptune-like worlds that whip around their host stars in a matter of hours or a few days. All of these are expected to be tidally-locked to their host star, meaning that they always present one face to their star. The next most interesting are similarly short-period planets, but those which orbit less luminous red dwarf stars and thus could be habitable. Finally, there will be planets orbiting less common, but still ubiquitous orange K-class stars. These planets are likely to rotate relatively slowly as they experience strong tidal forces, but not necessarily strong enough to cause them to lock. It is these three classes of world that will be the focus of this chapter.

The Climate of Tidally-Locked Planets: Assumptions and Expectations

Any planet that lies sufficiently close to its host star will lock to it. A young planet will experience tides within its hot, plastic mantle (and within any atmosphere and ocean it has). These tides are communicated between the planet and its star at the speed of light, which, being finite, takes time. Imagine then the planet orbiting its star but rotating niftily on its axis. As the star pulls on the planet it generates a tidal bulge. Given that light has a finite speed, the bulge will always be pointing towards the star in slightly the wrong direction because while the bulge is pulled one way, the rotation of the planet will carry it forward: the bulge will lead the pull of the star. With this happening, the star will pull on this in a direction pointing slightly backwards. This pulls on the bulge, slowing down its motion around the polar axis of the planet. After anything between a few tens of millions of years to a few hundred million years, the planet will have had its rotation slowed until one face permanently faces its host star. Now, the planet is still rotating on its axis, but now only so much as to rotate once per orbit. Like our Moon, the planet may librate, or appear to wobble, due to an eccentric orbit. It is thought that most tidally-locked worlds will have roughly circular orbits thanks to tidal locking.

What does this mean for its climate? Without going into detail just now, you would expect one side to be very warm, being permanently illuminated, while the opposing hemisphere, being permanently dark, was cold. The presence of even a Mars-like atmosphere will modify temperatures substantially. In 1998 Martin Heath and colleagues made some initial models which explored the effect of adding an atmosphere to the temperatures of planets orbiting red dwarf stars; and the subsequent impact on habitability. In the habitable zone, a planet without an atmosphere has a day side with temperatures approaching the boiling point of water. Meanwhile, temperatures on the dark hemisphere languish at less than −120 °C: pretty much what you find on the Moon. Add a Martian quota of atmosphere and the temperatures on the day-side fall to 50 °C, while the night-side warms to −70 °C. Add still more gas and the dayside maximum falls to a tropical 35 °C and the night-side rises to −20 °C. Finally, in more recent analysis by Yongyun Hu and colleagues, once you start modifying the amount of carbon dioxide and other greenhouse gases, the temperatures even out event more (as well as rise, overall).

The Structure of the Atmosphere of Jupiter-Like Worlds: Too Hot, Too Cold, or Just-Right

Before we look at the climate of tidally-locked planets we need to see how changing the temperature affects what gases are actually present in their atmosphere. Figure 10.1 shows the likely composition of clouds in the atmospheres of a planet with the composition

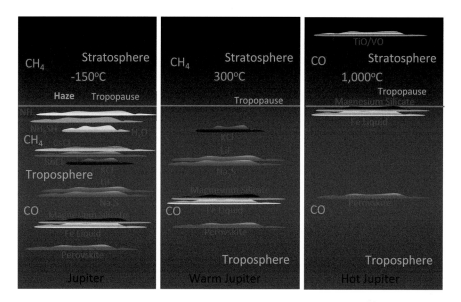

FIG. **10.1** Clouds and layers in the atmospheres of Jupiter-like planets. Jupiter, *left*, has visible clouds of ammonia (NH₃), Ammonium hydrogen sulfide (NH₄SH) and likely water (H₂O). Beneath this at higher temperature and pressure these substances are gaseous but more refractory compounds (those with higher boiling points) can condense to form clouds. These include Lithium fluoride (LiF); cesium chloride (CsCl); potassium chloride (KCl); sodium sulfide (Na₂S). At still deeper levels magnesium silicates and iron metal become liquids and gases. At the deepest levels, with the highest pressure a compound called perovskite is found as a gas. On Earth perovskite is only found beneath 1000 km of rock. As atmospheric temperatures rise (*center* and *right*) it becomes too hot at all levels for ammonia and water, and these are replaced by clouds from deeper down. Methane (CH₄) gives way to carbon monoxide (CO). This process continues as temperatures rise until it is hot enough for silicate clouds at the highest levels. Titanium and vanadium oxides may form clouds high in the atmospheres of the hottest Jupiter-like planets

of Jupiter, but heated to different extents by its parent star. Our cold Jupiter has icy clouds of ammonia and ammonium hydrogen sulfide, with water expected deeper down below the 5 bar level. Convention puts the upper lid on the troposphere (the tropopause) at 0.1 bars of pressure (Chap. 7). In the case of Jupiter this has a temperature of around –150 °C and this is the condensation point for ammonia at the pressures found there. If you lower the temperature to –220 °C, methane will form ice crystal clouds at this pressure, as is seen in the atmospheres of Uranus and Neptune (Chap. 8).

Conversely, raise the temperature to around 300 °C and now it is too hot for ammonia, water or ammonium hydrogen sulfide. Instead dusty clouds of halide compounds such as potassium or cesium chloride are formed. Hotter still and clouds of silicate dust and molten droplets of iron are found. It's not that water isn't found at these temperatures, rather that it is just too hot for it to condense to form clouds. So, in the first instance the types of clouds are determined by the available temperatures. Indeed, if you wanted to look for silicate vapor on Jupiter (but probably not clouds, per se) you'd have to drop down well below the 10 bar level until temperatures were at the appropriate temperature for these compounds to condense.

That's not quite all. In the cooler Jupiters the background gas (other than hydrogen and helium) is methane. This is because methane is only stable, chemically, at temperatures less than 1000 °C. As temperatures rise, methane will react with a variety of other oxygen containing compounds to form carbon monoxide. Chemical equilibria: the balance between different chemical compounds in a mixture of gases changes with pressure and temperature. In practice this means that methane picks up oxygen from silicates, water or other oxygen-containing compounds. This oxidizes the methane to carbon monoxide, while releasing the hydrogen. One of the benefactors of these reactions is oxides of iron. These chemically react with the hydrogen are then chemically reduced to iron metal. Therefore, in very hot Jupiters, methane reacts with available iron oxides to form carbon monoxide and liquid iron, which then rains out into the planetary interior. Hot Jupiters also show sodium, potassium and other metal halides such as sodium chloride in their upper atmosphere. The firs detection of this was made by Sarah Seager's group in the atmosphere of HD 209458b (Fig. 10.4) confirming the basic model.

One notable difference to this pattern of change was the detection of titanium and vanadium oxide clouds in the upper atmosphere of HD 209458b. This compound, common in the lower atmosphere of red dwarf stars, only appears at temperatures between 2000–2500 °C implying that the upper atmosphere is not only hot enough to host this chemical, but dense enough so that these compounds can condense in the first place. This clearly implies a strong temperature inversion in the middle atmosphere of this giant planet. These surprises aside, the general picture, which is emerging, matches expectations. Clouds of different compositions will form depending on the overall temperature and pressure.

If we increase the mass of these objects eventually we meet the brown dwarfs at 13 Jupiter masses and red dwarfs at around 75 Jupiter masses. Brown dwarfs make an interesting case and, yes, weather, has been observed on some of these by examining changes in the intensity and wavelength of radiation emitted. Such objects are important test-beds for understanding the overall structure of Jupiter-like worlds. All of these begin life hot. The only real difference between an infant Jupiter and a brown dwarf is the strength of its gravitational field. Both are born as hot, incandescent objects with broadly the same diameter. The increase in mass is broadly off-set by greater gravitational compaction. It is only when the core becomes hot enough to ignite does the diameter of the object increase again. Jupiter would cool more rapidly, having less internal energy to spare than a brown dwarf. The lower gravity in Jupiter primarily serves to puff-up its atmosphere, relative to a typical brown dwarf. The temperature gradient is somewhat less steep as a result, but otherwise Jupiter makes a pretty good model for an old and cold brown dwarf. Consequently, a young brown dwarf will have clouds of metal oxides (primarily titanium and vanadium oxide). As the brown dwarf cools it will go through the same series of changes we expect for Jupiter-mass planets at increasing distances from their host star. Clouds of oxides will retreat deeper inside the brown dwarf and become replaced by metal halide clouds. Over time the carbon monoxide gas, forming the background, will morph into methane as temperatures fall below 1000 °C. Such brown dwarfs are called T-class objects in an extension of the Hertzsprung-Russell diagram that is used to display stellar classification.

Finally, as temperatures fall lower, first water clouds, then ammonia ice clouds will form the uppermost cloud decks and the object will take on the veneer of our Jupiter and Saturn. Indeed, some very cool brown dwarfs have been found. The coolest brown dwarf (WISE J085510.83-071442.5 (W0855)) is colder than Mars with a cloud-top temperature of (−48 °C to −13 °C (225–260 K). Spectra taken and analyzed by Jacqueline Faherty (Carnegie Institute of Washington) suggest the planet has water clouds across half of its face. It still is far from an Earth-like atmosphere: aside from hydrogen and helium, the most abundant gas is methane, making it uninhabitable for humans.

Many other, hotter brown dwarfs also show indications of clouds of silicates drifting—or rather powering through their super-charged atmospheres. Brown dwarfs form a fascinating link between stars and planets, exhibiting weather as complex as any planet. This is infant territory at present, but the identification of growing numbers of nearby brown dwarfs is opening up another window on our universe's weather. The main difference between brown dwarfs and hot (or at least lukewarm) Jupiters is that the former are heated internally, while the latter are heated (primarily) by an external source. This should lead to differences in the dynamics of their atmospheres: brown dwarfs have convection-driven weather, while hot Jupiters have advection-driven weather, because they are hottest highest up.

This is an evolving picture—quite literally. Imagine the Solar System of 6–7 billion years hence. Jupiter will be heated to a few hundred degrees by the expanding Sun. Aside from steadily vaporizing its icy moons, the upper atmosphere will become too hot to allow first ammonia, ammonium hydrogen sulfide and finally water clouds. Heat will slowly percolate downwards from the upper atmosphere steadily evaporating Jupiter's cloud decks. After several tens of millions of years the upper atmosphere will play host to clouds of metal halides rather than ammonia. Deeper layers of silicate dust might be discernible. Further out, the methane clouds of Uranus and Neptune will also evaporate leaving deeper clouds of ammonia, ammonium hydrogen sulfide and water: our view of the Solar System's meteorology will be transformed. In essence if you want to see inside Jupiter, look at an alien hot Jupiter and this will reveal what lurks within its deeper layers. After the Sun expires and becomes a white dwarf, Jupiter will slowly cool off once more. Eventually, the planet will play host to methane clouds, before these, too freeze out and retreat to deeper, warmer layers. Only when the planet cools below the freezing point of hydrogen will the final changes occur. Jupiter and the other giants will truly liquefy throughout, before solidifying into immense icy balls. That will be a long, long way off into our universe's distant future.

Some time before our planets irrevocably alter, HD 209458b's will suffer the same fate. First its halide layer will be boiled away revealing deeper clouds of silicates or perovskite. This will be a brief transition before the planet as a whole is vaporized inside the outer layers of its expanding Sun.

There is a limit to how deep you can go as below some particular depth some of the materials that could form clouds will forever be too hot to do so. Moreover, below a critical depth the dominant hydrogen and helium gases morph into an ocean of liquid. At this point, meteorology gives way to oceanography and our experiment with a planet's atmosphere comes to an end.

What about the underlying driving force for the weather on giant planets? Well, this is a bit more controversial. For hot Jupiters, with temperatures above 500 °C, and an energy input measured in the tens of thousands of Watts per meter squared, it is likely that most of the energy needed to drive their weather comes from their parent star. Once the amount of energy falls to a few watts per meter squared then internal energy sources begin to play a more dominant role. On the Earth the surface heat flux released from the interior is around 0.087 W per meter squared, which is far less than the 240 W per meter squared it receives from the Sun (principally in the tropics). Jupiter releases considerably more energy from its steady contraction. This amounts to roughly 2.5 times the 14 W per meter squared captured by its planet's atmosphere. Here, the internal energy is more than sufficient to drive convection all the way from the planet's core to the top of the atmosphere.

Out at Neptune this world receives a dismal 0.7 W per square meter and the internal energy sources again dominate, driving convection. A hot Jupiter has an atmosphere which is hottest at its top—an extreme manifestation of our thermosphere and heat is readily conducted downwards making the atmosphere completely stratified to considerable depth—around 100 bars or ten times the depth of the water clouds on Jupiter. While convection ultimately delivers energy from the interior of our giant worlds, this is impossible in the atmosphere of a hot Jupiter where more than 100 times the energy available from the interior is delivered by its star. Winds are driven by the horizontal temperature contrasts between the day and night hemispheres, not by differences in temperature at height.

Finally, while Jupiter spins once every 10 h or so and Neptune once every 16 h, a tidally locked hot Jupiter will take at least twice this long to complete one revolution—and more typically it will take around 5–10 times as long. Consequently, the weaker Coriolis Effect these planets experience has a significant impact on how the atmospheres of hot Jupiter and hot Earth-like worlds work, as we shall now see.

The Mystery of Super-Rotation

The atmospheres of several bodies within the solar system show super-rotation: the movement of atmospheric gases in the same general direction as the planet or satellite rotates, but at a faster rate (Chaps. 5, 7 and 9). Super-rotation is a problem as how it isn't immediately obvious how you can power up a planet's winds so that they move faster than the underlying surface that drives their rotation around the planet's axis. The solution to this problem has come from various sources, including: observations of hot Jupiters; experiments based in over-sized round bathtubs; and observations of the atmosphere of our planet, particularly our tropical atmosphere. Within the tropics, rather out-of-sight of ground-based observations, are different forms of atmospheric wave. One of these, the Kelvin Wave is particularly relevant (Chap. 1). These Kelvin waves are the key to understanding how an otherwise sluggish easterly flow can be turned into a powerful westerly current that has profound impacts on the climate of tidally-locked exoplanets. For it is in the tropics that our atmosphere is heated most strongly and experiences the weakest Coriolis Effect, making it particularly analogous to the atmospheres of extrasolar worlds where the planet is tidally-locked to its star.

Before we examine how super-rotation is achieved (at least in some cases), we need to look a little more closely at how hot Jupiter planets operate.

How Do the Atmospheres of Tidally-Locked Worlds Move?

One might, naively assume that if a planet is tidally-locked to its star then air will be heated strongly on the star-facing side and simply rise by convection. This air will then blandly flow at altitude to the opposing, dark hemisphere where it will cool, descend and flow back across the surface to the star-lit side. This may be true for an unnaturally static planet—one that does not rotate on its axis. Radio and infrared observations of hot Jupiter worlds show that winds blow strongly around the equators of these worlds in sharp contradiction to any expectation based on the strength of their Coriolis Effect. How is this possible? We now look at how seemingly insignificant waves can rev up the atmosphere of a planet and drive super-rotation.

Kelvin waves are convection-driven eastward moving waves that move along the equator of the Earth—and obviously any other similar exoplanet. These waves are broadly symmetrical around the equator and have large wavelengths of 30–90° of longitude. Kelvin waves move at around 10–20 m per second somewhat faster than the Madden-Julian Oscillation (MJO) waves that were also discussed in Chap. 1. MJO waves can have wavelengths as long as the planet's circumference. Why mention these, aside from their importance in transporting energy in our tropics? Well, the odd property of these waves appears to lead, ultimately to the phenomenon of super-rotation seen in hot exoplanets and probably all planets that are tidally-locked to their stars. For while the predominant wind direction at the Earth's equator is easterly, Kelvin waves move from west to east: it is this that ultimately allows a hot Jupiter world to experience winds that are faster around the equator than the planet rotates.

How does this work? A few facts, first. Kelvin Waves, named after their theorist, Lord Kelvin, are alternating regions of low and high pressure that string along the equator. Where pressure is low there is enhanced convection (indeed, convection causes the pressure to drop). Air rising in these areas of low pressure ultimately cools, becomes denser and then falls back to the surface forming rear-flanking areas of higher pressure. Because of the underlying structure of these low and high pressure areas there is more convergence (coming together) of air at the eastern or leading edge of the low pressure than along its western edge. Winds and evaporation are stronger along the eastern edge, as well. Like the longer wavelength MJO waves (Chaps. 1 and 2), this leads to movement of the entire low pressure area (and rear-flanking high pressure area) towards the region of maximum uplift, pulling the wave from west to east.

Experiments done by Dargan Frierson (University of Chicago) confirm that it is the pattern of evaporation and wind that is the dominant driver of these waves. All of this is important because it sets up a pattern of west-to-east motion of mass and energy through the planet's tropical belt. On the Earth, this effect is largely, but not completely subordinate to the overall easterly trades so that although these waves move from west to east on monthly timescales, the predominant wind is still from northeast to southwest north of the equator, and south east to northwest to the equator's south (Chap. 1). On occasion these Kelvin waves do drive or at least contribute to a reversal of the prevailing trade winds, generating terrestrial superrotation which we know as an El Niño (Chap. 2).

How do Kelvin waves lead to the reversal of the prevailing winds and the development of super-rotation on tidally-locked planets? The key to solving this conundrum came from further modeling work by Adam Showman (University of Arizona) and Lorenzo Polvani (Columbia University). Together, they showed that this west-to-east motion can be amplified by disturbances generated by Rossby Waves moving further towards the Poles of these planets. This happens despite the contrary motion of the Rossby waves towards the west.

Where there is a strong difference in the temperature of the planet on its day and night side, this sets up large amplitude Rossby Waves in the middle and upper atmosphere (Chap. 1). For a hot Jupiter, or tidally-locked planet with one side permanently heated and the other side in permanent darkness, a large dome of warm high pressure is locked on either side of the equator around the point where most sunlight is received.[2] Mid-latitude westerly winds then flow upwards and around the feature. On the dark, anti-solar, side pressure is lower because the air is colder (Chap. 1). The westerly winds then flow back towards the equator (Fig. 10.2). This pattern of heating and cooling establishes fixed Rossby Waves in the planet's atmosphere around these fixed low and high pressure areas.

---

[2] You can see the same effect during an El Niño where two areas of high pressure flank the equatorial low over the central Pacific. On Earth, the surface low is anchored not by a fixed Sun, but by the pool of warm water that lies beneath it (Chap. 2).

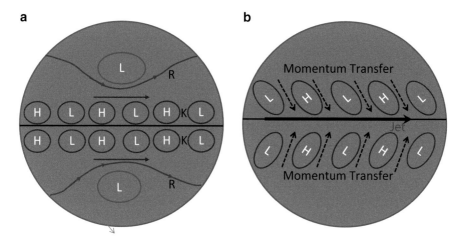

a

b

FIG. 10.2 The emergence of super-rotation in the atmosphere of hot Jupiters and super-Earth planets. Eastward propagating pressure waves form on either side of the equator (**a**). These features are known as Kelvin waves (K, *red circles* and *arrows*). North and south of these are westward propagating Rossby Waves (R, Chap. 1). The largest amplitude Rossby and Kelvin waves are then fixed by very uneven heating and by the equator, leaving shorter wavelength disturbances to move with the mean flow from west to east. Despite atmospheric momentum being lowest furthest away from the equator, this set up causes the combined Rossby and Kelvin waves to develop a northwest-southeast tilt north of the equator; with a south west-northeast tilt to its south (**b**). This allows the atmosphere to transport angular momentum from the poles towards the equatorial regions by directing more and more energy into an equatorial westerly jet stream. Once sufficient momentum has been transferred the Kelvin waves are modified by the developing westerly jet which prevents further acceleration. This leaves a self-sustaining westerly jet stream above the equator and light westerlies beneath it. In our Solar System, Venus (Chap. 5), Jupiter and Saturn (Chap. 7), Uranus (Chap. 8) and Titan (Chap. 9) all show super-rotating jets above their equators. Venus has an easterly flow; the others westerly. Super-rotation in the atmospheres of Venus and Titan is best explained by this mechanism; while Jupiter and Saturn, by different mechanisms involving convection and rotation

Normally this means that air flowing nearest the equator has the highest momentum and shouldn't be able to accept any more from air moving closer to the Poles, but the westerly (eastward-moving) Kelvin Wave motion to pick up momentum from the Rossby Waves further north and south. In their models, two sets of waves combine, forming tilted, chevron-like structures (Fig. 10.2). Although the eastward moving Rossby Waves would be expected to cancel out the westerly Kelvin waves, instead eddies and other disturbances generated by the Rossby waves can transport energy and momentum into the Kelvin waves lying at the equator and to quote the authors, "pump-up" the tropical westerlies. The chevron-like waves naturally direct momentum from the higher latitudes towards the east at the equator. Showman and Polvani show that even when you start with a static atmosphere, the interaction between the westward-moving Rossby Waves and eastward moving Kelvin Waves causes an acceleration of the equatorial westerly motion: in no time at all, super-rotation begins with a strong westerly jet stream overlying the equator (Fig. 10.2). This odd, and rather counterintuitive, phenomenon may explain super-rotation in the atmosphere of slowly-rotating Venus (Chap. 5), but probably not Jupiter and Saturn (Chap. 7), Uranus (Chap. 8) or Titan (Chap. 9). Here, other factors appear to be at work. At Venus momentum is transferred from the upper to the lower atmosphere, which undoubtedly also affects super-rotation on hot Jupiters (Chap. 5); while at Titan, Saturn's gravitational pull may affect or drive super-rotation through the action of the tidal bulge it creates. Tapio Schneider and Junjun Liu (Caltech) suggest that superrotation is driven by convection at the equator. This generates eddies and Rossby Waves (Chap. 1) on either side of the equator. This allows momentum to flow into the equatorial regions, accelerating winds into two prograde jets (Chap. 7). All of this is speculation at present.

Super-rotation ensures that on hot Jupiters the hottest conditions are found to the east of the region that is most strongly heated. Winds blow the hottest air towards the day-night terminator. In general, planets that rotate most slowly show the weakest displacement of the hot zone from the sub-stellar point. This means that for all tidally-locked worlds those that are located closest to their stars are most likely to show hot spots in their atmospheres that are located furthest from the sub-stellar point, while those furthest away, and rotate the slowest, will show maximum atmospheric temperatures in a lobster-claw-like pattern around the SST. Compare the rotation periods. For a typical hot Jupiter the planet takes at most a few days to rotate on its axis as it orbits its host star. Venus takes 243 days to complete the same trick, thus Venus experiences a far more leisurely Coriolis Effect than a typical hot Jupiter. These have a 1-6 day day-long orbit, while a super-Earth or Neptune orbiting a red dwarf in its habitable zone will have a 20–40 day-long orbit.

All of these models show that the maximum effects are found in the troposphere (1 bar) through to the lower stratosphere (100 millibar), while at higher altitudes (around the 10–20 milli-bar level) the hottest conditions are likely to be found where you expect them to be: under the SST. This is simply because the thin-ner air at this altitude is less able to hold and transport energy. Here, air flows in a thermal wind, much like in the middle atmo-sphere of Venus, in a relatively simple flow from day-lit side to night side. Unlike Venus, the illuminated and dark hemispheres are a permanent feature and do not rotate across the face of the planet: tidal locking prevents this.

In general, all tidally-locked worlds show the same, overall, pattern of atmospheric circulation. With one side heated and the other side not, air is forced into a fairly simple, generic pattern of airflow from warm to cool side. What is perhaps unexpected is that this is not accomplished by a single Hadley cell running from warm to cold. Even the modest rotation tidally-locked plan-ets experience is sufficient to modify air movement through the Coriolis Effect (Chap. 1). Thus although there is a broad ascent of air at the sub-stellar point, towards the anti-solar point at height, this is modified by the presence of a super-rotating westerly jet along the equator. Likewise, the return flow is diverted by the jet stream and by the large hemispheric lows and highs caused by uneven heating and planetary-scale Rossby waves (Fig. 10.2).

Hot Jupiters

What does this mean for the weather on a hot Jupiter? Well, there are two obvious effects that are illustrated in Fig. 10.3. The strong westerly flow around the equator brings cooler air from the east towards the sub-stellar point (SSP) where there is maximum stellar heating, and takes warm air eastwards over the day–night termina-tor towards the antistellar point (ASP). Depending on the strength of the wind and the amount of heating this displaces the hottest

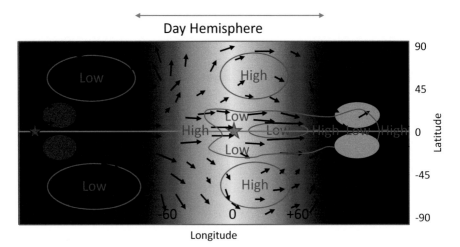

Fɪɢ. **10.3** The weather on a tidally-locked hot Jupiter. Models predict that on hot Jupiters the hottest conditions are found 60–120° east of the sub-stellar point, or SSP for short (shown by a *red* and *orange star* on the day hemisphere). A strong westerly jet stream (*equatorial black arrows*) blows air around the equator and displaces the hottest regions down-wind of the sub-stellar point. Where winds are particularly strong, the hottest regions, perhaps split by the jet, may be displaced all the way across the terminator into the night hemisphere (*pink circles*). Spitzer observations of HD 189733b, HD 209458b and Tau Boötes b appear to confirm this prediction. To the north and south of the equator at 30° east of the SSP, in the mid-to high latitudes, warm-cored areas of high pressure (may) bring more *settled* conditions—or form large Great Red Spot-like storms. In the night hemisphere, downwind of the anti-stellar point (or ASP, *black* and *grey star*), air descends and warms, generating a second, weaker warm spot (or spots) that are again split into two by the intruding westerly jet stream (*dark pink ovals*). Large areas of low pressure form on either side of the ASP and, again, these could be broad cyclonic storms. Across the entire planet, winds are thought to be pre-dominantly westerly. These patterns are for the 0.1 bar level and below. Higher in the stratosphere, winds are weaker with the hottest conditions found near the SSP. However, thermal winds blowing from day to night side still generate a weak warm spot where the air descends at either side of the ASP. Kelvin waves generate weak areas of higher and lower pres-sure along the equator. Equatorial winds move at hundreds of meters per second, dwarfing even the intensity of Neptune's

conditions eastwards, potentially over the terminator and into the dark hemisphere. Although this also may sound unlikely, VLTI, Spitzer and Hubble observations of various hot Jupiters, including HD 187933b and Tau Boötes Ab appear to confirm this pattern, with the maximum infrared emission occurring in a hot region far to the east of the area that is most strongly heated. Moreover, the observations confirm that the hottest region is shaped more like a lobster's claw than a sphere around the SSP because cooler air is being brought into the region by the westerly jet. There is no surface low pressure at, or near, the SSP simply because there is no surface. Air that is strongly heated has a high pressure and simply moves outward and away in a thermal wind towards the dark and cooler hemisphere. Here, pressure is lower because the air is colder. The overall pattern is then controlled by large, atmospheric Rossby and Kelvin waves, which produce super-rotation with strong west-to-east flow.

In the dark hemisphere, there is predicted to exist, a second zone where temperatures are relatively high. East of the ASP is a warm zone (or strictly speaking zones where air flowing from the sunlit hemisphere converges and descends into the planetary interior. This zone is again displaced eastwards by the action of the westerly jet. Observations of exoplanetary atmospheres are currently not sensitive enough to confirm or refute the existence of this region but will likely do so in the near future.

In general, planets that rotate most slowly show the weakest displacement of the hot zone from the sub-stellar point. This means that for all tidally-locked worlds those that are located closest to their stars are most likely to show hot spots in their atmospheres that are located furthest from the sub-stellar point, while those furthest away, and rotate the slowest, will show maximum atmospheric temperatures in a lobster-claw-like pattern around the SST. Compare the rotation periods. For a typical hot Jupiter the planet takes at most a few days to rotate on its axis as it orbits its host star. Venus takes 243 days to complete the same trick, thus Venus experiences a far more leisurely Coriolis Effect than a typical hot Jupiter—or even a super-Earth or Neptune orbiting a red dwarf in its habitable zone, with 1–6 day and 20–40 day-long orbits, respectively.

All of these models show that the maximum effects are found in the troposphere (1 bar level) and lower stratosphere (100 millibar level), while at higher altitudes the hottest conditions are likely to be found where you expect them to be: under the SST (1 millibar) with little displacement to the east.

Observations of HD 209458b, discovered in 1999, are particularly illuminating. This hot Jupiter orbits its Sun-like star every 3.5 days. Its mass is roughly 0.69 Jupiters but it is over twice Jupiter's volume. This contradictory pair of observations points to the trauma HD 209458b is experiencing. In its 0.7 million kilometer radius orbit the star is cooked by an intensity of radiation from its star over 1000 times that experienced by the Jupiter. Its troposphere has a searing 1000 °C heat that steadily increases with height. Spectral observations suggest that the stratosphere is relatively cloud free for most of its height. Here, measurements by Canada's MOST (Microvariability and Oscillations of STars telescope) show that the stratosphere grows hotter with altitude, reaching 2200 °C at the 33 millibar level. MOST also showed that the planet was very dark: its albedo was as low as 4 % meaning that 96 % of the visible radiation reaching it was absorbed. Compare that to Jupiter, which has an albedo of 52 %. This tells

Radius: 1.29 Jupiter Radii to 1 bar level 970°C. Hot silicate clouds likely below this level

Stratosphere: 1.32 Jupiter radii; 1,200°C but rising to 2,200°C 100-25mb contains dark cloud layer observed by HST

Exosphere: 3.1 Jupiter Radii, temp rising to 9,700°C 25mb-1μb

Hot hydrogen coma containing escaping gases that include carbon and oxygen

FIG. 10.4 The overall structure of the atmosphere of HD 209458b. The planet is embedded in a large, dispersed cloud of hydrogen and other gases that is being driven away from the planet by a combination of intense heating and a strong stellar wind. This is the coma and is analogous to a comet's outer cloud. Beneath this, within 3.1 Jupiter Radii lies the exosphere that constantly supplies gases to the expanding coma. Below this will lie a poorly defined thermosphere. Sarah Seager's group identified sodium in this layer, while Korey Haynes identified a layer of metal oxide clouds at 25 mb (stratosphere). Temperatures fall with depth until the troposphere, which is yet to be clearly observed but likely contains clouds of silicate dust from which a rain of molten rock likely descends. Layers not to drawn to scale

us that either HD 209458b has dark clouds or is relatively cloud-free and light is Rayleigh scattered by deep, translucent layers of hydrogen, much like the oceans of early Earth may have done (Chap. 2) (Fig. 10.4).

Moving upwards into the thermosphere, temperatures eventually reach over 9700 °C at its lid at the 25 millibar base of the exosphere. In 2001 Sarah Seager (currently at MIT) predicted that the atmosphere of this and other hot Jupiters would contain sodium. Subsequent spectral measurements made by Hubble detected this in the thermosphere, at pressures from 50 millibars to 1 microbar (one millionth of a bar). Separate work, carried out by Jeremy Richardson of NASA's Goddard Space Flight Center, showed that the lower atmosphere contained surprisingly little water vapor, which contradicted models. It might well be that the upper atmosphere has had all of its water broken down to oxygen and hydrogen through the action of the intense ultraviolet light the planet experiences in such a tortuous orbit. In 2009 JPL carried out further spectral analysis and this time identified water along with carbon dioxide and methane. This is an odd assemblage of incompatible gases: methane wouldn't normally be expected to coexist with carbon dioxide as the two will tend to react with another at high temperatures. By implication, there is a lot of dynamic atmospheric chemistry going on around HD 209458b that is caused by the rather hellish conditions.

What else do we know about HD 209458b? There are indications that at, or near to, the 25 millibar level there is a layer of dusty clouds resting above the clear hydrogen layer. Given the temperatures of 2000 °C (with a range of 260°), these are clearly not water. Instead a froth of vanadium oxide and titanium oxide seems likely. This makes HD 209458b rather similar to a cool red dwarf star at least in terms of its appearance. Titanium oxide also acts as a strong absorber of ultraviolet radiation, much like ozone in our atmosphere or titanium dioxide, which is used in commercial sun-creams. Like a sun cream, this layer absorbs energy which causes heating of the stratosphere. This partly explains why the temperature of this layer exceeds that of the troposphere, beneath.

Such a hot, absorbing layer is also seen in the hot Jupiter Wasp 33b. Hubble spied on this world and showed its stratosphere broils at 6000 °C. Research by Korey Haynes (George Mason University) and Drake Deming (University of Maryland) again made spectral measurements of Wasp 33b at different wavelengths, this time using water vapor as a tracer. In this planet's atmosphere, the lower stratosphere was a "cool" 1700 °C, but the layer immediately above it was twice as hot. Again, absorption of ultraviolet radiation by a titanium oxide "sunscreen" appeared to the driving force in generating this temperature inversion.

Returning to HD 209458b, deeper down, in the cooler troposphere, absorption patterns in the spectra suggest the presence of silicate clouds, which might bequeath HD 209458b with showers of molten rock. Imagine basalt falling out of the sky: perhaps Pele's hair drifting downwards through the hot, but otherwise fairly clear sky.

Yet, this is more than a little misleading. For measurements of carbon monoxide indicate that the atmosphere would hardly allow "drifting" in a leisurely sense. Carbon monoxide, again somewhat surprisingly, is one of the atmosphere's dominant gases after hydrogen and helium. Although only contributing a small percentage, overall, this gas is a particularly useful tracer of atmospheric motion. Not only does carbon monoxide emit and absorb radiation at useful wavelengths, revealing that the hottest part of the atmosphere lies eastwards of the sub-stellar point, in agreement with predictions, carbon monoxide emission also clearly shows that the atmosphere is in extreme motion. Doppler measurements made by the CRIRES spectrograph attached to ESO's Very Large Telescope show that at least at some levels, winds blow at nearly 7000 km per hour from the sunlit side to the permanently dark hemisphere. This intense wind is simply a response to the extreme difference in heating. Were HD 209458b to lack an atmosphere its dark face would freeze at less than –170 °C, while its sunlit side was over 1000 °C. With such an extreme contrast in temperatures, the fluid, dense atmosphere does what it can to even out the temperatures. The result is winds that blast at more than five times those of our windiest planet, Neptune (Chap. 8). Therefore, "drifting" might be something of a misnomer. Instead imagine a blizzard of molten basalt blasting downwards through the increasingly dense atmosphere, before it vaporizes deeper in the planets gassy interior.

In general most hot Jupiters are thought to have the hottest conditions highest up. The atmospheres are stable and stratiform in structure above the cloud tops because more and more radiation from their star is available to absorb. Various observations confirm this is true for most hot Jupiters. But the keyword is "most". Like any good theory (and set of observations) there is always an odd one out. In this case it is Tau Boötes Ab. Observations by the VLTI show that the upper atmosphere cools with height. The only way that this is possible is if the lower stratosphere contains materials that absorb energy extremely effectively. This would allow this layer to warm more than more transparent layers that were higher up. This set-up should allow the efficient transport of energy by convection. However, in the other hot Jupiters that have been observed, the upward transport of energy (if it occurs) will be through atmospheric gravity waves that we shall come to later in this chapter.

While the troposphere and stratosphere reveal unimaginably violent winds, HD 209458b's exosphere is a marvel of planetary destruction. The ellipsoidal exosphere extends over three Jupiter-radii from the planet's center (over 200,000 km) and reveals the scale of the torment this planet experiences. Somewhere between 100 million and 500 million kilograms of hydrogen gas is streaming away from this planet every second. Within this outward moving flow are heavier carbon and oxygen atoms that are being dragged away from the planet by the out-flowing hydrogen gas. This hydrodynamic drag helps strip away much of the heavier atoms that contribute to the mass of the atmosphere. This is the same process that is happening today above Titan, albeit at a far lower rate (Chap. 9). Mars (Chap. 6) also experienced this early in its history, with out-flowing hydrogen pulling away some of the planets carbon and oxygen. For both of these small worlds, the amount of gas that was shed is easily dwarfed by the vast outflow from this gas giant. Compared to the Earth, which loses around 1000 kg per second, HD 209458b is losing mass at a rate 100,000–500,000 times faster. Even with this prodigious rate, the massive planet will survive up until its parent star swallows it at the end of its main sequence life. However, it will survive in a much reduced state. These discoveries are described in Fig. 10.4.

How is HD 29458b losing so much mass? Obviously the extreme day-side heating is the underlying cause, but look again at the planet's dimensions. The mass is less than three quarters that of Jupiter but its volume more than twice as much. This shows that the planet's mass is far more dispersed than Jupiter, or even our fluffy Saturn. While Venus *only* gets hot enough to drive a thermal wind from day side to night side and this is sufficient to cool off the planet's atmosphere, this process is grossly insufficient to cool HD 209458b. Here, the planet is so strongly heated that it generates high pressures within the atmosphere and upper layers of the planet that blow the atmosphere off into space at hundreds of kilometers per second. The figure of 100 million to 500 million kilograms per second might well be conservative. Such hydrodynamic escape was discussed in Chaps. 5 and 6 but has little current impact on the atmospheres of our terrestrial worlds: Titan (Chap. 9) is a likely exception. In the case of HD 209458b it is stripping sizable quantities of the planet's mass over the course of a few billion years. Overall, if the current rate of mass loss is maintained, HD 209458b will lose only another 7 % of its mass before its host star becomes a red giant and destroys the planet. This adds to the similar amount the planet will have already lost since its formation. One might expect this rate to increase as the planet loses mass and its gravitational pull on its gases decreases. Conversely, the planet has a decent magnetic field which may help it retain gas—at least within its magnetosphere and thus slow the overall rate of gas loss. Is HD 209458b alone in suffering this state of affairs? Absolutely not: other hot Jupiters, such as Tau Boötes Ab, HD 198733b and 55 Cancri b also show gas escape revealing that this is a general phenomenon with hot Jupiter worlds.

Tidally-Locked Earths and Super-Earths: General Ideas

We can lump the hot worlds together: the hot Jupiters, hot Neptunes and hot Earth-like worlds. These all share the same principle atmospheric motion that is driven by strong heating on one side and cooling on the other. The only difference will be that the smaller super-Earths will have less voluminous atmospheres and thus be less efficient at transferring energy. Unless they have atmospheres much less massive than that of the Earth they should distribute heat in pretty much the same way. This will give rise to similar weather. The only exception to this rule will be where a smaller planet has a thinner atmosphere. This will allow the star to heat the surface as well as the atmosphere. Here, convection might dominate transport of energy from the day-lit to the night sides.

39 light years from Earth lies the red dwarf GJ 1132. In 2015 Berta-Thompson and his colleagues (MEarth-South Observatory) published their discovered of a planet orbiting this star. This Harvard University-led array of eight 40-cm-wide robotic telescopes is located in the mountains of Chile and is dedicated to scanning the heavens for Earth-like planets that orbit these little stars. GJ 1132b is a relatively small world, only marginally wider than the Earth. Consequently, with its density estimated to be similar to the Earth it is likely to be rocky in composition.

As the orbit of GJ 1132b takes only 1.6 Earth-days to complete it is well inside the region where temperatures exceed the boiling point of water. With its fairly substantial mass GJ 1132 should host a dense atmosphere. With cloud-top temperatures likely to lie around 230–250 °C, this world should resemble Venus, but spinning at over 151 times the speed of Venus it must have a much more substantial Coriolis Effect. Indeed, its Coriolis Effect is similar in strength to ours, so we can expect some unusual and distinctly non-Venusian weather. While Venus does show super-rotation higher in its atmosphere (Chap. 5), near the surface there is very little movement of air at the surface. By contrast, GJ 1132 would be expected to show a large-scale circulation that is dominated by a westerly equatorial jet and broad convection to the east of the SSP. With appropriate optics we might just be able to determine whether this is true in the very near future.

Continuing the trend from Hot Jupiter to tidally-locked Venus, what if we take our imaginary world to a location where the temperatures drop to the point at which liquid water becomes feasible? These tidally-locked worlds are only possible around red dwarf stars. For only these stars have a habitable zone, where temperatures are modest enough for liquid water in the region in which tidal locking will occur. What happens to the atmosphere of these worlds and how would this affect their weather and their habitability? As each of these worlds must orbit proportionately further from its star in order to be habitable—and this is only a few million kilometers—the orbital period is anything from 15 to 40 days. This depends on the mass of the red dwarf: larger periods are obviously associated with more massive and more luminous red dwarfs. Consequently, the Coriolis Effect is weaker for habitable Earth-like planets around red dwarfs than it is for hot Jupiters and other worlds orbiting Sun-like stars. Models show that there is a transition between broadly static and super-rotating atmospheres that occurs at different rotation rates, which is dependent on the size of the planet. With an Earth-sized world, once the rotation period increases beyond 5 days super-rotation slows and soon stops. At this critical point Rossby waves are ineffective at pumping energy into the tropics through eddies in the mid-latitude atmosphere. For planets with radii above that of the Earth, this transition moves to greater periods. A planet with twice the radius of the Earth will cease super-rotation at around 10 day orbital periods (and hence 10 Earth-day long rotations). Therefore, the more massive the red dwarf, and the more extended its habitable zone, the less effective super-rotation will be at redistributing heat around the planet.

Further refinements to this trend were made to these models by Ludmila Carone (KU Leuven, Belgium). In planets where the orbital (and rotation) period was greater than 45 days the equatorial super-rotation was steadily replaced by westerly jets at higher latitudes (around 60° north and south of the equator). Overall, there was a more direct flow of air from the sunlit hemisphere to the dark hemisphere at height. Indeed, as the rotation period was slowed further to 100 days (so, for habitable planets tidally-locked to the most massive red dwarf stars) the flow of air showed the least super-rotation and the greatest direct flow. In essence what is happening is that there is more and more flow from all directions into the thermal low pressure area at the surface. As this air rises and diverges it blocks the westerly flow around the equator at the tropopause. These planets experience the weakest Coriolis Effect so the weakest east-west drive with latitude. However, equatorial super-rotation became the dominant means of transporting air in planets with periods shorter than 25 days. Shorten the rotation—and hence move the habitable planet closer to its star—and the super-rotation increases.

When you continue the trend to even shorter orbital periods some odd things happen. Planets with periods less than 6 days oddly re-develop the same higher latitude westerly jets, but this time accompanied by equatorial super-rotation. When the rotation period is reduced to less than 4 days the mid-latitude westerly jets finally vanish and the equatorial jet dominates. On these worlds super-rotation is so strong that it disrupts the uplift of air east of the SSP and presumably cloud formation with it. On such planets the strength of the Coriolis Effect is close to, or the same as, the Earth's, but the difference in the pattern of heating drives the observed difference in the circulation.

In each case the super-Earth experiences the change from one regime to another at greater orbital periods than smaller, Earth-like worlds. This is primarily due to the greater radius, and hence circumference, of the planet. Larger worlds have to rotate faster than smaller ones in any give period—and Rossby waves can fit across the surface in increasing numbers or with greater wave-length with increasing surface area.

The only issue I have with these is that in the shortest period planets they used models where the planets had an orbital period measured in a few days, but insolation equivalent to the Earth. Now, in reality this can never occur. Any planet orbiting a red dwarf with such a short orbital period must be far closer to its star than the habitable zone and thus receive far greater amounts of radiation than the Earth does—unless it orbits an even dimmer *brown dwarf*, of course. That this aside, these models do suggest that the circulation on tidally-locked super-Earth like worlds might be much more complex and variable than we might first imagine.

Model 1: A Smooth, Unrealistic Planet with No Oceans

Now, let's take this information and make some worlds. If we first imagine a perfectly smooth, dry world (i.e., a completely boring and unrealistic one) then the atmosphere will move much like that around a hot Jupiter. You'll, therefore, find a lobster-claw shaped hot zone around the SSP (sub-stellar point) but extending to the east. A westerly jet stream will extend its influence all the way to the planet's surface driving warm air eastwards from the SSP across the day-night terminator. Above the SSP air will rise and blow towards the ASP (anti-solar point) predominantly from west to east. Around the SSP are two large areas of high pressure flanking the equator, while two areas of low pressure flank the ASP. The SSP will have low pressure where air is rising and the ASP will have high pressure where air is descending.

Let's spice it up with an ocean; then we'll add land and mountains to see what happens.

Model 2: Oceans on Super-Earths and Earth-Like Worlds

Life must have water—or at least some other solvent to transport heat, substances and to promote the chemical reactions necessary for life. If we stick with the universe's most abundant solvent, water, then we can begin to model our world. When we cover the speculative planet with oceans some interesting things start to happen that may ultimately have a bearing on the habitability of the planet.

When the first models of atmospheres were done around habitable, but tidally locked planets, it was thought that the only region in which the oceans would remain liquid was under the SSP. For here and only here would there be sufficient energy available to warm and liquefy H_2O. This led to "eyeball" like surfaces with open, ice free ocean only under the SSP. The rest of the globe would be frozen over, except that these models neglected one important thing: water is a fluid and will move. More recent models of potentially habitable super-Earths and Earth-like planets take into account both the efficient heat transport by water and the effect of planetary rotation on its motion. When you put all the information together you get oceans that move pretty much like the atmosphere above. Warm ocean currents move from west to east, carrying energy from the SSP towards the dark hemisphere. Concurrently, cold ocean currents can flow westwards around to the day side of the planet, cooling the atmosphere under the SSP. These motions both strengthen and are strengthened by circulation in the overlying atmosphere, which is predominantly west to east (Fig. 10.5). Therefore, conditions on a tidally-locked habitable planet resemble closely those in a hot Jupiter—only more moderate in terms of their temperature. Instead of an "eyeball", open water will exist, at least in a lobster-claw-like pattern, reflecting the motion of the ocean around the equator.

For the ocean, this circulation has two effects. Firstly, the ocean remains ice-free over a greater proportion of its surface as heat is more effectively transported across the planet. Secondly, there will exist two overlapping circulations within the ocean that should serve to transport nutrients effectively around the planet. This is vital if the oceans are to remain vibrant and potentially habitable. Detailed modeling work by Yongyun Hu and Jun

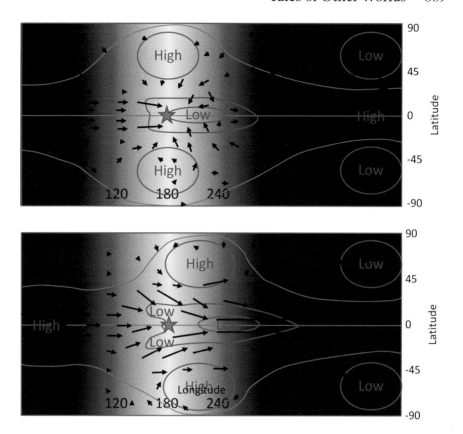

Fig. 10.5 Winds on tidally locked habitable planets generally blow from west to east, particularly near to the equator and at height. In these models of Yang et al. above slowly rotating planets air rises to the east of the SSP and generally descends elsewhere. Winds are very sluggish except west of the SSP. Nearest the surface winds are more strongly affected by heating so that winds converge on the SSP (180° Longitude at the equator) and diverge from the ASP. However, strong is a relative term. In the slowest rotating planets, the westerlies are 0.5–1 m per second (or 1–4 km per hour)—a very modest breeze. *Top*: 355 ppm CO_2; *bottom*, 200,000 ppm CO_2. In Timothy Merlis's models (not shown) a pair of relatively strong super-rotating westerly jets emerge at mid-high latitudes, above a region of westerly winds at the surface as planetary rotation is stepped up to 1 day. Meanwhile, at the equator the crescent shaped area of lowest pressure becomes more pronounced with the lowest pressure displaced 90° to the east by strengthening super-rotating winds

Yang (Peking University) suggests that oceans with depths greater than 2 km will have two circulations. In the top-most layer, above 2 km, warm waters move from the equator towards each pole.

This circulation is driven by the zone of equatorial westerly winds that drive water downwards at the equator then outwards to each pole, where the waters rise up once more. Beneath these hemispheric cells, if the ocean is deep enough, is a thermohaline-driven circulation. In this thermohaline cell, water moves downwards from the Poles towards the equator. The driving mechanism lies with the formation of ice. As sea water freezes, denser, salty water is displaced from the growing ice crystals because salt does not fit readily into the structure of water ice. This cold, salty water descends under its own weight towards the ocean floor. This generates a flow from the poles towards the equator: i.e., in the opposite direction to water driven by the westerlies in the atmosphere nearer the equator. If, however, the oceans are relatively shallow, this arm of the global ocean circuit will be broken.

What are the consequences for the climate, and how does changing the concentration of carbon dioxide, or the amount of stellar radiation affect the climate? Some of the effects confound expectations. In general increasing the concentration of carbon dioxide increases the area of open water. This isn't because the planet as a whole warms up, but rather because the atmosphere can hold more heat energy and thus transport it more efficiently from lit to dark hemispheres. While the day side barely warms at all, increasing the concentration of greenhouse gases, particularly carbon dioxide to around 2 % from an early-industrial, terrestrial 0.035 %, causes the dark hemisphere to warm substantially. Indeed, it causes the ice to melt across the entire globe. In terms of the atmosphere, the strength of the super-rotating westerly jet increases substantially and extends downwards towards the ocean beneath. In the atmosphere with early-industrial levels the atmospheric motion at the surface is only weakly from west to east, except closest to the western edge of the SSP. Once the carbon dioxide concentration is increased the increasingly deep overlying westerly jet stream drives a very noticeable westerly airflow around the equator, and ultimately it is this that causes the dark hemisphere to warm up.

If we then think back to the end-Permian mass extinction (Chap. 3), ocean circulation appears to have played a pivotal role in exterminating most of the planet's complex life. An already warm ocean became super-heated and stagnant, with microbial life filling its depths with methane and hydrogen sulfide. Ultimately, it was the hydrogen sulfide that killed off most of the ocean and possibly land-based life. For a tidally-locked planet, the active circulation described here, should minimize the risk of ocean anoxia (lack of oxygen) and euxinia (the build-up of hydrogen sulfide), which would likely cause the death of most aerobic (oxygen-loving) life forms.

Increasing carbon dioxide concentrations has other effects which might be surprising. You might expect more carbon dioxide and a warmer world, overall, might mean more convection and hence more clouds. Yongyun Hu and Jun Yang's work suggests instead that as carbon dioxide increases it is the effect of the increasingly strong westerly jet stream and ocean current that matters most. Instead of enhancing convection the strengthening westerly jet and ocean currents mix cooler air more thoroughly around the planet, along with cooler water. So, as was mentioned above, global temperatures are more even throughout. With less of a contrast in temperature convection is actually suppressed by increasing carbon dioxide. This is despite a higher tropopause (Chap. 5) which might favor taller convective clouds. In the end the planet is more cloud-free than if it has less carbon dioxide.

What about habitability? There are some contradictory outcomes here. You might imagine that a larger area of open water would allow more habitability. That might be true for a planet right in the middle of the habitable zone, but overall, more open water decreases the planetary albedo: the planet becomes darker. This was suggested for the early Earth as a mechanism to warm the early Earth with a dimmer Sun (Chap. 2). Now, the authors conclude that the end result is that with a darker planet, it will absorb more energy from its star and this, in turn, will cause the inner edge of the stellar habitable zone to move outwards away from the star. After all, a darker planet will be hotter at any given amount of stellar heating and this will lead to the planet overheating and experiencing a thermal runaway, sooner (Chap. 5). This narrows the habitable zone.

It must be stressed that this conclusion is still open to interpretation. Look for example at another piece of work by Jun Yang in the preceding year. Yang and colleagues concluded that convection under the SSP will produce enough cloud to cause negative feedback. What this means is that the extra cloud reflects incoming radiation and thus reduces the likelihood that the planet will over-heat. This prevents the dreaded thermal runaway and keeps the planet habitable. Therefore, in this work convection broadens the habitable zone because the clouds it creates are reflective. This work also moves the warmest area of the planet back towards the SSP despite the pervasive westerly jet stream around the equator. Now, in this set of published models oceans were included but they were not truly global in extent, with continents partly block-ing the movement of water around the globe. Thus, it is only on true water-worlds that the habitable zone will be affected.

Turn again to another study and something slightly differ-ent comes out. In the work of Jérémy Leconte and colleagues that was described in Chap. 5, thermal runaway is averted because of negative feedback as the planet warms. In their model, as tem-peratures rise the Hadley Cells strengthen. Although this leads to increased cloud cover near the equator, it causes a reduction in cloud cover elsewhere, where air descends, warms and can hold more water vapor.

In Yang's work reducing cloud cover is thought to have a destabilizing effect—reducing planetary albedo because oceans are darker—however, Yang's work suggests this is dependent on the type of planet. If you increase the amount of radiation a tidally-locked planet receives, then the amount of convection increases and the planet gets cloudier. This makes it more reflective. This has two effects. More cloud cover makes the dayside more reflec-tive and the night-side less: this evens out temperature contrasts, increasing habitability.

By contrast a non-tidally-locked planet, such as the Earth, experiences the opposite effect. Stronger heating, in Yang's model, means less of a temperature gradient and weaker Hadley cells. This causes greater warming as fewer clouds mean reduced reflec-tion of incoming radiation and more absorption of radiation by the planet's surface. Leconte's work suggests that on an Earth-like world greater warming leads to *stronger* Hadley circulation and (overall) *fewer* clouds, but those changes to albedo are off-set by increases in the ability of air to hold moisture. With less water escaping to the stratosphere, a runaway greenhouse is held off until the amount of radiation the planet receives increases to 375 W per meter squared—or approximately 100 W per meter squared more than the Earth currently receives from the Sun. This, broadly corresponds to an average surface temperature of 65–70 °C for an Earth-like planet. Leconte's work also shows that over the Horse Latitudes, where the air is far from saturated because it is warming and drying out, the atmosphere acts like a giant radiator and cools the planet down. Thus in the cloudless areas energy is lost and the planet can absorb more energy before it experiences a thermal runaway. More recent work by James Kasting, which was discussed in Chap. 5, also reaches the same conclusion. While Leconte's work only looked at the troposphere, Kastings included more water vapor and examined what happened in the stratosphere. At similar temperatures to those seen by Leconte, the stratosphere fills with water vapor and the planet dries out. These two pieces of work suggest that planets can remain habitable for longer, with greater amounts of heating before they expire.

Moreover, Leconte's work reaches the opposite conclusion of Yang: increasing cloud cover in their model increases the greenhouse effect and leads to a faster runaway. Here, the water droplets and ice crystals in clouds absorb more energy than they reflect and, therefore, increase warming. How come this is so different? Well, it's down to the type of clouds in the models. Yang has mostly cumulus and cumulonimbus convective clouds, while Leconte favors an increase in cirrus and cirrostratus because they suggest under hotter conditions more high level clouds form, while lower clouds are evaporated. Cirrus and cirrostratus are dominantly absorbing rather than reflecting, meaning that as they increase in abundance, the greenhouse effect increases in sync. Think back to Chap. 2 and the 9/11 attacks in the US. When planes were grounded, there was a 1.8° increase in the diurnal (day–night) temperature range—primarily because more radiation was escaping at night, with fewer plane contrails to absorb it. Despite less reflection in the daytime, the dominant effect was absorption of outgoing radiation at night, not more reflection during the day.

All of this is rather confusing. It comes down the role of clouds. If their dominant effect is to reflect energy as Yang concludes, then habitable zones are extended where tidally-locked planets have limited ocean circulation. If Leconte is correct then increasing cloud will have the opposite effect—but that this depends on the type of clouds a tidally-locked planet will form. In Leconte's models decreasing cloud will stabilize the planetary temperature and limit the chances of a thermal runaway because, overall, there will be less cirrus and cirrostratus-like clouds. At the moment, the best we can do is speculate (Figs. 10.6 and 10.7).

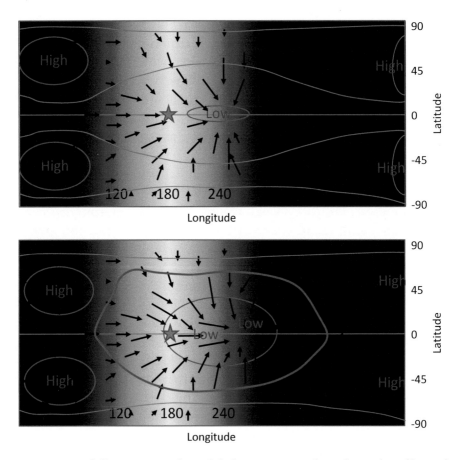

FIG. 10.6 A different pair of models by Jun Yang that show the effect of slowing the Earth's rotation to once every 197 days (*top*) and moving it to the current location of Venus (*bottom*). With a slow rotating Earth, in roughly its current orbit, a large area of low pressure tracks the slowly moving Sun across the face of the planet, flanked by areas of high pressure to it rear (0/360°, approximately). When the amount of radiation is increased to that of modern day Venus, in Wang's models, the Earth remains habitable because enhanced convection, linked to the larger surface low pressure area, produces more clouds and these reflect more sunlight. Where Yang simply moves the current Earth, with its 24 h rotation period, the Earth becomes uninhabitable as there are no longer enough clouds to reflect the higher level of solar energy causing the Earth to overheat. Leconte, on the other hand would have an uninhabitable Earth. See text for details

 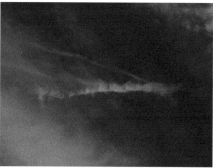

Fig. 10.7 The types of clouds hot planets manufacture appears critical to their fate. If they produce large, convective, cumulus and cumulonimbus clouds, like those above left, they will reflect much of the incoming radiation and remain cool and habitable for longer. However, should they produce mostly high, ice clouds, such as this unusual example of cirrus (*above, right*) or the cirrostratus ice cloud above it, then they will preferentially trap more heat at night time than they release during the day. This will cause the planet to overheat more quickly than if it had no cloud at all. Photographs by author

In another set of models, Yang puts another effect into play: rotation (Fig 10.6). In their models where the Earth is made to rotate slowly (roughly similar to modern Venus, with a 197 day rotation period) the Earth takes on an aspect similar to a tidally-locked planet. There is a strong area of uplift, low pressure and convective clouds tracking the Sun as it slowly moves across the globe, much like the SSP on tidally-locked worlds. If you then move this slowly rotating Earth to the location of Venus the Earth surprisingly remains habitable. This is again down to how you view the role of clouds. In Yang's models, clouds are reflective and moving the Earth towards the Sun makes it, naturally hotter. This enhances the formation of convective clouds which causes a negative feedback. The Earth is then shielded from the additional radiation, keeping it habitable. Yang goes further and runs another simulation but this time with an Earth that rotates as fast as it currently does. In this model the Earth soon overheats as it cannot produce sufficient cloud cover to keep the Earth cool. With a fast rotating Earth (as it is now) convective clouds remain fixed to a narrow equatorial belt and unlike Leconte's models, cannot emit or reflect enough energy to keep cool. A thermal runaway ensues. This is all to do with the types of clouds planets form at high temperatures. Where icy cirrus and cirrostratus clouds dominate, the planet is destined to boil dry. Large, fluffy cumulus, filled with relatively large water droplets are found, lead to planets that will reflect more energy and stay cool for longer.

Until we can accurately model the type and distribution of clouds on tidally-locked worlds, we really won't understand how they will affect climate. How this will combine with the effect of cloud cover on ice cover and ice cover on evaporation from oceans remains in the realms of speculation, but at least we are beginning to see how tidally-locked planets will begin to function and hence make some accurate models of weather and climate on these worlds.

One outcome from Yang's models of fast and slow-rotating Earths is we can again look at why Venus overheated. Yang suggests that if the early Venus rotated quickly, this would have led to an early thermal runaway, making Venus uninhabitable early on. However, if it had rotated slowly then it could have been cooler for longer and thus, remained habitable even as the Sun slowly brightened. One must consider the influence of the Sun's gravity on Venus here. Although this might have slowed Venus early on, it seems more likely that Venus only slowed to its current, sluggish (retrograde) rotation after its atmosphere overheated and filled with sufficient carbon dioxide that the Sun *could* slow it down.

All these models show a few things. For one they are still not completely consistent. More than this, they indicate how a wealth of different factors will likely influence how a planet's climate adapts to distinctly unearthly conditions, such as slow rotation or increased stellar radiation.

Tidally-Locked Earths and Super-Earths: Towards A Realistic Planetary Model

Let's begin to put all of the above information together and combine it with what we know about our Solar System's worlds. We've got some issues with the effect of clouds, but we can model the rest of the atmosphere and oceans with a fair degree of accuracy (or *likely* accuracy). Now, we'll take a lead from Yang's 2013 paper, which modeled the circulation with varying amounts of carbon dioxide and the presence or absence of land-masses (Fig. 10.8).

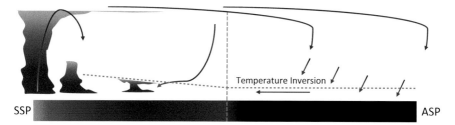

SSP   ASP

Temperature Inversion

Fɪɢ. 10.8 The general pattern of air flow across a tidally-locked (or very slowly rotating) planet. Air rises at the SSP then flows at height to the dark hemisphere. There is uplift within 15° of the SSP and general, slug-gish descent elsewhere. This leads to warming and the formation of a temperature inversion across most of the planet. In turn, this limits how high clouds can rise by convection (*left*) so that only at the SSP is convec-tion strong enough to produce thick enough clouds for precipitation. Bear in mind this model is for a perfectly smooth planet. This pattern of circu-lation is superimposed on the general west to east flow around the planet

Here, as we've seen, without land-masses, air on a tidally-locked Earth-like planet, or super-Earth, will super-rotate around the equator with Hadley Cells that flank a large area of warm low pressure east of the region that is most strongly illuminated. Warm air flows from this equatorial region into each Hadley cell and moves north and south at height. Air also flows to the coldest region, east of the ASP, predominantly eastwards, again at height. Air then descends and returns, across the Polar Regions, from the ASP to the region east of the SSP. Air also flows in the surface westerly, super-rotating flow from the ASP towards the SSP. This completes the global circuit in the troposphere. The whole process is driven by heating at the SSP and cooling at the ASP. Put land in the way and the land disrupts the flow in two ways. First of all, land exerts greater friction than water and cannot be shifted by it. Therefore, the surface westerlies are broken up. Moreover, land heats and cools more rapidly than seawater (it has a lower heat capacity). Land also dries out meaning that it contributes less moisture to the atmosphere above it, which also allows it to heat up more rapidly than sea water. This is the driving mecha-nism for the Monsoon (Chap. 2). Consequently, air tends to have a lower pressure over sun (star)-heated land than sea water. The con-verse is true where land cools down. Here, land cools the fastest.

This generates higher surface pressure. What does this mean for the climate? Well, for a tidally-locked planet the arrangement of continents (if there are any) is crucial to whether the planet will be habitable.

On Earth the continents are both formed and shuffled around by the process of plate tectonics. If an Earth-like world is less than six billion years old there is a good chance that this process will operate. On larger, super-Earths plate tectonics may or may not work. Studies suggest that anything up to ten times the mass of the Earth will operate this process, but that as the planet grows larger plate tectonics may gum up faster, once the mass exceeds 2–3 times that of the Earth. The reason is continental crust. Larger planets will be hotter inside and will brew out their granite faster (see also Chap. 5). Granite, the stuff of continents, is too light to subduct and gradually builds up on the planet's surface. This eventually chokes off the recycling of the crust which is needed to operate the process.

Where plate tectonics operates the arrangement and position of the continents will change over periods of tens to hundreds of millions of years. This, as we saw in Chaps. 2 and 3, changes the movement of the atmosphere and hence planetary climate. In extreme cases plate tectonics can cause the climate to shift radically across the entire globe, potentially turning a balmy planet into a freezer. There is yet another way in which plate tectonics can conspire with the fundamental properties of a planet to alter its climate: this is angular momentum. On a rotating globe equatorial regions spin the fastest around the center of mass, while Polar Regions spin the least. In this situation bulky, massive continents are arranged in their most stable configuration when they lie closest to the equator. Move them to the Poles and it's like putting a toy car on a spinning top: it will topple over. On Earth, this process has been identified and is known as true polar wander (or TPW for short; Chap. 2). Instances of TPW are relatively rare as the process requires a particular arrangement of continents— and a suitable mass of them. On planets with few continents this process may not happen at all. Increase the mass of these and the planets spin may become unstable. On its own, this process can accelerate a continent over tens of degrees of latitude in only a few million years–around ten times faster than normal plate motion.

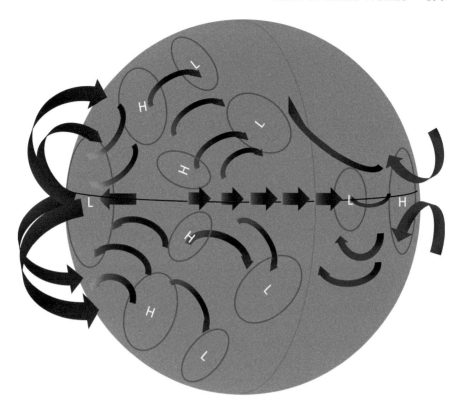

Fɪɢ. **10.9** An early attempt at reconciling different pieces of information on the likely circulation on tidally-locked exoplanets that was used in "Under a Crimson Sun". Some features such as the low pressure near the SSP and the super-rotating westerly flow are correct. However, the other flanking lows and highs are probably not. Nice, imaginative try, however. As the next figures show the circulation for aquaplanets is likely much simpler, although changing the parameters used in models has some significant changes. Despite the inherent simplicity of the circulation above water worlds, the introduction of landmasses can have significant effects making these planets more like the one proposed here

For a tidally-locked planet this process is likely to become rather cyclical because the likely most stable configurations of the continents will be either grouped facing their star or grouped facing away from their star. This is a reflection of the current situation with our Moon. The Maria face us while the highlands, predominantly, face away. This, as it turns out is the most stable configuration with mass centered facing towards or away. But, why might the process be cyclical? Well, think about plate tectonics.

This will periodically assemble supercontinents and these will experience the largest pull as they are the largest collection of mass. However, super-continents are ephemeral: neighboring subduction zones pull on the edges of these continental beasts, while hot spots undermine their interiors. Ultimately, this causes them to fall apart, before closing ocean basins bring them back together once more. When assembled in random locations, true polar wander will bring them back in line with their star; then disassembly will scatter them once more, only for the process to repeat.

Taking all this into account, let's assemble most of the continents nearest to the SSP and then see what a likely consequence for the planet's climate is. Figure 10.10 illustrates this, taking into account what we already know about super-rotation and the effect of the fixed heating. I've placed the SSP over something that looks awfully like Africa. Indeed, for the sake of argument I used the Earth of the future: Amasia (Chap. 2). This way you can relate some of the climatic features of this model to those of the Earth. The same arrangement of continents will be used in the subsequent scenarios.

What of the climate? To keep things straightforward equatorial Africa was chosen as it's a relatively small continental mass compared to the assemblage further north and will heat very dramatically under the intense illumination at the SSP. This will draw in moisture from surrounding oceans generating persistent rainfall and a wet, tropical climate, particularly along its windward coasts. To the north of this region, the current Sahara desert conditions might persist, as air flowing into this region predominantly comes across a long fetch of land and will have dried out. On the continent's north-western shores onshore westerly winds will bring precipitation leading to something more like a Mediterranean climate. Along the eastern Atlantic shoreline a cooler, maritime climate, much like the present day one would be found. Australia would have a cold climate, but it may be kept temperate by super-rotating westerly winds. If these winds are blocked or sufficiently subdued by the arrangement of the continents Australasia will freeze over; most likely with an ice cap developing as weak westerlies deliver snowfall from the neighboring warmer oceans.

Turning our attention to North America and South America, the Amasian organization places most of these across the day-night

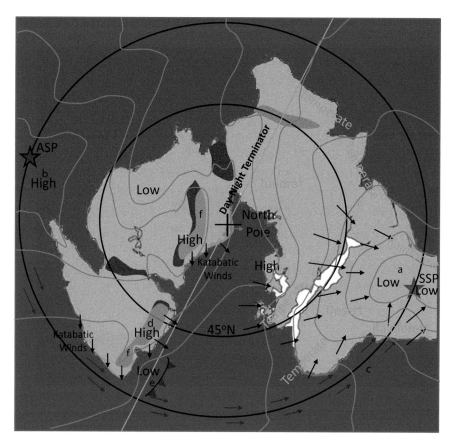

Fig. 10.10 Transporting Amasia to a tidally-locked Earth around an M-class dwarf. The SSP has been placed over Africa and the ASP over the future Panthalassa. Rotation is slowed to once every 30 days (720 h), which is appropriate for an Earth-like planet in an orbit around a mid-M-class red dwarf, were it to be habitable. Features: The large thermal low nearest the SSP is distorted by the continent which warms faster than neighboring ocean (**a**); pressure is highest east of the ASP (**b**); weak equatorial westerly winds carry cool air to the SSP from the ASP and warm air away (**c**); high pressure zones for over mountains where westerlies cross them (**d**) and low pressure in their lee (**e**). Mountain ice caps are white (**f**). *Light black arrows* indicate regional winds caused by surface features

terminator so that they are cold, dark deserts. On the north-eastern side of North and South America some mountain belts are retained to add some flavor to the climate. These mountains are produced by subduction and continental collision (Chap. 2). As these have predominantly cold climates, but are adjacent to potentially warmer oceans, some ice caps that are fueled by moisture from the adjacent seas have been added. What is interesting about dong this is that the ice caps will likely shed very cold, dry winds onto the neighboring oceans. This is a result of both the prevailing light westerly winds which are likely at these latitudes, but also because the very cold air is denser than surrounding air and will sink rapidly under its own weight. Thus locally, particularly where glaciated or other valleys empty out into the ocean, there will be particularly strong cold winds called katabatic winds. These local or regional features will lower temperatures along the affected coastlines and also generate a lot of shower activity as cold air pushes out over warmer water. In Europe one such valley wind is called the Mistral, which blows up to hurricane strength down the Rhone valley when conditions are suitable, particularly during the early spring.

Mountains have another important effect on the overall circulation of the atmosphere. They divert westerly winds north and south. As winds approach the windward face they are compressed and develop a northward (cyclonic) spin. At the summit of the mountains, winds diverge (spread out) and develop the opposite (anticyclonic) spin. This leads to the formation of a semi-permanent ridge of high pressure aloft, with airflow bending to the south in the mountains leeward down-slope. Further east, as the air moves back down to the leeward side of the range, it develops the opposite (cyclonic) spin to that on the ridge, bending to the north. This forms a semi-permanent trough in the atmosphere. As we saw in Chap. 1 this is a Rossby wave that generates further waves downstream. Since these waves develop in the predominantly light westerly or north-westerly mid-latitude winds, at this longitude, they will then move towards the south-east and interact with the northern fringes of the large, warm cored (or thermal low) over Central Africa.

In terms of the climate, on our M-class Amasia, will likely then have a low pressure area anchored to the east of the mountains that fringe North and South America (Fig. 10.9). In combination with cold air pouring off the continents onto the ocean, one would expect storms to develop here, which would likely drift east towards the European side of our super-continent. At the surface these might manifest as frontal systems (Chap. 1), bringing steady precipitation to the European flank of our tidally-locked planet. If they were to continue further east or south, they might stir up dust storms over drier areas, much as they do on Mars (Chap. 6). How these incursions of colder air would affect the low pressure area over Africa is hard to say. They might well peter-out over southern Europe or they may, like Martian storms intensify bringing severe dust storms into the tropics or drive the formation of more persistent storm cells over the equatorial regions east of the SSP.

Overall, the climate is likely to be extremely varied on such a world. Yes, one hemisphere is likely to be a tad chilly, with temperatures permanently below freezing, but a rich variety of geographical locales will produce a wide variety of weather. All that can be said is once you have that style of weather it is fairly likely to persist. For without a proper day-night cycle and a fixed angle of illumination and heating the weather is not likely to change much from day to day. Your best chance for weather is on the windward slopes of the mid-latitudes, where onshore winds and possible frontal systems would likely bring day-to-day, or week-to-week variation. The temperature will depend on your geometric distance from the SSP and whether you are on its western or eastern side of the SSP. East of the SSP is where the warmest conditions will be found—but not necessarily the driest for converging moist air will generate copious convective showers and storms. Had I chosen to put the SSP in the heart of Eurasia or North America, the overall climate would have been much drier. Rainfall would have been on the ocean-facing coastlines as warm, moist air converged across these. The interior of the continents would likely be dry, hot desert by comparison, depending on the strength of heating and the distance from the SSP to the nearest shoreline.

The End of Weather: Atmospheric Collapse and Atmospheric Erosion

We conclude our tour of the tidally-locked planet with two problems that could make habitability rather precarious: atmospheric collapse or erosion. Collapse occurs when the temperature of the atmosphere falls below that need to condense (or re-sublimate) its major component. In the case of the terrestrial planets the most important gases would be water and carbon dioxide, for without these greenhouse gases global temperatures would fall lower still. The critical temperature for our world and Mars is around –80 °C for a fairly unrealistic concentration of carbon dioxide. This is an average temperature and far below the average temperature achieved even on Mars (Chap. 6). Thus while carbon dioxide can freeze out over the Martian Poles it never becomes cold enough for the entire atmosphere's mass to freeze out: instead on Mars, it circulates between polar stores.

Atmospheric collapse was considered to eliminate habitability on most tidally-locked planets (or our tipped-over world) where one hemisphere would, potentially, become cold enough for water vapor and carbon dioxide to freeze out. It turns out that for a planet in its star's habitable zone in most cases the atmosphere around a potentially habitable planet needs to be rather thin (and therefore not very habitable) before collapse happens. How thin? In the late 1990s work by Joshi suggested it could be as low as 30 millibar—or three times the average surface pressure on Mars. Robin Wordsworth took another look at this and found it could happen for pressures as high as 140 millibar—or one tenth the surface pressure on Earth. Low by our standards, rather a lot of organisms on Earth would survive this pressure.

Wordsworth's results made the atmospheres of tidally-locked planets appear less stable than previous work. This was a result of changing the effectiveness of carbon dioxide as a greenhouse gas at low pressure. At such low pressures carbon dioxide has broader windows (narrower absorption bands) which allow more energy to escape the planet over its dark hemisphere (Chap. 3). At pressures of less than 0.14 bar, carbon dioxide leaks enough energy to allow collapse on planets that receive the same stellar heating as the Earth does at present. Therefore, although we shouldn't worry that relatively large red dwarf planets, such as GJ 581d, might suddenly lose their habitability, it does pose a threat to Mars-like planets. With a thinner or less stable atmosphere these could lose some critical support when gases on the permanent nightside become severely chilled. Even when we consider our Mars, the atmospheric pressure has decreased at a fairly high rate over geological time, thanks to various escape mechanisms (Chap. 6). On a tidally-locked planet, the atmosphere might suddenly change from habitable to inhabitable over a few million years. This would happen once the overall pressure had fallen below the cut-off; and this would be despite the presence of a sizable amount of green-house gases that were still present.

While, atmospheric collapse is not a significant problem for most planets in orbit around red dwarfs, atmospheric erosion is. This is all a question of time. In our Solar System the only planets and moons that show significant loss of atmospheres are Mars and Titan. The latter has large enough reserves for this not to matter at present (Chap. 8). What about a planet orbiting a red dwarf? Such planets are only a few million kilometers from their Sun and are subject to a greater bombardment of charged particles from their star's stellar wind. Although red dwarfs typically sport limited stellar winds (less than one hundred trillionth of a solar mass of gas per year) when they are young this rate is perhaps a thousand times greater. Calculations by Jesús Zendejas (Max Planck Institute for Astrophysics) suggest that if this persists for more than one billion years then any planet without a sizable magnetic field, in its red dwarf's habitable zone, could lose much, or all, of its atmosphere.

What this means depends on a number of factors. Do tidally-locked planets lack magnetic fields? That is hard to say, but we can look at Mercury, which has a low mass and rotates very slowly (much more slowly than most habitable red dwarf planets). Clearly, Mercury still sports a decent field. Now, although most red dwarf worlds might well retain a protective shield—or an atmosphere like Titan that is replenished from deeper stores—these stars have lifetimes measured in trillions of years. It seems fairly likely that over time any planet residing in the habitable zone of a red dwarf will, ultimately lose its atmosphere to space long before its parent star dies.

The biggest threat to such worlds will probably happen after six billion or so years. For it is here that plate tectonics is likely to fail, ether because the planet has become too cold or because the surface has become clogged with granite and is too buoyant to support the subduction needed to stir convection within its planet's core. Here, the loss of the magnetic field will most likely instigate a very slow process of attrition that whittles the planet's atmosphere down over tens of billions or (less likely) hundreds of billions of years. Gas giants and ice giants will most likely retain most of their mass because they have a higher gravitational pull and because they have more mass to lose. However, Earth-like worlds, or more massive "super-Earths" will likely lose most of their atmosphere long before their central star dies. Life, it seems, is not eternal for any world. Remember, once the density has fallen to suitably low levels, atmospheric collapse might complete the job of converting a habitable planet into an icy desert like Mars.

Earth-Like but not Tidally-Locked

Instead of placing our planet near an M-class red dwarf star, we're going to put it around a slightly more massive orange K-class star. If it is to be habitable, then it must lie further from its more luminous host. This places it outside the region around the star where gravitational forces will be strong enough to cause tidal locking. It's year is still relatively short at 130 Earth days (in an orbit of 0.4 AU). The length of year is based on some reasonable assumptions about planetary mass, tidal-locking radius and stellar mass.[3] The planet rotates relatively slowly (3 Earth days) so that the Coriolis Effect is more like that experienced by Mars than our Earth. We've also given it a tilt of 23.5° so that it has seasons. What's the weather like?

On this world the atmosphere circulates in a manner that is somewhat like that seen on our planet. With slower rotation the tropical Hadley cells reach further across the globe, bringing a tropical climate closer to the Poles. Hadley cells span roughly 30° on either side of the equator on our planet. On Mars, which has a comparable Coriolis Effect to the planet we've made here, Hadley cells extend to 60° north and south of the equator. Smaller Ferrel cells are induced on the Polar side and extend almost the full way to each pole. Finally, like the atmosphere of Venus at each Pole there is a quieter area where cold air descends towards the planetary surface.

During the winter months the arrangement of continents around the northern polar regions will lead to strong internal cooling as land has a lower heat capacity than water (Chap. 2). This leads to strong outward blowing winter monsoon winds to the south of these land masses (Chap. 2). Over what is the North Atlantic, an

---

[3] See Jérémy Leconte's 2015 paper in the journal Science for more details. This is listed in the references.

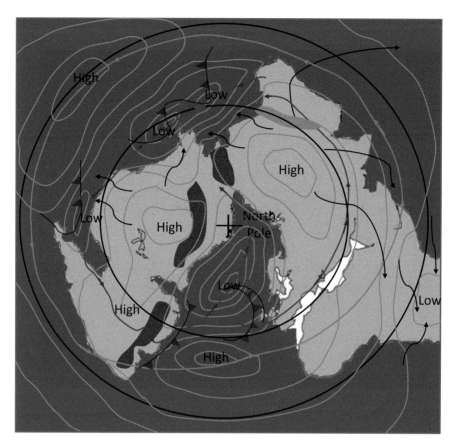

Fig. 10.11 The surface circulation during the northern winter on an Earth-like world with a Mars-like Coriolis Force. On this map the sub-tropical jet stream (*orange*) and polar-front jet stream (*purple*) are indicated. Large areas of cold-cored high pressure dominate the continental interiors driving cold, dry winter monsoon winds out-wards. The Ferrel cell over the northern oceans is strengthened because of the larger temperature contrast between the cold air and the ocean underneath. Summer monsoon winds blow southwards over Africa towards its southern half (*black arrows*). The Weaker Coriolis Effect means that air blows more strongly across pressure lines (isobars) than it does on our Earth

invigorated polar front jet stream extends stormy weather across our alien Europe. Similar storm belts lie along the southern margins of the other continents bringing seasonal wind and rain. This is shown by cold and warm fronts anchored to deep, cold low pressure areas (Fig. 10.11). An anchored Rossby Wave in the jet stream generates storm systems east of what is South America.

A second branch of the Polar front jet brings seasonal storms across what is our alien Southern Europe and northern India. Meanwhile, the weather inland is predominantly cold, windy but dry.

Zooming in on the northern side of our alien South America is a small narrow in the mountains. Here, we might see gap winds. One of the finest examples of these is the Chivela Pass winds in Central Mexico. These occur in the winter and early spring when cold fronts penetrate far to the south of the United States' southern shores. These can arc around the Gulf of Mexico, frequently south of a strong blocking anticyclone which is situated over North America (Chap. 1). When these cold fronts curve back westwards, their progress is blocked by the Sierra Madre Mountains. However, at Chivela Pass there is a clear route east towards the Pacific basin. The cold front carries denser air to its rear, which ponds behind the mountain barrier—except at Chivela Pass. After filling the Gulf of Mexico to a certain depth, the cold air pours like a waterfall, through Chivela Pass and out into the Pacific. This generates fierce easterly winds along the western coast of Mexico that fan outwards for over 150 km into the Pacific. The whole process makes an aneurysm-like blob on the otherwise blocked surface cold-front. This aneurysm balloons westwards through the Chivela Pass and out into the Pacific, carrying with it the otherwise stalled cold-front. On our world, the gap winds can advance eastwards into our future Atlantic-like basin and generate both storm-force winds and vigorous convection where they push over warmer waters.

The summer circulation (Fig. 10.12) is dominated by large warm-cored thermal lows over the continents that bring strong summer monsoon climate to the shores and mountains of our world. Again, because of the weaker Coriolis Effect and the arrangement of the landmasses, the Monsoon regime extends much further towards the North Pole. All of the alien Americas, North Africa and Asia are under its influence, bringing warm, summer rains. To get an idea of climate think of Central China or the South Western States. Cold and relatively dry winters followed by a short mild spring, then a hot and fairly wet summer. Weaker circulation around the North Pole generates frontal storms that afflict mainly the most northerly latitudes. The polar front jet stream organizes the polar circulation with Rossby waves partly anchored by the

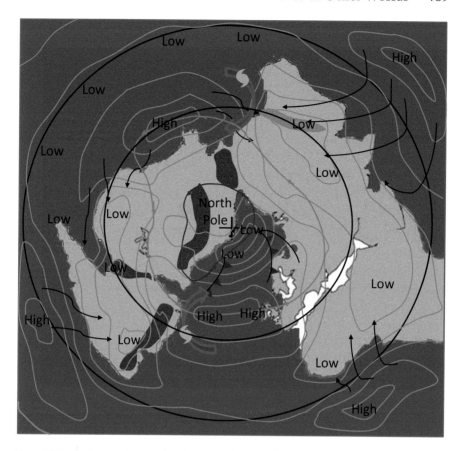

FIG. 10.12 The surface circulation during the summer on an Earth-like world with a Mars-like Coriolis Force. Tropical Hadley Cells extend further towards the north of the equator than on Earth. The northern extent is shown by areas of high pressure centered at 50–60° N. Large areas of continent north of the equator heat strongly during the summer and generate broad areas of surface low pressure. These will be overlaid with easterly jet streams. These lows drag in strong monsoon winds (*black arrows*). A small Ferrel Cell is indicated nearer to the North Pole, indicated by the presence of surface fronts. Tropical storms are indicated by whirlpool-like symbols along with their likely tracks (*red, arrows*). Polar Front jet stream (*purple*); sub-tropical jet (*orange*)

arrangement of mountains and oceans (Figs. 10.11 and 10.12). This will lead to storm formation along what is northern Canada (and indicated by a cold front on Fig. 10.12).

Despite the grouping of the continents hot deserts are likely limited in extent due to the much stronger monsoon circulation that brings moisture further inland. Much like the Miocene climate of Earth, areas that are currently desert will most likely be open savannah with summer rains (Chap. 2).

The southern hemisphere is largely open-ocean on our world. The only land area (southern-central Africa) would experience a monsoon climate, with hot, wet summers and cool winters. However, unlike the northern continents, greater exposure to the mid-latitude westerly winds in the winter would likely bring some rain, much as it does to South Africa today. Over the southern ocean, unimpeded westerly winds will bring a fairly steady mild and wet climate, with relatively little seasonal variation. Overall, the climate would most resemble the Eocene climate of Earth with limited Polar chill and a broad tropical climate that was mostly fairly wet (Chap. 3).

Tipped Over Worlds

Now let's go a bit crazy and tip our planet over, so that its rotation pole is analogous to Uranus's. What would an Earth-like planet experience now? Sadly, this arrangement isn't stable and tidal forces would soon align the planet's equator with that of its star, making it much like the planet in Fig. 10.10. However, we'll take the planet and place it around another K-class star at a suitable distance and let it Uranian climatic sensibilities get to work. What will the weather be like?

To keep things in perspective, we've stuck to our "uber-Earth" with its super-continent organized around the North Pole. This makes it possible to make a comparison with the other model planets and with contemporary Earth. The orbit is still 130 days long in an orbit four tenths the radius of Earth's, just as it was for our planet in the section above. The spin is kept at a third of ours to give it a weaker, but still substantial Coriolis Effect. The main difference is heating over the North Pole for roughly 42 days (the northern summer); heating over equatorial regions for its 42-day long spring and autumn; and heating over the South Pole during the northern winter (southern summer).

Overall, strong heating on one hemisphere should lead to the formation of a strong, surface thermal low over the North Pole or South Pole during their summers (Figs. 10.13 and 10.14). However, the relatively short seasons may prevent them from becoming as

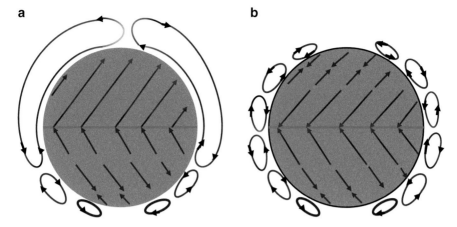

Fig. 10.13 The likely, overall circulation within the atmosphere of a tipped Earth-like planet. Such a world has winds that blow from hemisphere to hemisphere for half of the year. During the northern summer, a strong area of low pressure will dominate the North Pole with winds blowing from southwest to north east towards it. These winds begin their life south of the equator at the opposite end of a large Hadley cell, bending from south-easterly to south-westerly as they cross the equator. Above the equator should lie a strong easterly jet stream, which persists all year round. Observations of Uranus suggest a westerly jet stream is likely over the mid-high latitudes in the southern hemisphere A similar jet is found over the mid-high latitudes in the northern hemisphere during its summer—at least on Uranus. This flow completely reverse during the northern winter. During spring and autumn (**b**). Sunlight strikes equatorial regions most strongly giving rise to a circulation like that found on Earth during its spring or autumn. The weaker Coriolis Effect on our imaginary world might limit the number of cells to two per hemisphere, rather than the three shown here. Any land will alter this flow both by warming and cooling at a faster rate than neighboring ocean and because of greater friction. Vertical extent of cell is exaggerated

strong as I suggest in the figures. That aside, this band of story weather will migrate towards the equator from the south during the autumn and away from it to the north during the hemisphere's spring. Hadley cells will dominate the overall circulation, but this will be complicated by the spin of the planet, which adds a significant Coriolis Effect. Winds would, therefore, be south easterly to the south of the equator, then bend to form a south westerly flow north of the equator. Winds would converge on the surface thermal low over the northern continents that were closest to the North Pole. For here, heating would be strongest and air would

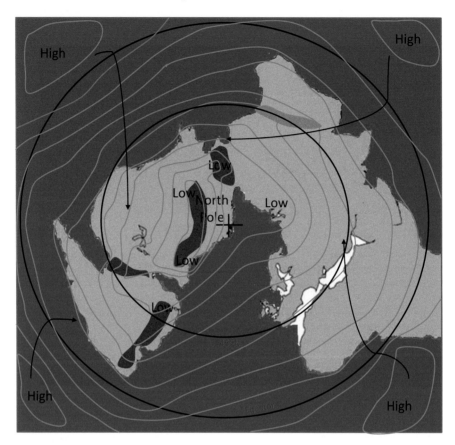

Fig. 10.14 Summer circulation in the northern hemisphere of a tipped-over planet? Strong surface heating generates a large area of low pressure will develop over the northern continents, with a focus or foci on elevated landmasses. Winds will circulate clockwise around this region but cross lines of equal pressure more strongly than on Earth. Winds are south easterly south of the equator and south westerly to its north because of the Coriolis Effect (*black arrows*). Above the equator an easterly jet stream will persist all year round. As autumn comes on pressure will rise over the poles and the zone of low pressure retreat equator-ward. Although this model seems reasonable it doesn't match Uranus… More details in text

be the most buoyant. With such a strong temperature contrast between the sunlit and dark hemispheres and the modest Coriolis Effect, winds might be rather strong. However, remember that this is not true for tidally-locked planets where winds are expected to peak at less than 50 km per hour: a modest gale. Moreover, surface features such as coastlines, mountains or forests would further weaken surface winds due to friction.

Thus, in the model, illustrated in Fig. 10.14, there is a relatively simple circulation with winds bringing warm air from the southern to northern hemisphere. There are two problems with this. Firstly, the best match for such a world would be Uranus. However, Uranus has a very different pattern of circulation (Fig. 8.2). Yes, it has a weak easterly jet over the equator, but winds are westerly around each Pole. There are two possible explanations: for one Uranus has a much stronger Coriolis Effect, with a rotation of around 17 h; secondly, Uranus's weather is probably driven by heat from its interior, rather than from the Sun. The first factor is definitely the case, while the second is still uncertain. However, a lack of solid surface which would heat up under its Sun and the two factors suggested here, might allow for the difference.

Secondly, it isn't clear how the weaker, but still significant Coriolis effect would influence the circulation on such a topsy-turvy planet: in principle there should be 2–3 different cells per hemisphere with the spin we've given the planet, here, much like the Mars shown in Fig. 6.5. Circulation in the cold hemisphere might, therefore, be broken up into multiple cells (as is suggested in Fig. 10.13) rather than simply stream from Pole to Pole.

Weather-wise, obviously all seasons would be severe. Summers in the northern hemisphere during its summer would be brief but hot, and likely very wet along coastlines and over mountains. The continental interiors would be hot and dry. The transition to winter would be dramatic as winter itself. The surface low pressure would peak in strength roughly a month after the Pole received its maximum heating due to thermal inertia: the lag in heating and cooling caused by the structure of the material. However, not long after the low pressure reached its strongest, the retreating Sun would drag the northern margin of the Hadley cell back towards the equator. Reaching it 33 days later, the Sun would drag the region of maximum precipitation back across the equator—again with a lag of a couple of weeks. Meanwhile the northern Hadley cell would uncouple from the North Pole, with its high pressure edge also retreating south towards, but perhaps not moving beyond 60° N. However, while the southern Hadley cell was shoved further south and eventually eliminated, the northern cell would extend across the equator.

An increasingly large Hadley cell would extend southwards towards the South Pole as this warmed up. To the north of the

Hadley cell, over the North Pole a large, cold cored low pressure area might form, bringing first rain then snowfall on western coasts and some distance into the interior. Finally, this too might then be displaced south to around 70–80° N while a small cold-cored high pressure area takes up residence over the North Pole, persisting until spring arrived 33 days later. However, equally, during the northern winter, the pressure may just be predominantly high across the entire hemisphere.

Higher up, as on Earth a westerly jet stream would likely develop above the northern edge of the Hadley Cell—the analogy of the sub-tropical jet stream. Meanwhile, as the zone of maximum heating moves south towards the South Pole, pressure would rapidly fall until the southern thermal low appeared bringing its dose of summer rainfall.

One interesting feature that we mustn't forget is the absence of a continent over the South Pole. We've "lost Antarctica" from our alien Earth leaving open ocean. This will have profound implications for our Uranian (or terran) mirror-world, for oceans heat and cool more slowly than land. The brevity of each season (33 days) will likely preclude the formation of polar ice caps, although snowfall is a certainty over the planet's northern territories. Open ocean means a vigorous circumpolar circulation, probably at all seasons, and even with strong heating in the summer. Thermal inertia in the oceans—the resistance of a material to change its temperature—may even prevent a warm-cored summer low from forming over the southern hemisphere. This would have radical implications for the movement of air north and south of the equator during the southern summer. Moreover, pressure may remain stubbornly low in a belt around the South Pole with westerly winds circulating around it. There is no contemporary analogy that we know of.

What of the equatorial regions? While the Poles experience nights and days 65 Earth-days long, the equator will have an illuminated part of the day that is never shorter than half the spin period: 72 h. Therefore, the "shortest day" will have approximately 36 h of daylight. However, within the 130 Earth-day-long year that our planet has, the Sun-lit portion of the day increases twice to a maximum value of approximately 144 h, when the Sun is overhead each Pole. The equatorial climate is the most stable, with two rainy seasons per year that are associated with the biannual crossing of the doldrums. This is the same as on Earth, except that on this topsy-turvy world "the doldrums" move Pole to Pole rather than Tropic to Tropic with the overhead Sun (Chap. 1). Roughly overhead, but in the summer to the south of the North Polar region will lie an easterly jet stream: easterly, because the coldest air is going to lie to the south of the Pole. This jet will retreat equator-wards during the autumn but likely never reach the Pole (and disperse) because of inertia: the air won't have time to heat strongly at height before the brief summer comes to an end. The jet will be strongest along the northern march of the Sun; then weaken when warmer conditions lie to its north in the early autumn. Shortly after the autumn equinox the easterly jet will die out as it approaches the equator, with little temperature contrast on either side of this geographical divide. Later, as autumn continues in the north, a westerly jet stream may emerge during the remainder of the southern spring. This will follow the Sun's continued march south towards the southern Pole. With the coldest air to the jet's north, but with the jet over the southern hemisphere, this should also be of the easterly type.

Atmospheric Gravity Waves

As a final thought, we will look at the effects of mountains on the transport of energy and on the weather within a planet's atmosphere. Figure 10.10 showed one impact of mountains. Where they are very cold, either all year round or seasonally, cold air can drain down valleys and onto adjacent warmer surfaces. This not only generates locally strong and potentially damaging winds, but it will lead to the development of showers and so-called polar lows where underlying warmer water triggers convection in the overriding cold air. In Chap. 1, and here, we saw how mountains can also trigger horizontal Rossby Waves within the atmosphere that have a critical influence over the weather and climate. Such impediments can also generate vertical waves in the atmosphere, as can storms, which have an equally critical role in the formation of atmospheric "weather".

There are other waves which move in the vertical plane which have regional influence, as well as acting as agents that deliver energy from the troposphere to the lower ionosphere. These are atmospheric gravity waves. Although the name may conjure up images of colliding black holes and neutron stars, these "gravity waves" are simply vertical motions within the atmosphere that propagate like ocean waves. Atmospheric gravity waves are initiated in three, broad settings: motion across mountains; motion initiated by thunderstorms; or finally frontal motion. Figure 10.15 illustrates one of these. This atmospheric gravity wave had a wavelength of several kilometers and moved, broadly, from south

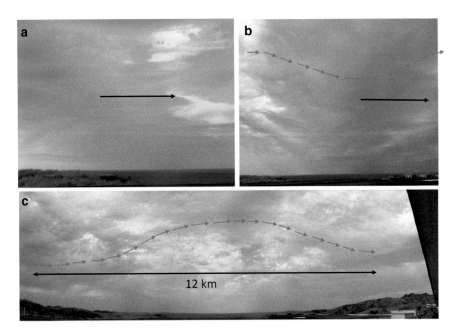

Fig. 10.15 The passage of a long wavelength gravity wave across the Cornish coast. A series of troughs and crests passed from south to north, roughly perpendicular to an advancing cold front out to the west (photographs (a) through (c)). *Small, blue arrows* indicate air movement through the waves. Bands of thick altocumulus castellanus gave thundery downpours interspersed with warm sunshine, as successive wave-fronts moved overhead in an upper layer of unstable air. The lowest layer in the troposphere was stabilized by the cool ocean, beneath. *Single-headed arrows* indicate direction of motion of the waves, while the double-headed arrow indicates approximate wavelength. Photograph (c) is a panorama of three photographs and is slightly distorted in the middle by camera motion. Photographs by author

to north ahead of the cold front that likely initiated it. In this wave an upper layer of warm unstable air, undulates up and down over a layer of drier, more stable air. Clouds were thickest in the troughs of the wave. Air is blowing in the general direction of the wave, but the wave is propagating more rapidly through this layer than the wind is blowing. This takes energy horizontally along the boundary as well as vertically through the atmosphere above and below the boundary that separates both layers.

Two additional, well-characterized forms of atmospheric waves are associated with mountains and weather fronts: these are illustrated in Fig. 10.16. In the top half of the figure, waves develop as faster, less dense and more humid air advances over slower moving, cooler, denser and drier air. Instabilities at the interface cause ripples, much like the wind blowing over water. This forms lines of altocumulus clouds perpendicular to the

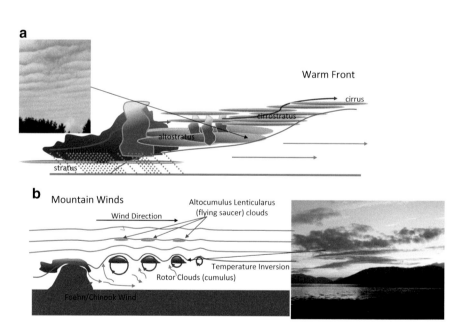

Fig. 10.16 Gravity wave clouds caused by air masses moving at different speeds at different altitudes. In (**a**), warm air advances over colder air ahead of a warm front and produces lines of altocumulus cloud. In (**b**) air moves across mountains and undulates. In each case waves have a wavelength measured in tens to hundreds of meters. In (**a**) undulating air above the temperature inversion forms clouds, and in (**b**), roll clouds form beneath it with altocumulus above. Photographs by author

direction of air flow. In the lower half of the figure, air crossing mountains is forced to ascend rapidly. Where there exists a temperature inversion, the air on either side of the inversion is driven into a series of waves. These waves form roll clouds, where cooler, but moist air periodically rises above its dew point causing condensation against the inversion. A few thousand meters above the inversion, in the warmer air, altocumulus clouds can form at higher elevations. Each of these lies above the lower roll clouds, with cirrus and cirrocumulus forming higher still if the waves extend high enough into the troposphere. Thus the entire depth of the wave is illustrated by clouds.

Toshitaka Tsuda (Kyoto University) analyzed how such waves propagate through the atmosphere. Until the 1970s it was thought that west-east winds would strengthen with height all the way into the lower thermosphere (Chap. 1). However, observations showed that winds not only decreased in strength above 70 km but that they often reversed direction. It has been successfully shown that this attenuation and reversal of these mid-atmospheric winds on Earth is due to the breaking of gravity waves. Such waves are launched by motion across mountains or by storms. By the time they reach the mesosphere they have wavelengths measured in tens to hundreds of kilometers. Tsuda used ground-based radar to monitor the propagation of such gravity waves that were associated with frontal storms. The waves measured by Tsuda penetrated the polar front jet stream and continued unabated into the mesosphere. Waves that began with a vertical wavelength of 1–2 km in the upper troposphere grew until they were 10–15 km in size at 60–70 km up in the mesosphere. At this height the waves began to break apart as a combination of wind-shear and convection tore the waves apart. Convection is caused by the steady drop in temperature in the mesosphere with height that makes this region of the atmosphere unstable.

In the stratosphere, winds blow towards the west during the summer (an easterly jet stream) and to the east in the winter (a westerly jet). Gravity waves, on the other hand move in all directions. However, this summertime easterly flow captures westward propagating gravity waves and prevents them rising higher into the atmosphere (Fig. 10.17). However, eastward expanding waves can

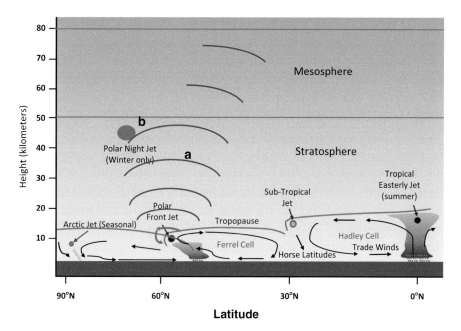

FIG. 10.17 Atmospheric gravity waves deliver energy to the mesosphere from the troposphere. Atmospheric gravity waves (a) are launched by thunder or frontal storms, or by motion across high mountains, where air is forced to rise rapidly through stable upper layers. In the troposphere and lower stratosphere waves can spread evenly east and west. Gravity waves deliver energy through the lower stratosphere, growing in amplitude as they rise. Stratospheric jet streams can block upward motion in the direction opposing the motion of the jet (b), but they continue to around 80 km elsewhere

rise higher until they break in the mesosphere. Here, these breaking waves deposit energy into the layer at an altitude of 60–90 km and cause it to accelerate from west to east, driving the overall circulation. Measurements showed an acceleration of 7–13 m per second where gravity waves hit this layer. In the winter the opposite effect occurs with westward propagating waves reaching the mesosphere causing the winds to reverse direction. Thus, phenomena that are seemingly insignificant on a planetary scale, such as tropospheric storms or mountains that barely penetrate the depth of the troposphere, can in fact alter the motion of the atmosphere upwards to the edge of space.

If all of this still sounds rather academic, then look at Saturn. Measurements by Cassini show the clear signature of atmospheric waves affecting the planets equatorial winds and stratospheric temperatures. A clear downward propagating pattern of temperature changes and alterations to the strength of the planet's jet streams is best explained by upward moving waves from the troposphere. As these interact with easterly and westerly jets they cause the jets to move downwards, towards the tropopause. Indeed, at present Saturn is the only planet we know of, other than the Earth, that shows this pattern. It would be interesting to find out is this kind of energy transfer lies behind super-rotation of Saturn's equatorial atmosphere (Chap. 7). It certainly seems to tie with observations of Venus, where the transport of momentum from the top to the bottom of the atmosphere seems to lie behind super-rotation, here (Chap. 5). Our understanding of this phenomenon will undoubtedly change once Juno starts its mission at the King of worlds; and when better measurements come in from extrasolar planets. As with the thunderstorms we looked at in Chap. 4, there exist a plethora of connections between events on the ground and the phenomena we observe at the top of our atmosphere and this must be true for all planets. Weather is a phenomena not bound to the clouds below.

Atmospheric gravity waves and Earthquakes

After the March 11th 2011 magnitude 9 earthquake, instruments at Japan's National Institute of Information and Communication Technology detected concentric waves radiating outwards 300 km above the Tohoku quake epicentre. The magnitude 9 quake and accompanying tsunami generated atmospheric gravity waves that extended upwards to the Ionosphere, where they altered its structure and conductivity. This observation opens up an interesting avenue for researching geology on planets such as Venus, where the surface cannot be observed. Strong ground shaking should be evident through the waves this generates hundreds of kilometres overhead.

Weather, Unbound: Brown Dwarfs, Stars, Galaxies and Galaxy Clusters

When we think of "weather" we naturally think of planets. Essentially this is true, but a number of meteorological phenomena have found themselves applied to more esoteric objects some hundreds of thousands of light years across. Here we find the idea of weather reach its somewhat illogical conclusions.

Thinking first of brown dwarfs, the idea of weather is fairly understandable. For although these objects begin their lives as hot as the coolest, red dwarf stars, they soon cool down, once their supply of deuterium fuel has been exhausted. For most of these objects this happens in less than ten million years, leaving a cooling orb. Like the hot, young Jupiters, these objects go through a succession of orderly changes in their atmospheres as they cool. Initially, with temperatures above 2000 °C, the atmosphere is filled with metal oxides, drifting in an atmosphere of carbon monoxide, hydrogen and helium. As temperatures fall, these compounds fall out until they have precipitated to deeper, hotter levels where they can become gaseous once more. This weather falls deeper and deeper within the brown dwarf with time. As it does so other gases reach their condensation points and emerge. First, compounds like perovskite, then metal halides. When temperatures fall below 1000 °C the carbon monoxide and hydrogen react to form water and methane and the Brown dwarf morphs from an L to a T-class object.

Beyond 100 million years of age the brown dwarf cools down to a few hundred degrees and its red hue vanishes beneath a sea of cloud. Initially, these are composed of metal halides: compounds such as sodium and cesium chloride. But as temperatures fall below 300 °C or so, water vapor clouds take prominence and the brown dwarf finally takes on the aspect of Jupiter. Perhaps one billion years old, this dwarf has bands of water clouds in a sea of methane, hydrogen and helium. Much older still and the water vapor cools until it can form ice clouds. This is the status of the coldest brown dwarf known (see earlier in this chapter). Over time, the water vapor and ice clouds will be replaced by ammonia ice clouds the methane ice clouds and aside from the stronger gravity, will leave an object indistinguishable from our Jovian world.

What of stars and weather? In a loose sense, yes, they do have it. Stars obviously have atmospheres that are affected by the energy generated beneath their torrid surfaces. Cooler stars (red dwarfs, orange dwarfs and yellow dwarfs) have fairly strong magnetic fields that are generated within their outer layers. This leads to a variety of interesting phenomena such as prominences and flares. Now, although these are not standard "weather" the Sun does occasionally generate something we might recognize on Earth. For example in September 2015 the Sun generated a tornado—a rather large one more five times higher than the diameter of the Earth. Now, as tornadoes are generated from the top–down and this was clearly generated from the solar photosphere, the term fire devil might be more appropriate (Chap. 6). Convection, generated from the solar surface combined with a generous twirl from twisted solar magnetic fields, whisked very hot plasma into the characteristic shape of a tornado. The temperature, based on its spectrum, was a rather fearsome 2.8 million degrees Kelvin.

The whole sequence lasted 2 days and delivered another batch of energy from the solar interior to the corona above. How the Sun accomplished this process has been something of a conundrum for astronomers: energy is apparently transferred from the cooler photosphere to the hotter corona above. However, the reality is a little more complex: energy is in fact transferred from the solar interior to the corona through a combination of magnetic waves and shock waves. The "tornado" was simply another manifestation of this energy transfer process, with its rapidly spiraling structure carrying energy from the interior to the hot corona. Similar "twisters" had been seen before but on this occasion the Solar Dynamic Observatory (SDO) had a beautifully clear observation of the event over its entire lifetime, giving astronomers the opportunity to see how such storms develop and deliver their cargo of energy to the sun's outer atmosphere.

Solar Wind and Stellar Gales

The Sun is obviously hot and all hot objects radiate energy. Liquid and gaseous objects also release particles because there will always be a population of particles that have enough energy to overcome the force of gravity. In the case of the Sun, the outflow of gas from its surface blows at hundreds of thousands of miles an hour (greater than or equal than 400 km per second). The solar wind is divided into two components: the first is known as the slow wind, with slow being very much a relative term. Wind speeds are 400 km per second with a temperature of 1.4 million to 1.6 million Kelvin. This wind blows primarily from the Polar Regions and from regions called coronal holes, where magnetic field lines extend from the solar surface into interplanetary space. The gases have a composition matching the corona and essentially represent gases boiling away from it. Gases are fed into the coronal holes by small-scale magnetic fields anchored in the Sun's outer convecting region. When these field lines reconnect they launch particles which can then flow directly away from the Sun.

The second component of the solar wind is far more variable in nature and blows at much higher speeds from the Sun's equatorial regions (within 35° of the equator). These winds carry material from the photosphere and include gigantic bursts called coronal mass ejections. Although wind speeds are higher—up to 750 km per second—and their temperature is lower, at around 800,000 K. These fast winds are twice as dense as the slow wind and have a chemistry more broadly matching the photosphere. Because these winds tend to owe their existence to violent explosions, from prominences, through flares to coronal mass ejections, they are very turbulent and carry a lot more momentum than the steadier slow winds.

Would you then feel the Solar wind blowing gently (or fiercely) on your face, assuming you wouldn't die from exposure to the vacuum of space? Not really. The term vacuum is important. Now, obviously you can't have a wind in a vacuum: that's an oxymoron. However, the gas is so rarified that even close to the Sun where the wind is densest, the pressure the wind exerts is only 1-6 billionths of a Pascal. Recall the Earth's atmospheric pressure is around 100,000 Pa's at the Earth's surface. That's a difference of 15 orders of magnitude. Hence, the term vacuum is often applied. Yes, strictly speaking no part of the volume of space is a true vacuum, but with a handful of particles per cubic centimeter you'd be hard-pressed to think of it as anything else.

As for other stars, as all stars are by definition hot, they all generate a wind. The red dwarfs generate the weakest winds, with outflows similar or weaker to those of the Sun. More massive stars generate proportionately stronger winds. These winds blow faster—perhaps ten times faster, and carry up to a million times more mass. Stars also vary the amount of mass they lose over time. Very young stars tend to be more magnetically active than older stars and various processes happening within their rapidly rotating magnetic fields tend to launch stronger winds. So a red dwarf that releases less than a trillionth of its mass every year when a few billion years old may have released 1000 times this amount per year when it was first formed. We encountered the problems this may cause for any planet that was orbiting closely earlier in the chapter. Likewise, the Sun, and indeed all stars, expand when they age. Red dwarfs expand the least and the smallest red dwarfs do this in the middle of their lives. More massive stars expand greatly—perhaps a few hundred times their original size. With their surfaces far removed from their center of mass the pull of gravity is lessened. Moreover, older stars are brighter and the higher luminosity blows gases more effectively. Therefore, both effects mean that aging red giant stars can blow vast amounts of mass away from their surfaces. Now, although the speed is reduced to only a few tens of kilometers per second, the much greater mass (up to a few tens of millions of times that of the Sun) have considerable effects. The star is effectively blown away leaving only its core.

When the Sun dies and its wind finally subsides, will mark the beginning of the end of interplanetary weather. For then, the remaining planets will cool and their atmospheres progressively rain out into their interiors. Trillions of years from now, each planet that remains orbiting the cold, dark Sun will see one gas after another condense then freeze. The process of atmospheric collapse will mark the ultimate end of weather in our solar system and presages the end of weather in the universe as a whole.

For now, the Sun's hot breath impacts us in two ways. For one it slowly erodes the Earth's atmosphere but also contributes to geomagnetic storms. For when those coronal mass ejections hit our planet's magnetosphere, the change in the Earth's protective field causes part of it to collapse into our ionosphere. Thus, the continual attrition of our atmosphere parallels the steady loss of mass from the Sun. Everything loses out in the end. However, each moment of loss is marked with aurorae that light up the atmosphere's of all the planets in the solar system bar Mercury. Pluto? Well, its thin atmosphere might just glow green-blue when the solar wind blows more strongly. We will need to wait for the continual download of New Horizons data to complete to see if this is apparent above its hazy blue firmament.

So, when your power supply fails in a geomagnetic storm, just look at what the planet and its Sun lay on for compensation. We may love our artificial sodium-neon glow; however, the blue-green fluorescence of aurora has a far greater majesty than anything we can create.

Galactic Winds

Galaxies that contain active star formation exhibit galactic winds. In our galaxy these winds are modest and only blow gas from the disc into the halo. This gas is expelled from the disc by a combination of strong stellar winds generated by massive stars and through the actions of supernovae. These winds form as hot bubbles within the cooler gas of the disc. As this gas expands upwards and downwards away from the galactic plane, it pushes some cooler disc gas ahead of it.

As this gas reaches the halo of the galaxy it cools and rains back into the disc. In the loosest sense, this movement of hot and cold gas might be thought of as weather. However, if you are dissatisfied with this analogy then look at starburst galaxies, such as M82. Here, furious rates of star formation generate winds hundreds of times stronger than our galaxies puny breeze. Such winds propel gas out of the galaxy and much of it into intergalactic space. Chandra x-ray observations, coupled to Hubble observations reveal cool streamers of gas shredding and blasting outwards away from the galactic core.

Elsewhere in the universe, giant elliptical galaxies hide a dark secret: a central super massive black hole. These dark lords reveal their presence through their eating habits. The galaxies they inhabit reside in a deep well of hot gas (below). This gas can cool and stream into the core of the central cluster galaxy. As it plunges in towards the central black hole, it gathers into a disc around the central behemoth. Although 99 % of the gas is ultimately consumed, magnetic fields within the disc can propel gas outwards in long, thin jets at 90° to the disc: these jets point along the magnetic poles of the black hole and its disc. This gas pummels outwards at velocities between 60 and 99 % the speed of light where it slams into and heats gas in the galaxy cluster's surrounding medium. This turmoil generates regions where the temperature and pressure within the gas changes over relatively short distances. Perhaps we could think of these as fronts?

Cold Fronts in Galaxy Clusters

Taking this analogy somewhat further and explicitly using the term "front" are "meteorological features" that are seen in the centers of entire clusters of galaxies. It's been known for a couple of decades that galaxy clusters contain very hot x-ray emitting gas and that in some of these clusters there appear to be regions, hundreds of thousands of light years long, where the temperature of the gas changes by millions of degrees over relatively short distances (hundreds of light years). We might well be pushing the boundaries of what could reasonably be called weather, but these structures do bear some resemblance to terrestrial cold fronts. For example, as we cross the front temperature and pressure falls and there is a large difference in the direction gas is flowing. Don't expect thunderstorms or even a slight drizzle. Instead (if you wait long enough – a few tens of millions of years) there will be a reduction in the amount of x-radiation and cosmic rays that you are exposed to.

How do such cold fronts form? Simon Ghizzardi (Istituto di Astrofisica Spaziale e Fisica Cosmica) and colleagues analyzed a large number of galactic clusters, containing up to a few thousand galaxies, each orbiting the cluster's mutual center of gravity. As they sweep through the cluster they interact with very hot gas that permeates the cluster in a number of ways. Some gas-rich galaxies violently collide with the hot cluster gas, generating shockwaves that speed through it. Gas that is compressed by the advancing shock cools as its density increases, simply because the denser gas can radiate heat more efficiently than the remaining rarified material. These are the cold fronts – in most cases. The vast majority of clusters that show cold fronts have gas-rich galaxies plowing into and merging with the cluster.

A few clusters don't show collisions but still show the presence of active cold fronts near the cluster core. In these clusters, the galaxies are rather more sedate and not interacting as strongly with one another. Here it appears that the fifty percent or so of clusters with cold fronts have a very steep temperature gradient from the outside to the center of the cluster. The hottest gas lies furthest from the core of the cluster where gas is least dense. Near the center gas is colder and denser: cold still means temperatures of a few million degrees, but this is much less than the 20–30 million Kelvin gas found further away. The cold fronts form near the edge of the coolest gas. Not precisely determined, it looks like cold fronts form here when small galaxies approach the dense, cold core. As they pass their gravity pulls on the gas causing it to slosh back and forth near the edge of the hotter, surrounding gases. The advancing boundary of this sloshing gas is a cold front. Does that mean there is a warm front where the cold gas is retreating back towards the cluster core? Although this use of "front" is somewhat unconventional, it is certainly not unreasonable and pushes our concept of what weather is to scales tens of times larger than an entire galaxy.

Conclusions

Understanding the atmospheres of exoplanets is perhaps the most exciting area of meteorology there is. For not only does there now exist a vast array of complex and highly diverse planets to examine, but there are also clear impacts on the understanding of our weather. Exoplanets are the testing beds for our ideas on how our planet's atmosphere works. Even the most extreme planets, the hot Jupiters, have a direct connection to the atmosphere and weather on our world. Their super-rotating atmospheres mirror, in the extreme, the tropics. In each case eastward moving Kelvin waves drive super-rotation. On earth this is a very modest affair, with bursts of westerly winds moving against the prevailing easterly trades, or on occasion assisting in the development of El Niño patterns. On Hot Jupiters, with a far stronger push, Kelvin waves drive super-rotation measured in hundreds of kilometers per hour.

More modestly, the effects of atmospheric tides on hot Jupiters and other less strongly heated worlds are reflections of atmospheric processes on Earth, Venus and Mars. Although our current telescopic resolution is insufficient to probe frontal boundaries on planets hundreds of light years away, soon that may be possible.

Then we can get a comprehensive picture of how these planets' atmospheres work and truly be able to give an accurate weather forecast for a planet we formerly could merely dream of. For example, the aging Kepler 452b paints a view of our planet in a billion or so year's time. Although we cannot quite probe its weather yet, its age and location suggest its atmosphere is likely evolving from one able to support life to one that is as hot and as torrid as modern day Venus. As we obtain more and more extensive libraries of worlds we can map weather and climate in space and time.

While we extend what weather we can see, we also expand the definition of what might reasonably be called weather. We are more than happy to talk about the weather on Jupiter, but what about a brown dwarf? After all these take on the aspect of Jupiter with time, so where will you draw the line? Happy with brown dwarfs having weather, then what about stars, groups of stars in clusters or entire galaxies and groups of galaxies? If weather is simply the motion of gases, then the universe has weather on all scales from objects the size of Pluto through to clusters of galaxies that extend over tens of millions of light years.

This chapter gives only the simplest introduction to what is becoming an increasingly comprehensive picture of exoplanetary weather. With a bit of dreaming and a lot of hard graft planetary scientists are developing increasingly sophisticated views of worlds hundreds or thousands of light years beyond our cosmic shore. And while the atmospheric scientists dream of worlds still out of sight, cosmologists see weather on scales that we can only imagine. This is an amazing time.

References

1. Visscher, et al. (2010). *Atmospheric chemistry in giant planets, brown dwarfs, and low-mass dwarf stars III. Iron, Magnesium, and Silicon*. Preprint available at: arXiv:1001.3639 [astro-ph.EP].
2. Heath, M. J., Doyle, L. R., Joshi, M. M., & Haberle, R. M. (1998). *Habitability of planets around red dwarf stars*. Preprint, available at: http://www.as.utexas.edu/astronomy/education/spring02/scalo/heath.pdf.
3. Marley, M. S., Ackerman, A. S., Cuzzi, J. N., & Kitzmann, D. (2013). *Clouds and hazes in exoplanet atmospheres clouds and hazes in exoplanet atmospheres*. Preprint, available at: http://arxiv.org/pdf/1301.5627v1.pdf.
4. Loadders, K., & Fegley, B. (2006). *Chemistry of low mass substellar objects*. Preprint, available at: http://arxiv.org/ftp/astro-ph/papers/0601/0601381.pdf.
5. Hu, Y., & Yang, J. (2014). Role of ocean heat transport in climates of tidally locked exoplanets around M dwarf stars. *PNAS, 111*(2), 629–634.

6. Holmström, M., Ekenbäck, A., Selsis, F., Penz, T., Lammer, H., & Wurz, P. (2008). Energetic neutral atoms as the explanation for the high-velocity hydrogen around HD 209458b. *Nature, 451*, 970–972.

7. Showman, A. P., Cho J. Y-K. & Menou, K. *Atmospheric circulation of exoplanets.* Retrieved from http://www.lpl.arizona.edu/~showman/publications/showman-etal-exoplanets-review-revised.pdf.

8. Liu, B., & Showman, A. P. (2012). *Atmospheric circulation of hot Jupiters: Insensitivity to initial conditions.* Preprint, available at: http://arxiv.org/pdf/1208.0126v2.pdf.

9. Showman, A. P., Menou, K. & Cho, J. Y-K. Atmospheric circulation of Hot Jupiters: A review of current understanding. Retrieved from http://www.lpl.arizona.edu/~showman/publications/showman-etal-santorini-paper-submitted.pdf.

10. Holmström, A., Ekenbäck, F., Selsis, F., Penz, T., Lammer, H., & Wurz, P. (2008). Energetic neutral atoms as the explanation for the high-velocity hydrogen around HD 209458b. *Nature, 451*, 970–972.

11. Swain, M. R., Vasisht, G., & Tinetti, G. (2008). The presence of methane in the atmosphere of an extrasolar planet. *Nature, 452*, 329–331.

12. Schinder, P. J., Flasar, F. M., Marouf, E. A., French, R. G., McGhee, C. A., Kliore, A. J., Rappaport, N. J., Barbinis, E., Fleischman, D. & Anabtawi, A.(2011). Saturn's equatorial oscillation: Evidence of descending thermal structure from Cassini radio occultations. *Geophysical Research Letters, 38*, Article Number: L08205. http://onlinelibrary.wiley.com/doi/10.1029/2011GL047191/full.

13. Leconte, J., Forget, F., Charnay, B., Wordsworth, R., & Pottier, A. (2013). Increased insolation threshold for runaway greenhouse processes on Earth-like planets. *Nature, 504*, 268–271.

14. Yang, J., Boué, G., Fabrycky, D. C., & Abbot, D. S. (2014). *Strong dependence of the inner edge of the habitable zone on planetary rotation rate.* Preprint, available at: http://arxiv.org/pdf/1404.4992v1.pdf.

15. Yang, J., Cowan, N. B., & Abbot, D. S. (2013). *Stabilizing cloud feedback dramatically expands the habitable zone of tidally locked planets.* Preprint, available at: http://arxiv.org/pdf/1307.0515.pdf.

16. Tsuda, T. (2014). Characteristics of atmospheric gravity waves observed using the MU (Middle and Upper atmosphere) radar and GPS (Global Positioning System) radio occultation. *Proceedings of the Japanese Academy, Series B, 90*, 12–27. Preprint available at: http://www.ncbi.nlm.nih.gov/pubmed/24492645.

17. Faherty, J. K., Tinney, C. G., Skemer, A. & Monson, A. J. (2014). *Indications of water clouds in the coldest known brown dwarf.* Preprint available at: http: //arXiv:1408.4671v2.pdf.

18. Wordsworth, R. (2014). *Atmospheric heat redistribution and collapse on tidally locked rocky planets.* Preprint, available at: http://arxiv.org/pdf/1412.5575v1.pdf.

19. Zendejas, J., Segura, A. & Raga, A. C. (2010). *Atmospheric mass loss by stellar wind from planets around main sequence M stars.* Preprint available at: http://arXiv.org/pdf/1006.0021v1.pdf.

20. Hummocky terrain and volcano collapses. Retrieved from http://volcanoes.usgs.gov/observatories/cvo/cascade_debris_avalanche.html.

21. New Horizons public access data. Retrieved from https://www.nasa.gov/mission_pages/newhorizons/main/index.html.

22. Carone, L., Keppens, R., & Decin, L. (2015). *Connecting the dots II: Phase changes in the climate dynamics of tidally locked terrestrial exoplanets.* Preprint, available at: http://arxiv.org/pdf/1508.00419v1.pdf.

23. Leconte, J., Wu, H., Menou, K., & Murray, N. (2015). *Asynchronous rotation of Earth-mass planets in the habitable zone of lower-mass stars.* Preprint, available at: http://arxiv.org/pdf/1502.01952v2.pdf.

24. Ghizzardi, S., Rossetti, M., & Molendi, S. (2010). Cold fronts in galaxy clusters. *Astronomy and Astrophysics, 516*, A32. Preprint available at: http://arxiv.org/pdf/1003.1051v1.pdf.

25. Tsugawa, T., Saito, A., Otsuka, Y., Nishioka, M., Maruyama, T., Kato, H., et al. (2011). Ionospheric disturbances detected by GPS total electron content observation after the 2011 off the Pacific coast of Tohoku Earthquake. *Journal of Earth Planets and Space, 63*, 875–879.

Glossary

A

Acetonitrile   A simple compound of carbon nitrogen, oxygen and hydrogen, produced through the action ultraviolet light on nitrogen, methane and water or carbon monoxide.

Advection   The horizontal transport of energy by wind.

Aerosols   Small particles that are produced by varying processes in planetary atmospheres or in interstellar space. These include spray from oceans or falling water feldspar dust from dry areas and chemicals released by plants or plankton in the oceans. Terrestrial aerosols also include bacteria viruses and other biological materials.

Altocumulus   A type of mid-level (approximately 3000–5000 m in altitude) cumuliform cloud composed of very cold water droplets. These may be arranged in rows or streets lying in the direction of the wind or at 90° to it.

Altocumulus castellanus   A type of altocumulus cloud associated with unstable mid- level air masses. These may produce showers or develop into high base storms.

Altostratus   Flat stratiform clouds composed of water droplets that typically obscure but do not completely block out the sun. These often form ahead of warm or occluded fronts.

Ammonia   A compound of one nitrogen and three hydrogen atoms found abundantly in the outer solar system's planetary atmospheres and in interstellar clouds and comets.

Ammonium Hydrogen Sulfide   A noxious compound of ammonia and hydrogen sulfide found within the atmospheres of the giant planets. Ammonium hydrogen sulfide smells of rotten eggs.

© Springer International Publishing Switzerland 2016
D.S. Stevenson, *The Exo-Weather Report*, Astronomers' Universe,
DOI 10.1007/978-3-319-25679-5

Anthropocene   The terms given to the latest geological period beginning around 1700 and characterized by our propensity to alter the atmospheric chemistry, temperature and climate, as well as significantly alter surface and deep-level geology through industrial processes.

Anthropogenic Global Warming   The term given to the outcome of putting additional greenhouse gases into a planet's atmosphere - causing more outgoing energy to become trapped and the temperature to rise.

Anticyclone   A region where air flows in a clockwise direction in the northern hemisphere and anticlockwise in the southern hemisphere. Anticyclones are characterized by calm conditions on Earth with higher surface pressure. They are formed when air descends from greater altitudes. They may have a warm or cold core. Those with warm cores extend through the entire troposphere while cold -cored high pressure areas eventually give way to regions of lower than normal pressure at height.

Atmospheric Tide   The bulk motion of a large portion of a planet or its satellite's atmosphere caused by gravitational pulling (Titan) or day-night differences in heating (Venus Earth and Mars).

Aurora (aurorae/auroras)   Luminous displays associated with the ionization of gases in the high atmosphere. Charged particles from the Sun arrive in waves during solar storms and these cause realignments in the magnetic field around the Earth (or other worlds). These reconnection events cause a cascade of particles that sweep down into the upper atmosphere blasting electrons off neutral atoms. As these fall back onto their charged particles they emit light. The blue-green color is predominantly caused by oxygen and nitrogen ions collecting electrons and emitting light.

B

Bacteriophage   A type of virus that infects bacteria.

Ball lightning   A rare form of lightning that involves some form of organized and roughly spherical ball of plasma that is generated (most likely) by some mechanism involving lightning discharges.

Ballooning (spiders)   The process by which (mostly) small spiders can take off using small lengths of silk to catch the wind and/or pick up electrical charge which causes them to repel from the underlying surface.

Bergen frontal model   The original model of frontal storm systems with an advancing warm front, closing cold front and intervening warm sector. Developed in the wake of the First World War this model successfully describes most frontal low pressure areas that are found in mid to high-latitudes.

Blocking (blocking anticyclone)   A block is an area of warm-cored, high pressure that often becomes stranded on the cold side of the polar front jet stream. Such high pressure areas then block the normal west-east progression of frontal low pressure areas and give rise to periods of steady and predominantly dry weather.

Blue jets (associated with lightning)   Upward directed jets of luminous gas that connect thunderstorms with the mid to upper stratosphere.

C

Charge exchange (escape mechanism)   An atmospheric escape mechanism, whereby a fast moving ion exchanges its charge with a less massive neutral atom, such as hydrogen or oxygen. In the process the ion becomes electrically neutral and is accelerated. With enough kinetic energy it then escapes into interplanetary space. This escape mechanism is important in the upper atmosphere of Venus, for example.

Chinook (Foehn/Föhn)   A warm wind that blows down the lee of mountains and (in winter and Spring) readily melts lying snow. The Chinook is formed when warm, moist air cools on the windward side of mountains and latent heat is released as moisture condenses and precipitates. As this, now dry, air descends in the lee of the mountains, it warms at a faster rate than it cooled on the windward side, because it is moisture-free. It is also warmer, through the earlier release of latent heat. It thus reaches lower elevations considerably warmer than it would otherwise have done.

Cirrus   A type of high (typically above 5000 m), fibrous ice cloud.

Cirrostratus   A fibrous ice cloud that forms translucent sheets.

Cold front   A dividing line at the Earth's surface where cold air advances into and displaces upwards warmer air. Cold fronts fall into two classes—Kata and Ana—with the former dominated by descending, warming air and the latter by ascending and cooling air.

Cold high   An area of high pressure typically produced by air cooling and rapidly descending. These occupy only the lowest 15,000 ft (4000 m) or so of the troposphere. Above, they gradually morph into areas of cold low pressure.

Cold low   A mid-latitude or sub-tropical area of low pressure that is cold throughout its depth and has low pressure throughout.

Conduction   The process of heat transport that that involves the exchange of particle energy through collisions or the direct exchange of electrons between atoms that lie close to one another.

Convection   The transport of energy by the bulk motion of matter in a liquid or a gas that is driven by density differences.

Coriolis effect   The outcome of any fluid (including gases) moving on a rotating surface. In the context of planets with atmospheres, particles that move away from the equator towards either pole are moving faster than the ground underneath and must curve to the east. Particles moving from either pole to the equator are moving slower than the underlying surface and curve towards the west. The Coriolis Effect is thus dependent on two factors the speed (or angular velocity) of the planet's rotation and its size. Small planet and those which rotate most slowly experience the weakest effect.

Cryovolcanism   The release of icy fluids and gases from vents or fissures on the surface of a cold, icy body. These can be made of ammonia, water, methane or any other moderately abundant substance that is above its melting point.

Culminate   Whereby a star or other celestial body reaches the highest altitude relative to the observer when it crosses the imaginary line between the observer's southern horizon and the point immediately above his or her head. (Observer's northern horizon if they are in the southern hemisphere.)

Cumulus   A bulbous cloud that is produced by convection and is composed mostly of water droplets.

Cumulonimbus   The tallest type of cloud with a typically low base. The cloud is composed of water droplets through the lower half of its bulk, but ice particles in its middle to upper levels. Such clouds give rise to hail, heavy showers or thunderstorms on all planets we observe them. On Earth and warm to hot planets, cumulonimbus clouds may have high bases where the air feeding them is relatively dry.

Cyanide   A compound of carbon and nitrogen (with one other element), which is highly toxic to aerobic organisms but is a key component in the non-biological formation of parts of DNA and RNA. Found in Titan's atmosphere.

D

Derecho   Literally "horizontal wind" these are violent, gale to hurricane force winds that blow outwards from the base of violent thunderstorms. These storms are largely self- sustaining, once they are initiated, as long as there is a continued supply of unstable air to drive the formation of thunderstorms.

Dry line   A type of atmospheric frontal boundary between warm, moist air and warm, dry air that forms in a few geographical locations on Earth. In each there is a source of warm dry air (preferably at some elevation) to the west of the source of moist air. Consequently, dry lines most commonly form in places like SW USA, eastern Spain and Argentina. They are central to the formation of tornadoes.

Dust devil   A type of swirling, convection-driven storm where the air is dry and the column of swirling dust is generated from the ground upwards.

E

Earthquake lights (EQLs) A blanket term given to lights of various sorts which appear to be associated with earthquakes. However, although some lights—such as those erupting from the ground—are clearly linked to tectonic strain, others have more tenuous associations which may be coincidental.

Electromagnetic pulse (EMP) A burst of radio waves associated with strong electrical discharges—or nuclear explosions.

El Niño A pattern of weather that emerges once or twice per decade and is associated with the warming of the eastern and central Pacific Ocean surface waters. Changes to the flow of air in the tropical Pacific are followed by a realignment of mid-latitude weather patterns. In most (but not all) years El Niño tend to develop in the autumn and persist into the following spring.

El Niño Southern Oscillation (ENSO) The blanket term covering the movement of weather patterns from normal trade-wind dominated tropical airflow to El Niño conditions and then (often) to El Niño's opposite pattern, the La Niña.

Elves (associated with lightning) A very faint, outward propagating pulse of light and other electromagnetic radiation associated with an EMP (above) which accompanies a Sprite (below).

Eocene The geological period from 55 to 34 million years ago that was associated with predominantly high temperatures and a higher than present atmospheric level of carbon dioxide gas.

Euxinia Stagnant and largely anaerobic conditions, usually with an abundance of hydrogen sulfide.

Exoplanet Any planet outside the solar system.

Exosphere The region at the top of a planet's atmosphere where gases can escape through different mechanisms.

Equatorial Rossby Wave (ERW) A pairing of alternating high and low pressure areas which move from east to west on either side of the equator. Unlike eastward moving Kelvin waves that appear as fast moving changes in wind direction, on either side of a region of enhanced rainfall, ERWs appear as pairs of high pressure leading pairs of low pressure with rainfall enhanced at the leading edge of the low pressure regions.

F

Fall-streak   (see Virga).

Ferrel cell   A pattern of tropospheric circulation in the mid to high latitudes of the Earth and Mars (but also apparent in the atmosphere of Venus). This motion is induced by Hadley cell circulation above the tropics. Air descending in the cool limb of each Hadley cell displaces air along its poleward side causing it to move in a counter flow towards the pole, along the planet's surface.

Fire Devil   A nasty incarnation of a dust devil, where strong heating above the ground, within a large fire, induces the formation of a swirling column of plasma. This can ignite fires elsewhere.

Fire Tornado   Fortunately, a very rare phenomenon on Earth, where very large fires generate first cumulonimbus clouds, which then generate tornadoes. These vortices then entrain fire and other hot debris within their circulation, causing significant additional damage and posing a major threat to fire-fighters. The most closely documented example struck parts of Canberra in 2003.

Foehn (Föhn) wind   (see Chinook).

Front   A generic term that arose in the light of the trenches in the First World War. Fronts mark the dividing line between air masses with different properties, such as temperature, pressure and humidity.

G

Gap wind   A strong wind that blows through mountain valleys. In a classic case a front becomes trapped behind mountains, but some of the air it leads can flow through valleys. A classic example occurs in the Chivela Pass in Mexico where cold fronts moving south across the Gulf of Mexico come up against the Sierra Madre mountains. Cold air eventually floods through a gap—the Chivela Pass—and out into the Pacific Ocean.

Gigantic jets   As the name suggested these enlarged versions of the blue jets (above) reach the mesosphere from the top of thunderclouds, rather than petering out in the stratosphere. The exact composition is unknown but they are believed to share some similarities with lightning.

H

Hadley cell   The largest circulatory cell within the troposphere of terrestrial planets. This cell (or rather pair of cells, in most cases) is a convection cell driven by solar heating near the equator. Air warms, becomes less dense and rises. From here it expands out towards either pole. The Coriolis Effect (and increasing density) eventually limits the poleward advance. Air then descends to the surface and either returns to the region of rising air or blows out across the surface towards the poles.

High pressure   A region of higher than normal pressure caused by subsidence of air from aloft.

Horse latitudes   A belt of high pressure that forms the poleward limit of the Hadley Cells. Here, air is relatively calm and the weather cloud free. On Earth, the Horse Latitudes are advancing towards the poles at a rate of around 60 km per decade, in a process almost certainly caused by anthropogenic global warming.

Hydrodynamic escape (including hydrodynamic drag)   A dynamic process of atmospheric escape. Gases are heated strongly within the atmosphere of a planet, either by stellar heating or tidal heating. This creates sufficient pressure to drive the gases out into space. Typically, light gases such as hydrogen are most strongly affected. However, these can drag other, heavier gases, such as nitrogen or oxygen, with them. This is hydrodynamic drag.

I

Impact erosion   The removal of part or all of a planet's atmosphere following a collision or collision with asteroids or comets.

Insolation   The term given to the total amount of radiation a planet receives from its host star.

Isobar   An imaginary line connecting points of equal pressure.

Isotope   Isotopes are versions of elements that have differing numbers of neutrons.

J

Jeans escape   The name given to the process whereby particles have a range of kinetic energies. Those with sufficient energy can overcome the pull of gravity and escape the atmosphere. Jean's escape is therefore proportional to the mass of the atom or molecule and its kinetic energy.

Jellyfish clouds (see Virga)   Fall-streak or Virga is a precipitation of ice, typically from cirrocumulus or altocumulus clouds that evaporates before reaching the ground.

Jets (associated with lightning)   See blue jets and gigantic jets, above.

Jet stream   The generic term given to relatively narrow bands of high wind speed within the atmosphere of a planet. These may lie at varying altitudes and move at speeds varying from 10 to several hundred meters per second. They are associated with sharp differences in pressure and temperature.

K

Katabatic wind   A fast moving type of mountain wind that descends rapidly under its own weight from very cold upland regions. These winds pour into valleys and can attain storm-force. If they cross warmer ocean waters they can pick up moisture and develop squalls.

Kelvin wave   A fast-moving westerly wave in the otherwise easterly trade winds. Easterly winds are fastest on its leading edge, while strong westerly winds dominate to the rear of the advancing wave and rainfall band. These waves and Equatorial Rossby Waves (above) are crucial drivers of super-rotation on hot, tidally locked planets. Kelvin waves are also found in the oceans and have similar roles in directing or reversing the direction of ocean circulation.

L

La Niña   The colder sibling of the El Niño, where waters across the eastern Pacific become colder than normal as a result of stronger trade winds. While the eastern and central Pacific Ocean are colder than normal, the western Pacific is warmer than normal.

Little Ice Age (LIA)   A period of climatic variability extending from 1200 AD to 1900 AD. In Western Europe the peak of the period coincided with the frequent freezing of the Thames in London. However, despite claims to the contrary the LIA was a regional phenomenon with different regions experiencing cold while others, simultaneously, experienced enhanced warmth.

Loess   Fine, wind-borne silt-like deposits that are produced by glaciation and then blown away from their source by regional winds. Loess deposits can be hundreds of meters thick. Perhaps the most extensive are found in central China where they contribute to the coloration of the Yellow river.

M

Madden-Julian oscillation   A periodic, long-wavelength atmospheric disturbance that is found in the tropics. These waves move from west to east, crossing the Indian then Pacific oceans. As they do so they first enhance then reduce convective rainfall over periods 60–90 days long.

Maunder minimum   The somewhat infamous period during the 1600s where there appear to have been few significant and possibly no sunspots. Many pieces of work suggest a link to the Little Ice Age; a link that falls when the global temperatures are looked at, rather than focusing on those of Western Europe.

Medieval Warm Period (MWP)   Another regional climatic event where temperature in the North Atlantic basin, and southern Greenland in particular, rose by 1–3 °C compared to the 1960–1990 average. This excursion in temperatures appears to be linked to a prolonged La Niña in the Pacific, which simultaneously cooled and dried out Meso-America. The tail end of this period is associated with China's coldest temperatures of the last 1000 years.

Mesosphere   The region in a planet's atmosphere above its stratosphere, where temperatures fall again, with height. Above the Earth the mesosphere lies at 60–90 km above the surface and is associated with a variety of unusual meteorological effects such as noctilucent clouds, Elves, Sprites and aurora.

Metallic hydrogen   When hydrogen is compressed to 3 or more million atmospheres -300 billion Pascals (and this is still not confirmed) it should behave like an alkali metal. At present it is assumed to be present in the interiors of Jupiter and Saturn but the conditions needed to produce it have yet to be reproduced on Earth.

Mei-yu/Baiu front   A seasonal frontal band that marks the intersection of the mid-latitude westerly winds over Asia and the advancing or retreating edge of the summer monsoon winds that are blowing in from the Pacific. The northern side of the front has cool, drizzly weather, while thunderstorms and showers predominate along its southern edge. Frontal lows may develop and move northeastwards into the Pacific Ocean. A similar frontal feature is occasionally visible along the eastern seaboard of the United States where tropical air moves north along the coast in the summer and impinges upon cooler westerly winds blowing from the continental interior.

Milanković cycle   A series of overlapping cycles that affect how much solar radiation reaches the planet at different latitudes over time. Cycles are related to the shape of the Earth's orbit; where the Earth is in its orbit during the northern winter and the amount of tilt in the axis of the planet. Each of these cycles varies in a predictable manner and evidence for its influence in climate extends at least as far back as the Eocene. Here, Milanković cycles may have driven the catastrophic release of greenhouse gases causing massive global warming. More contemporaneously they are largely responsible for the cycles of glaciation and thaw we call ice ages. There is some uncertainty over which cycles drive glaciation and why this appears to have changed in the last million years.

Momentum   A physical property of moving matter which is the product of mass and velocity. Momentum is always conserved. Angular momentum is a slight variant which takes into account the motion of a mass around a center of rotation, with the distance the mass lies from the center of rotation of equal importance.

Monsoon   Seasonal winds that blow from cooler to warmer regions. On Earth the summer monsoon blows warm, moist air into the heart of Asia, bringing life-sustaining rains. On Mars the southern summer monsoon brings dust storms.

N

Night-glow   A pervasive but dim blue-green glow that is detectable in the visible and ultraviolet portions of the spectrum. It is produced by the recombination of oxygen (and nitrogen) ions with electrons high in the atmospheres of the terrestrial planets (except Mercury, for obvious reasons).

Noctilucent clouds   Rare, high (60–80 km) ice clouds that form when the mesosphere is at its coldest.

Non-thermal escape mechanisms   A group of mechanisms whereby particles obtain sufficient energy to escape the gravitational clutches of their planet. These may involve chemical reactions, collisions between ions and atoms, or the collision of cosmic ray particles with gas atoms and molecules high in the thermosphere and exosphere. These mechanisms are "non-thermal" because the particles involved do not have a continuum of energies that would be found in a warm gas.

O

Ozone   A molecule consisting of three oxygen atoms. Electrons are shared between them in an unstable organization that makes the molecule particularly suitable for absorbing a wide range of ultraviolet wavelengths that might otherwise harm living organisms on the ground beneath. Ozone, however, is toxic if inhaled and is a major pollutant that is produced by combustion engines.

P

Paleocene-Eocene Thermal Maximum (PETM)   A period of several tens of thousands of years in duration where the temperature of the Earth rose by 6–8 °C on average, but perhaps 10 °C at the poles. The underlying cause was the release of large quantities of carbon dioxide and methane - two potent greenhouse gases. The underlying reason for the release may have been volcanic activity in the North Atlantic or the action of Milanković cycles.

Paleoclimate   Any prehistoric climate.

Permian   The geological period lasting from 298 to 252 million years ago.

Permian mass extinction   An abrupt period lasting 60,000 years or so at the end of the Permian where 90 % of complex species became extinct. The underlying cause was a period of volcanism-induced global warming. Temperatures likely rose 10–15°C. The effect of heat was almost certainly exacerbated by the release of copious hydrogen sulfide from warm, anoxic ocean waters.

Permian-Triassic boundary The narrow divide between the Permian and Triassic periods where the mass extinction of life occurred and its subsequent recovery began.

Polar cell The portion of the Earth's or another planet's atmosphere on the polar side of the Ferrel cell.

Polar front jet stream The semi-continuous ribbon of high velocity air that streams in an easterly direction. The polar front jet stream is the most visible manifestation of the Earth's and Mars' Rossby waves that buckle and twist it as it courses around the globe. The jet marks the interface of the mid-latitude Ferrel and Polar cells.

Prograde In the context of this book prograde motion is the rotation of a planet in the same direction as the planet orbits its star (or a satellite orbits its planet). Thus the Earth and most other planets orbit the Sun in a prograde manner (same direction as the Sun rotates on its axis) and rotate on their axis in this manner as well. (See, also, retrograde.)

Q

∎ ∎

R

Radial velocity method (of planet finding) As a planet orbits its star it pulls on it. This the star moves back and forth in its orbit. This, in turn, causes its spectral lines to Doppler shift backwards and forwards in manner proportional to the mass and orbital period of the planet.

Retrograde A planet that orbits its star in the opposite direction to the rotation of the star. Or a planet that rotates on its axis in the opposite direction to its direction of motion around its star. In terms of atmospheres these would blow around their planet's axis in the opposite direction to the planet rotated. Except in specific regions or at specific seasons this is not seen.

Rossby wave   A series of intertwined waves of differing wavelengths which form in different locations around the earth and other worlds. Rossby waves may be equatorial (above) or they may form very visible structures in the Polar Front Jet Stream (above). Rossby waves have eastwards and westward directed components, with the waves that have the shortest wavelengths moving in an easterly direction and bearing frontal storms in the mid-latitudes.

S

Shapiro-Keyser model   A revised model applying to some frontal storms whereby the normal undercutting cold front is ripped in two by a descending jet of cold air. The more equatorial portion of the front then slices into the warm sector at speed, while the central low is isolated on the polar side of this feature. The warm front wraps around the core of the central low and warm air is rapidly bulldozed to altitude, causing the low to explosively deepen. Several notable historical storms, such as the 1979 Fastnet storm in the UK, or the 1993 "superstorm" in the US were of this type.

Solar wind stripping   A non-thermal mechanism of atmospheric loss whereby the atmosphere is progressively removed through the action of particles in the solar wind colliding with and removing them in the planetary exosphere.

Sprites (associated with lightning)   Transient, red luminous phenomena seen above the anvils of some thunderclouds following the discharge of positively charged lightning bolts to the ground.

Sputtering   Another non-thermal process, whereby energetic cosmic ray particles collide with atoms and molecules within a planet's atmosphere. These may transfer sufficient kinetic energy to overcome their planet's gravitational pull.

Sub-tropical jet stream   The ribbon of high velocity westerly winds that marks the temperature belt edge of the Hadley cells in the terrestrial atmosphere. The sub-tropical jets overlie the Horse Latitude high pressure belts.

Stratosphere   The region in a planet's atmosphere immediately overlying the troposphere where temperatures rise with altitude. On Earth the rising temperatures are caused by the absorption of ultraviolet radiation by ozone. This process reverses at greater height where the density of gases is much lower. Here the stratosphere gives way to the mesosphere.

Sumatra (storm)   A type of tropical line squall that forms near eastern Sumatra during the months of April through to November, during the southwesterly monsoon. Sumatras bring relatively narrow bands of thunderstorms and strong winds as they move north eastwards towards the coastal Malaysia.

Superbolt (lightning)   A rare (five in ten million strikes) version of lightning that comes from the anvil portion of a cumulonimbus cloud. Such strikes last ten times longer than conventional negative strikes and may be associated with diverse electrical phenomena such as ball lightning, sprites and elves

Supercell   A type of long-lasting rotating thunderstorm that may generate tornadoes, strong horizontal winds and extensive hail. These storms usually form in the spring and summer in a few restricted locations such as Tornado Alley in the US, eastern and central Europe, Argentina and southern Australia.

Super-rotation   The phenomenon whereby part of the atmosphere moves faster in the same direction (has a greater velocity) than the underlying surface or liquid interior. The phenomenon is still fairly poorly understood and more than one mechanism may underlie it. However, super-rotation is yielding to improved modeling and its origin may well be understood on hot planets that are tidally-locked to their parent star.

T

Temperature   A measure of the average kinetic energy of particles within a substance. The greater the average temperature, the greater the object's temperature. However, temperature is a measure of the overall particle energy and many particles in an object may have higher or lower kinetic energy. This explains how gases can sublimate from solids when the temperature is far lower than the boiling point.

Thermal escape mechanisms   A group of mechanism through which gases can escape a planet's clutches that are linked to their overall temperature. The higher the temperature the greater the overall kinetic energy of the particles and hence the greater likelihood that the particles will have a velocity in excess of the planet's escape velocity.

Thermal (warm) low   A region of lower than average pressure at a planet's surface, where air is rising because it is warm and less dense than surrounding air. Thermal lows morph into upper level high pressure areas because the air is warm and the particles moving quickly. Therefore, they exert greater pressure than the surrounding air.

Thermal wind   Also known as a thermal tide, this is the flow of air from the warm to the cold side of a planet caused by uneven heating. Such flow is detectable in all of the terrestrial planets that have an atmosphere and is particularly significant in the hot atmosphere of Venus and the tenuous atmosphere of Mars. Although detectable as pressure waves within the Earth's atmosphere, the thermal tide has a limited role compared with other atmospheric processes.

Thermosphere   The region near the top of all planetary atmospheres where cosmic rays and stellar radiation heat the very tenuous gases to high temperatures. The density of gas is so low that the concept of hot is largely lost, despite temperatures measurable in hundreds or thousands of degrees Celsius.

Tornado   A swirling column of air that is generated by converging and spinning air within a thunderstorm. In most cases the storm is a rotating supercell. The tornado is visible in part because the column contains a focus of low pressure which lowers the condensation point and also because such vortices kick up debris from the ground.

Trade winds   Belts of easterly winds that span the bulk of the tropics on the earth, Venus and Mars. These winds move towards the equator from areas of high pressure and blow towards the region of lowest pressure underlying the overhead Sun. Trade winds should be present on all planets with moderate or greater spin and a modest input of heat from their star.

Transit-timing variation method (of planet finding)   If a planetary system hosts more than one planet the gravitational pull of each planet affects the orbital period of each. If one of these planets is visible in its orbit by a transit, then the presence of a second or third unseen planet (one that does not transit) can be inferred from the effect it has on the time the observed planet takes to transit the star.

Tropics   An arbitrary name given to the region on a planet's surface that lies between the most northerly and southerly extent of the overhead Sun. This is dependent on the tilt of the planet so that, for example, on Jupiter the tropics are from 3° north and south of its equator. This means that the cloud tops of Uranus are all at some point within its tropics over the course of its 84 year orbit.

Tropical (equatorial) jet stream   The tropical (easterly) jet stream is a seasonal feature formed over Asia and Africa when the land north of the jet is hotter than the air to its south during the northern summer.

Tropopause   The narrow transition region between the troposphere where temperatures fall with altitude and the stratosphere where they rise with height.

Troposphere   A lowest region within a planet's atmosphere, where temperatures fall with increasing altitude. The troposphere tends to show the most dramatic weather, because it is here where gases are densest and transport heat energy most efficiently.

Turbulence   Short wavelength disturbances that transport energy rather chaotically through a combination of advection and convection. Turbulence can occur when warm air overlies cold air or when the temperature gradient in a gas or liquid is too great for organized convection to occur. Turbulence will also appear where air is forced over or around obstacles.

U

V

Virga   (See fall-streak)

W

Warm high   A region of higher than surrounding pressure where the air is relatively warm. These high pressure areas extend throughout the depth of the troposphere and may block the motion of other pressure features in the polar front jet stream. By definition tropical high pressure areas are all warm cored.

Warm low   (see Thermal low)

X

Y

Z

Index

© Springer International Publishing Switzerland 2016
D.S. Stevenson, *The Exo-Weather Report*, Astronomers' Universe,
DOI 10.1007/978-3-319-25679-5

Printed in the United States
By Bookmasters